Also in the Variorum Collected Studies Series:

MENSO FOLKERTS
Essays on Early Medieval Mathematics
The Latin Tradition

PAUL KUNITZSCH
Stars and Numbers
Astronomy and Mathematics in the Medieval Arab and Western Worlds

RICHARD LORCH
Arabic Mathematical Sciences
Instruments, Texts and Transmission

GEORGE MOLLAND
Mathematics and the Medieval Ancestry of Physics

EDWARD PETERS
Limits of Thought and Power in Medieval Europe

GILES CONSTABLE
Culture and Spirituality in Medieval Europe

WESLEY M. STEVENS
Cycles of Time and Scientific Learning in Medieval Europe

GAD FREUDENTHAL
Science in the Medieval Hebrew and Arabic Traditions

VERN L. BULLOUGH
Universities, Medicine and Science in the Medieval West

Y. TZVI LANGERMANN
The Jews and the Sciences in the Middle Ages

BRUCE S. EASTWOOD
The Revival of Planetary Astronomy in Carolingian and
Post-Carolingian Europe

EDWARD S. KENNEDY
Astronomy and Astrology in the Medieval Islamic World

JULIO SAMSÓ
Islamic Astronomy and Medieval Spain

VARIORUM COLLECTED STUDIES SERIES

The Development of Mathematics in Medieval Europe

Menso Folkerts

Menso Folkerts

The Development of Mathematics in Medieval Europe

The Arabs, Euclid, Regiomontanus

Routledge
Taylor & Francis Group
LONDON AND NEW YORK

First published 2006 by Ashgate Publishing

Published 2016 by Routledge
2 Park Square, Milton Park, Abingdon, Oxon OX14 4RN
711 Third Avenue, New York, NY 10017, USA

Routledge is an imprint of the Taylor & Francis Group, an informa business

ISBN 9780860789574 (hbk)

British Library Cataloguing in Publication Data
Folkerts, Menso
The development of mathematics in medieval Europe : the Arabs, Euclid, Regiomontanus. – (Variorum collected studies series)
1. Euclid – Influence 2. Regiomontanus, Joannes, 1436–1476
3. Mathematics, Medieval
I. Title
510.9'4'0902

US Library of Congress Control Number: 2005932704

VARIORUM COLLECTED STUDIES SERIES CS811

Contents

This volume contains xii + 340 pages

Publisher's Note

The articles in this volume, as in all others in the Variorum Collected Studies Series, have not been given a new, continuous pagination. In order to avoid confusion, and to facilitate their use where these same studies have been referred to elsewhere, the original pagination has been maintained wherever possible.

Each article has been given a Roman number in order of appearance, as listed in the Contents. This number is repeated on each page and is quoted in the index entries.

Preface

In this volume there are 12 articles on mathematics in Europe from the twelfth century to about 1500. It may be considered as a continuation of the volume, also in this series, "Essays in Early Mediaeval Mathematics" (2003), which treated the same subject from the eighth to the eleventh century. Of the twelve articles seven have been simply reprinted (items II, IV–VI, VIII, X, XI), two are reworked translations from German (items I, VII), one has been radically revised (item III), one has not yet appeared in print (item IX) and one is published here for the first time (item XII). The main themes of the volume are the reception of Euclid in the Latin West and the contributions of Johannes Regiomontanus (1436–1476), the most important Western mathematician of the fifteenth century.

The greatest influence on Western mathematics in the period under consideration were the translations from Arabic into Latin, made for the most part in the twelfth century. Accordingly, the first item in this book is a survey of the mathematical texts which were made available through the translations and which formed the basis for later developments.

The Hindu-Arabic numerals and calculation with them were among the things transmitted through the translations. In item II the earliest Western treatises on this subject are described. The core of this article is a study of the oldest known Arabic writing on the new method of calculation, which al-Khwārizmī wrote in the first half of the ninth century. The Arabic text of this work is lost. We know its content only because it was translated into Latin. This translation has apparently not survived in its original form, but a text that is evidently very close to it has been discovered by the author of this volume. It is analyzed here in item II.

In item III the history of Euclid's principal work, the *Elements*, in the Latin Middle Ages is considered in some detail. This text was translated three times from Arabic in the twelfth century. There were also translations of it from Greek in late antiquity (Boethius) and in the twelfth century. The translations gave rise to a large number of reworkings and commentaries written between the twelfth and fifteenth centuries. Some of these compilations incorporated material from contemporary and older mathematical writings. Research in the last few decades has radically altered our understanding of the medieval Euclid. Item III is an up-to-date account of our knowledge of it. Part 2 of this article comprises lists of manuscripts of the *Elements* in Arabic and Latin: all known manuscripts of the Arabic translations and all known manuscripts of the Latin texts, both translations from Greek or Arabic and reworkings or commentaries.

The double translation of the *Elements* – from Greek into Arabic and from Arabic into Latin – led in the Western Middle Ages to obscure passages in the production of the text. In item IV the most important of such obscure passages are presented. Discussions of passages from Euclid combined with Aristotelian ideas led to some of the most interesting scientific achievements of the Middle Ages (e.g. ideas about the continuum and about the mathematical relation of velocity, force and resistance). Descriptions of these developments are integrated into this article.

Items V–VIII are on the contributions of Johannes Regiomontanus to mathematics. Item V gives a survey of his early mathematical achievements. Already in these years, which he spent in Vienna (1449–1461), he had found some new results, not only in algebra and trigonometry, but also in number theory and the determination of the volumes of barrels, the so-called "art of gauging" (*Visierkunst*). Most of these results he wrote down in manuscripts which were until recently unknown or have not been sufficiently studied. Regiomontanus made great efforts to recover Greek mathematics in its purity, unobscured by the translations into and out of Arabic. As a good mathematician he was able to detect and correct mistakes and obscurities in the transmitted texts. In this task the friendship of Cardinal Bessarion was important: he invited Regiomontanus to come to Italy and allowed him free use of his important collection of Greek and Latin manuscripts. Here Regiomontanus developed his skill in Greek. His role in the transmission and transformation of Greek mathematics is the subject-matter of item VI. In item VII it is shown how Regiomontanus tried to improve the text of Euclid's *Elements* by means of methods resembling those of modern textual criticism. Item VIII shows that Regiomontanus also played an important role in the tradition of mathematical problems. He loved to set his correspondents mathematical problems and was able to solve even complicated number-theoretical, algebraic and geometrical problems. He was an important figure in transmitting mathematics based on Arabic sources from Italy into the lands north of the Alps.

Items IX–XI are case-studies for the knowledge and application of Euclid's *Elements* by Italian mathematicians of the thirteenth and fifteenth centuries. Item IX is about Leonardo Fibonacci, the most important mathematician of the early thirteenth century. In his *Liber abbaci*

(1202, reworked edition finished 1228) and in his other mathematical works he shows a surprisingly good understanding of the *Elements* for his time and also of the content of other important Greek and Arabic mathematical writings. Piero della Francesca (ca. 1410–1492) and Luca Pacioli (ca. 1445–1517) were important Renaissance mathematicians with strong connections to art and perspective. In items X and XI it is demonstrated that they, too, profited from the *Elements*; and it is shown which texts they knew and used.

The last contribution (item XII) treats the development of algebra in Germany in the second half of the fifteenth century and the beginning of the sixteenth. Like the *maestri d'abbaco* in Italy, the *Rechenmeister* in Germany made important contributions to the development of algebra. In particular, much of the modern symbolism (*inter alia*, the sign for the unknown and the signs + und –) was developed in Germany at this time. Again, Regiomontanus, though not a *Rechenmeister*, played an important role. Since this development is little known outside the German *Kulturgebiet*, and since new facts have come to light in the last few decades, it was thought appropriate to give here an account of the present state of knowledge on the matter.

It is a pleasure to thank Richard Lorch (Munich) for helping translate the German articles into English, Paul Kunitzsch for his constant help and encouragement, and finally Dr. John Smedley for including the book in his series and for his kindly advice.

MENSO FOLKERTS

Munich
June 2005

Acknowledgements

Grateful acknowledgement is made to the following for kindly permitting the reproduction of articles originally published by them: Cambridge University Press (II); Prof. Helge Kragh, Chief Editor of *Centaurus*, Aarhus (IV); Österreichische Akademie der Wissenschaften, Vienna (V); Brill Academic Publishers, Leiden (VI: exceptional permission); Franz Steiner Verlag, Stuttgart (VIII); Fondazione Piero della Francesca, Sansepolcro (X); Prof. Enrico Giusti, Florence, editor of the volume *Luca Pacioli e la matematica del rinascimento. Atti del convegno internazionale di studi Sansepolcro 13–16 aprile 1994*, Città di Castello 1998 (XI).

I

Arabic Mathematics in the West*

In this paper I should like to present a survey of the mathematical texts that were translated from Arabic (and Greek) into Latin in the twelfth and thirteenth centuries. In Spain the Muslims were gradually pushed back in the *reconquista*, beginning in the eleventh century. With the fall of Toledo (1085) the Christian world came into possession of an important centre of Arabic scholarship. The Iberian peninsula was subsequently invaded by Western scholars thirsty for knowledge. Under the patronage of bishops and lay rulers there arose a great activity in translating Arabic works into Latin – especially in Toledo, Barcelona, Segovia and Pamplona. Through these translations Greek scholarship was to a great extent reintroduced in the West. In addition, Arabic scholarly works, some of which contained material from India, were also translated. After the meagre intellectual fare of the early Middle Ages, this increase of available knowledge induced an enormous stirring of scholarly activity in the West. The newly founded universities, especially the faculties of arts, were greatly influenced by this mass of new material. From the point of view of the historian of science this translation activity may be considered as the most important factor in the development of science in the western Middle Ages. It led to a renaissance of learning that may be compared with the later, better known, Renaissance[1]. In the main the texts translated in the twelfth and thirteenth centuries were scientific and philosophical writings; philology, poetry and history were mostly left aside. Many of the translators are known by name: the most important were Adelard of Bath, Hermann of Carinthia, Robert of Chester and, above all, Gerard of Cremona (1114–1187).

The present account is confined to mathematical texts translated at this time. For practical reasons, I here limit the term "mathematical" to

*Revised English translation of "Arabische Mathematik im Abendland unter besonderer Berücksichtigung der Euklid-Tradition", in O. Engels, P. Schreiner (ed.), *Die Begegnung des Westens mit dem Osten. Kongreßakten des 4. Symposions des Mediävistenverbandes in Köln 1991 aus Anlaß des 1000. Todesjahres der Kaiserin Theophanu*, Sigmaringen 1993, pp. 319–331.

[1]For details see Haskins, 'Studies', – who incidently coined the term "Little Renaissance".

its narrow sense: I will treat only those topics which are today counted as mathematics. In particular this means that the mathematical part of astronomy is not be considered here, though it must be admitted that, historically considered, this choice is rather artificial[2]. Incidentally the term "Arabic mathematics" is to be understood as referring to the Arabic language and has no ethnic implications. Thus Arabic mathematics comprises both translations from the Greek and genuine Arabic writings – these being often greatly influenced by Greek and Indian works. We may note that no Indian mathematical works in Arabic translation were further translated into Latin – Indian methods and ideas were known through works originally written in Arabic. My limitation to the twelfth and thirteenth centuries excludes Arabic texts only known to the West much later, from the sixteenth century on.

Which works were translated in this time into Latin? I should like to divide the material into the following groups: arithmetic, algebra, geometry, geometry of the sphere and trigonometry, conic sections. A special section will be on Euclid.

1. Arithmetic

The treatise by al-Khwārizmī (first half ninth century) on arithmetic is the oldest known treatment in Arabic of calculation with Hindu-Arabic numerals. Of this text no Arabic manuscript is known. What we have are two manuscripts which appear to present revisions of a Latin translation made in the twelfth century. Until recently only one of these two manuscripts was known and this was not complete[3]. The second of these manuscripts, which transmits the whole Khwārizmī text, has been recently discovered and published[4]. It is not known who translated the arithmetic. The work treats, in the Indian fashion, the decimal place value system and the six basic operations (addition, subtraction, dou-

[2]We note that Fuat Sezgin makes the same practical distinction between mathematics and astronomy in his *Geschichte des arabischen Schrifttums* (Sezgin, 'GAS 5': mathematics; Sezgin, 'GAS 6': astronomy). There is not yet a similar encyclopedic survey of Latin writings, but one is being compiled by H. L. L. Busard and the present writer and it will come out shortly.

[3]Cambridge, UL, Ii. 6.5. Editions: Boncompagni, 'Trattati'; Vogel, 'Algorismus'; Allard, 'Calcul Indien'; Folkerts, 'Älteste lateinische Schrift', pp. 28–88 (MS **C**). Analysis of the text in the Cambridge manuscript by Juschkewitsch, 'Über ein Werk'.

[4]New York, The Hispanic Society of America, HC: 397/726, f. 17r–24v. Edition with German translation and commentary by Folkerts, 'Älteste lateinische Schrift' (MS **N**). This edition presents both the Latin texts (**N** and **C**) in facing columns. See also the English summary, reprinted in this volume as item II.

bling, halving, multiplication, division) and these are followed by calculations of fractions (including sexagesimal fractions) and the extraction of square roots[5].

Two influential Latin books were based upon the translation of al-Khwārizmī's arithmetic: the so-called *Liber Ysagogarum Alchorismi* and the *Liber Alchorismi.* It is not certain who wrote these two treatises, but in some manuscripts the *Liber Alchorismi* is ascribed to a "magister Johannes" or to "magister Johannes Yspalensis"[6]. Both these derivative texts were written in the twelfth century[7]. From the twelfth century on there are numerous Latin texts on arithmetic which were based on al-Khwārizmī's treatise and the two reworkings. These treatises were used on the universities to teach how numbers were written in this system and how they could be used for calculation. They were called "algorismus treatises", a word derived from the name al-Khwārizmī, though this derivation was soon forgotten[8]. As far as we know, no other arithmetic treatise was translated from Arabic.

2. Algebra

As with arithmetic, the Arabic tradition in algebra begins with al-Khwārizmī. His *al-Kitāb al-mukhtaṣar fī ḥisāb al-jabr wa-l-muqābala* contains both pure algebra and applications of algebra: a short chapter on mercantile calculations, a section on problems on mensuration and a long section on legacies. The purely algebraic section teaches the solution of linear and quadratic equations in geometric form; six types of equations are defined and solved[9].

The *Algebra* of al-Khwārizmī was translated twice in the twelfth century, on both occasions without the parts on mensuration and legacies[10], by Robert of Chester in 1145 in Segovia[11] and later by Gerard of Cre-

[5] For further details, see item II in this volume.

[6] See Allard, 'Calcul Indien', p. 62.

[7] The author of the *Liber Ysagogarum Alchorismi* could be Adelard of Bath or Petrus Alfonsi (whoever he was). According to Burnett, 'Algorismi vel helcep', it is possible that not Johannes Hispalensis, but Johannes Ocreatus was the author of the *Liber Alchorismi.*

[8] The modern term "algorism" (sometimes Grecised to "algorithm") is in turn derived from *algorismus.*

[9] A good analysis of the content is to be found in Juschkewitsch, 'Geschichte der Mathematik', pp. 204–214. See also Sezgin, 'GAS 5', pp. 228–238.

[10] A list of manuscript is given by Hughes, 'Medieval Latin Translations'.

[11] Edited by Karpinski, 'Contributions'; new edition by Hughes, 'Robert of Chester's Latin Translation'.

mona[12]; further there is a reworking, of which the authorship remains a matter of dispute[13].

The *Algebra* of al-Khwārizmī influenced the *Kitāb fī l-jabr wa-l-muqābala* written shortly afterwards by Abū Kāmil (ca. 850–930)[14]. It is written at a higher theoretical level than al-Khwārizmī's algebra. It consists of three parts: pure algebra, in which, again, linear and quadratic equations are treated; a section on the pentagon and decagon, in which the sides of the polygons are determined by algebraic methods; a section on inderminate quadratic equations and on problems of recreational mathematics. An Arabic manuscript of Abū Kāmil's Algebra has been published in facsimile[15]. Abū Kāmil's Text was translated into Latin at an unknown date and into Hebrew in the fifteenth century. Only one manuscript is known of the Latin text (Paris, BnF, lat. 7377A, f. 71v–97r.). This was edited in 1993 by R. Lorch and J. Sesiano[16].

Until recently the three parts, into which the treatise falls, were considered separate treatises. Of the first part there is an edition of the Hebrew translation, some articles on the content and Sesiano's edition of the Latin translation[17].

The Latin translation of the section on the pentagon and decagon was edited by R. Lorch, and there are earlier translations into Italian, German and English[18].

The third section shows that Abū Kāmil in solving indeterminate equations is not in the tradition of Diophantus[19]. The only manuscript

[12]Edited by Libri, 'Histoire', pp. 253–297; new edition by Hughes, 'Gerard of Cremona's Translation'.

[13]Edited by Kaunzner, 'Lateinische Algebra'. See also Kaunzner, 'Frühe lateinische Bearbeitung'. The author could be William de Lunis.

[14]For Abū Kāmil see Sezgin, 'GAS 5', pp. 277–281.

[15]Istanbul, 'Book of Algebra'.

[16]Lorch, 'Abū Kāmil'; Sesiano, 'Version latine'.

[17]Levey, 'Algebra'. German translation by Weinberg, 'Algebra'. The Latin translation is in Paris, BnF, lat. 7377A, f. 71v–93v. See Karpinski, 'Algebra' (with edition of parts of the Latin text); complete edition in Sesiano, 'Version latine', pp. 325–420. For the content see also Juschkewitsch, 'Geschichte der Mathematik', pp. 220–227.

[18]Edition of the Latin translation in Lorch, 'Abū Kāmil', pp. 234–252. The Latin text is in Paris, BnF, lat. 7377A, f. 93v–97r. For the content see Lorch, 'Abū Kāmil', pp. 216–222, and Suter, 'Abhandlung'. Suter translated the Italian version, made by Sacerdote from the Hebrew (Sacerdote, 'Trattato'), into German. English translation by Yadegari / Levey, 'On the pentagon and decagon'.

[19]See Sesiano, 'Méthodes', and Schub / Levey, 'Indeterminate Problems' (with English translation).

of the Latin translation breaks off after a short passage in the middle of the second problem[20].

The name of the translator and the date of the translation are still uncertain[21]. Abū Kāmil's Algebra had great influence on the algebra in the West, especially on Leonardo Fibonacci, as is shown in his *Liber abbaci* and his *Practica geometriae*.

Another important algebraic work, which was translated from Arabic into Latin in the twelfth century, is the *Liber augmenti et diminutionis*[22]. The authorship is still unclear: it could have been Abū Kāmil, but Abraham ibn Ezra (ca. 1090–1167) has also been suggested[23]. In this work it is shown how linear equations with one unknown or systems of linear equations with two unknowns may be solved with the help of the rule of double false position[24].

A work, which until recently has not appeared in the standard histories of mathematics, is the *Liber mahameleth*, a large collection of mathematical problems with an introduction on the mathematics needed to solve such problems. This treatise, which has come down to us in three manuscripts[25], was written by Johannes Hispalensis (fl. ca. 1135–1153) in Spain[26]. It is not a translation from the Arabic, but based on Arabic sources[27].

3. Geometry

Apart from Euclid's *Elements*, the most important geometric texts translated from the Arabic were the writings of Archimedes. In his fundamental *Archimedes in the Middle Ages* in five volumes Marshall Clagett has not only fully described the fate of Archimedes' own works in the West (with editions and English translations), but also followed every trace of

[20] Paris, BnF, lat. 7377A, f. 97r. This short Latin text has been edited by Sesiano, 'Version latine', pp. 448–451.

[21] See the different opinions of Lorch and Sesiano (see note 16). Lorch considers Gerard of Cremona, but sees too many inconsistencies. Sesiano dates the translation with the manuscript, written in the fourteenth century.

[22] Edited by Libri, 'Histoire', pp. 304–376.

[23] See Sezgin, 'GAS 5', pp. 396–398.

[24] See Juschkewitsch, 'Geschichte der Mathematik', pp. 214–216.

[25] Paris, BnF, lat. 7377A, f. 99r–203r; Padua, Biblioteca Capitolare, D.42, f. 1r–86v; Paris, BnF, lat. 15461, f. 26r–50r.

[26] See Sesiano, 'Survivance'.

[27] Jacques Sesiano has prepared an edition. In the meantime, a study of the text, with a complete edition and French translation, has been prepared by Anne-Marie Vlasschaert in a Ph.D. dissertation at the University of Louvain-la-Neuve.

Archimedean methods in other treatises of the Latin Middle Ages[28]. Here I can only make a short report on the most important results. Apart from the translation of almost the whole Archimedean corpus from the Greek in 1269 by William of Moerbeke, only two of Archimedes' works were known in the West before the fifteenth century: the *Mensura circuli* and an incomplete text of the *De sphaera et cylindro*. The *Mensura circuli* was translated in the twelfth century twice – first, apparently, by Plato of Tivoli and then by Gerard of Cremona. Gerard's version became widely known and it was several times reworked: of the first theorem, on the relation between the surface and the circumference of a circle, Clagett found no fewer than seven different reworkings[29]. Based on this theorem was a tract entitled *De quadratura circuli*, by an anonymous author writing about 1340. This author also used theorems from Archimedes' *Spirals*, which he had evidently known from Moerbeke's translation. In the same tradition are still further works on squaring the circle written in the fourteenth and fifteenth centuries. Of Archimedes' *De sphaera et cylindro* we have fragments of a translation, apparently by Gerard of Cremona[30].

Another important geometrical writing translated, also by Gerard of Cremona, from the Arabic is the *Verba filiorum* of the Banū Mūsā (ca. 850)[31]. This work is in the Archimedean tradition – it contains two propositions of the *Mensura circuli* and several theorems from *De sphaera et cylindro* –, but is critical of the approach of the ancients. *Inter alia*, the *Verba filiorum* presents Hero's formula of the area of a triangle, the determination of the two mean proportions between two given quantities and the trisection of the angle. The work was immensely influential in the West; it was used by Leonardo Fibonacci and Roger Bacon, among others.

Also translated by Gerard of Cremona was Euclid's *Liber divisionum*, which treats the problem of cutting a given figure with a straight line so that the resulting sections have a given form or so that their areas stand in a given ratio. Like the Greek original, this translation has been lost. But there are traces of it in several mathematical treatises, such

[28] Clagett, 'Archimedes', vol. 1–5. For translations from the Arabic volumes 1 and 5 are the most important.

[29] Clagett, 'Archimedes', vol. 1, p. 5.

[30] Clagett, 'Archimedes', vol. 1, Chapter Six.

[31] Edited by Clagett, 'Archimedes', vol. 1, chapter 4. See also Sezgin, 'GAS 5', pp. 246–252.

as Leonardo Fibonacci's *Practica geometriae* and Jordanus de Nemore's *Liber philotegni*[32].

There are a few other, shorter, texts on theoretical geometry translated by Gerard[33]. These strongly influenced some passages in later writings, especially in the *Liber de triangulis Iordani*, which is an anonymous redaction of Jordanus' *Liber philotegni*, though considered a genuine work by Jordanus de Nemore until Clagett's edition[34].

Of greater influence than these short texts are three tracts on mensuration, also translated by Gerard. In the manuscripts they are always found together and always together with the text *Liber augmenti et diminutionis*. The three texts belong to the *misāḥa* tradition, as the Arabs called the determination of areas. One of these texts, called in the manuscripts *Liber Saydi Abuothmi*, may have been written by Abū ʿUthmān Saʿīd (d. 979 in Toledo)[35]. Another of these texts is called *Liber Aderameti*; its author has not been identified[36]. Equally unknown is the identity of the Abū Bakr, the author of the third text, the *Liber mensurationum*[37]. This last work treats the application of algebra to geometrical problems. It is particularly remarkable that the methods used by the author for determining areas and volumes are procedures known from Babylonian mathematics.

Many of the problems in the *Liber mensurationum* are also to be found in the *Liber embadorum* (Book of measurements), which was translated by Plato of Tivoli in 1145[38]. The text from which it was taken was a book on practical geometry written by Abraham bar Ḥiyya (Savasorda) in the year 1116. In this work there are not only theorems on plane and solid figures – in particular the determination of areas and volumes –, but also a chord table and a section on the geometric solution of quadratic equations. In his text he also solved geometric problems algebraically.

---

[32]See Clagett, 'Archimedes', vol. 5, p. 581, note 1.

[33]They are to be found in Paris, BnF, lat. 9335, the most important single manuscript with mathematical texts translated by Gerard from the Arabic, and sometimes in other manuscripts. For details, see Clagett, 'Archimedes', vol. 5, pp. 595–603.

[34]The relevant fragments and passages of longer works are presented by Clagett, 'Archimedes', vol. 5, pp. 591–603.

[35]See Sezgin, 'GAS 5', p. 387. Edited by Busard, 'Vermessungstraktate', pp. 169–171.

[36]See Sezgin, 'GAS 5', pp. 394f. Edited by Busard, 'Vermessungstraktate', pp. 171–174.

[37]See Sezgin, 'GAS 5', pp. 389–391. Edited by Busard, 'Liber mensurationum'.

[38]Edited by Curtze, 'Liber embadorum'.

The work was used by Leonardo Fibonacci, among others, in his *Practica geometriae.*

4. Geometry of the sphere and trigonometry

On the geometry of the sphere the most important mathematical works of antiquity are the *Sphaerica* of Theodosius (fl. ca. 100 BC) and the *Sphaerica* of Menelaus (fl. 98 AD). Among Theodosius' sources are the more descriptive treatises of Euclid (*Phaenomena*) and Autolycus (*De sphaera quae movetur* and *De ortibus et occasibus*). Of these, Autolycus' *De sphaera quae movetur* was translated from the Arabic into Latin by Gerard of Cremona[39].

The medieval transmission of Theodosius' *Sphaerica* has been described in Lorch, 'Transmission'. We give here only a short summary: one of the two Arabic translations of the Theodosius work was translated into Latin by Gerard of Cremona; a second Latin text was compiled, pretty clearly by Campanus. Both Latin texts were well known in the West, as is clear from the number of manuscripts and the citations by later authors[40].

With Menelaus' *Sphaerica* the matter is not so clear: it is lost in Greek; at least two Arabic translations appear to have been made and there were several redactions; one of the latter, lost in Arabic, was translated into Latin, again by Gerard of Cremona[41].

In the Middle Ages the various arcs and angles of spherical astronomy were determined by applying either Menelaus' theorem or special theorems on spherical triangles. As the name implies, Menelaus' theorem appears in Menelaus' *Sphaerica* (Proposition III.1 in Gerard's version); but the chief instrument of transmission was Ptolemy's *Almagest* (2c. AD). This was translated three times: once from the Greek, possibly by

[39] Edited by Mogenet, 'Traduction latine'. For the transmission into Latin, see also Mogenet, 'Autolycus', especially pp. 181–184.

[40] Richard Lorch has prepared a critical edition of the Arabic together with the Latin translation by Gerard.

[41] For the transmission of Menelaus' *Sphaerica*, see Björnbo, 'Studien'; Krause, 'Sphärik', pp. 1–98; Lorch, 'Thābit ibn Qurra', pp. 327–342. The Latin translation by Gerard has not been edited. Björnbo started to make an edition, the results (in manuscript) are in his *Nachlaß* in the Royal Library in Copenhagen.

Hermann of Carinthia, and twice from the Arabic, by Gerard of Cremona[42] and (in commented form) by an unknown translator[43].

Menelaus' theorem was also treated in a separate treatise, usually called *De figura sectore*, by Thābit ibn Qurra (d. 901). This was translated three times into Latin: perhaps by Hermann of Carinthia, perhaps by Plato of Tivoli, and certainly by Gerard of Cremona[44]. Thābit's work was the subject of several comments in Latin. Together with a proof by Maslama (Maslama al-Majrīṭī, d. ca. 1007) of a case of Menelaus' theorem not treated by Ptolemy, it inspired Campanus' *De figura sectore*, a much-copied work that influenced, for instance, Richard of Wallingford in the fourteenth century[45].

Menelaus' theorem was expressed by means of compound ratios. Ptolemy's *Almagest* and Thābit's *De figura sectore* were consequently two of the works by which compound ratios were known in the West. Influential treatments of the combinatorial parts of Thābit's book are to be found in two short tracts ascribed to Jordanus de Nemore and Campanus[46].

Further material on Menelaus' theorem and compound ratios and related matters is to be found in Aḥmad b. Yūsuf b. Ibrāhīm Ibn al-Dāya al-Miṣrī (ca. 900; in Latin: Ametus filius Josephi), "Book on Ratio and Proportion" (*De proportione et proportionalitate*), translated by Gerard of Cremona[47]. Also by Aḥmad and also translated by Gerard is the short

---

[42]See Kunitzsch, 'Überlieferung'.

[43]The title of the unique manuscript of this text, Dresden, Sächsische Landesbibliothek Db 87, f. 1–71, is *Phylophonia Wuttoniensis Ebdelmessie*. The manuscript has only the first four books.

[44]The Gerard version was edited by Björnbo, 'Thabits Werk'. The other two translations are edited in Lorch, 'Thābit ibn Qurra', pp. 124–153. For the transmission of Thābit's *De figura sectore* see ibid., pp. 30–37.

[45]For an edition of the Campanus text see Lorch, 'Thābit ibn Qurra', pp. 426–442. For Richard of Wallingford see North, 'Richard of Wallingford', vol. 1, pp. 21–178; vol. 2, pp. 23–82; vol. 3, pp. 3–14. For derivative Latin material see Lorch, 'Thābit ibn Qurra', pp. 402–425.

[46]Both texts have been edited by Busard, 'De Proportionibus'. The *Incipits* are: *Proportio est rei ad rem determinata ...* (Jordanus); *Proportio est duarum quantitatum eiusdem generis ...* (Campanus).

[47]On Aḥmad see Sezgin, 'GAS 5', pp. 288–290. The text has been edited by Schrader, 'Ametus'. For the composition of ratios in this text, see Koelblen, 'Combinatoire'.

De arcubus similibus, on similar arcs, a discussion of some ideas in Euclid, *Elements*, book 3[48].

Theorems of spherical trigonometry, relating to spherical triangles, were developed in the East about 1000 AD by a group of mathematicians including Abū 'l-Wafā', Abū Naṣr Manṣūr and al-Bīrūnī. A collection of such theorems was included by Jābir ibn Aflaḥ (12c., Seville) in his *Iṣlāḥ al-majasṭī* ("Improvement of the *Almagest*"), a work translated by Gerard of Cremona. Jābir ("Geber" in Latin) also presented explicit rules for plane trigonometry implicit in Ptolemy's procedures in the *Almagest*. Thus "Geber" became the principal source of trigonometry in the West – though later works such as Levi ben Gerson's *De sinibus, chordis et arcubus* (as it is known in the anonymous Latin translation) no doubt also made their contribution. Both plane and spherical trigonometry were systematized by Johannes Regiomontanus in his *De triangulis omnimodis libri V*, written mostly in the years 1462–1464.

There are several tables of sines and *umbrae* (tangents / cotangents) and also *canones* (instructions for use of tables) of obviously Arabic provenance. These tables and *Canones* are often associated with the name of al-Zarqāllu ("Canones Azarchelis")[49].

5. Conic sections

The most important Greek work on conic sections, the *Conica* of Apollonius, was translated into Arabic in the ninth century[50]. Only a fragment (the beginning) is known of the translation into Latin by Gerard of Cremona. It is included in the introduction to the Latin translation of Ibn al-Haytham's *De speculis comburentibus*[51]. John of Palermo, who lived for some time at the court of Emperor Frederick II, seems to have translated a short treatise on the hyperbola from Arabic into Latin[52]. Other traces of the doctrine of the conic sections are to be found in translations of the optical work of Alhazen (Ibn al-Haytham), in the derivative *Perspectiva* of Witelo, and in other works on optics. All texts of this sort can be found in Clagett's fundamental work[53].

[48] Edition of the Arabic and Latin texts by Busard / van Koningsveld, 'Liber de arcubus similibus'.

[49] See Curtze, 'Urkunden', who edited some of the texts.

[50] See Sezgin, 'GAS 5', pp. 136–143.

[51] Edited by Heiberg, 'Apollonii ...', pp. LXXV–LXXVIII, and by Clagett, 'Archimedes', vol. 4, p. 4, note 5.

[52] See Sezgin, 'GAS 5', p. 398. For an edition of the Latin text, see Clagett, 'Tract on the Hyperbola', and Clagett, 'Archimedes', vol. 4, pp. 33–61.

[53] Clagett, 'Archimedes', vol. 4.

6. Euclid's "Elements" and his "Data"

The most important mathematical text in Greek antiquity and in the Middle Ages was Euclid's *Elements*. The transmission of this work into Latin and the subsequent compilations and commentaries have been investigated intensively in the last half century, especially by M. Clagett and H. L. L. Busard. Only a few remarks are given here, since the subject is treated in detail elsewhere in this volume[54].

The *Elements* was translated three times from Arabic in the twelfth century and once, anonymously, from Greek. The translators from Arabic were Adelard of Bath, Hermann of Carinthia and Gerard of Cremona. More influential than these translations were the compilation of Robert of Chester (about 1140) and the elaborate text by Campanus (shortly before 1259), which took over most of the definitions and enunciations from Robert and added some material from other texts.

Besides the *Elements* itself, three commentaries on it were also translated from Arabic in the twelfth century: that of al-Nayrīzī to books 1–10, which was translated by Gerard of Cremona[55], and those of Muḥammad ibn ʿAbd al-Bāqī (?) and Pappus on book 10 alone. The commentary of Muḥammad ibn ʿAbd al-Bāqī appears to be lost in Arabic[56]. Pappus' commentary is lost in Greek. It has been preserved in Arabic, but only the beginning of the Latin translation is known[57].

Another of Euclid's mathematical writings, the *Data*, became known in the West through a translation from the Arabic. In a way, the *Data* can be considered as an extension of the *Elements*; it treats questions of the theory of ratios, geometry and algebra, which, as in Book II of the *Elements*, appears in geometrical form. The translation of the *Data* from Arabic was made by Gerard of Cremona – it appears in his list of his writings made by his pupils –, but seems to be lost. But we do have a translation of the *Data* – from Greek. It was made in Sicily shortly before 1160[58].

---

[54] *Euclid in Medieval Europe* (item III).

[55] There are three known complete manuscripts. Editions by Curtze, 'Anaritii commentarii', pp. 1–252, and by Tummers, 'Albertus', pp. 121–190 (only book 1). See Sezgin, 'GAS 5', p. 284.

[56] Edited from only one manuscript by Curtze, 'Anaritii commentarii', pp. 252–384; critical edition by Busard, 'Commentary on Book X'. The authorship is not clear: Sezgin suggests Abū Yūsuf Yaʿqūb b. Muḥammad al-Rāzī as possibly the author (Sezgin, 'GAS 5', pp. 388f.).

[57] Edited by Junge, 'Fragment'. See Sezgin, 'GAS 5', p. 175.

[58] Edited by Ito, 'Medieval Latin Translation'.

7. Final remarks

Through these translations from Arabic into Latin the West came to know the most important works of the Greek mathematicians, though some authors were not translated, notably Pappus, Diophantus and Hero. In addition, mathematics from India and from the Arabic world also became known, especially treatises on calculation with the new Hindu-Arabic numerals and on algebra. That this new knowledge brought an unprecedented impetus to mathematics and science in the West, is well known. In the following centuries this new knowledge was digested and formed the basis for further developments. This is not the place to go into details: here I can only make some general remarks.

The translation of the arithmetic of al-Khwārizmī and derivative texts led ultimately to the transfer of calculation from the medieval abacus to paper and written symbols developed from the numerals of the Indians; in this tradition are the much-copied *algorismus* treatises of the thirteenth century, by Johannes de Sacrobosco and Alexander de Villa Dei, and also the vernacular writings of the *maestri d'abbaco* in Italy and the *Rechenmeister* in South Germany. Arabic algebra made its influence in the West both direct through translations and indirect, especially through Leonardo Fibonacci's *Liber abbaci*, a summary of the mathematics that included material which Leonardo acquired in his journeys in the Orient. The geometric methods of the Greeks, above all the method of exhaustion, exerted a strong influence of the mathematics of the high and late Middle Ages; all this is described in magnificent detail in Marshall Clagett's five-volume *Archimedes in the Middle Ages.* Especially important were the translations of Euclid's *Elements.* They not only induced reworkings and commentaries, but also encouraged the development of various branches of mathematics and also of non-mathematical disciplines. I may cite here not only the theory of ratios, which on the one hand became important for mathematical astronomy, and on the other hand led to the quantitave treatment of physical problems, such as the relation between motive force, resistance and velocity, that did not disagree with the accepted doctrine of Aristotle. Another example are questions of continuity which led to discussions on the continuum which had a surprising similarity to the set-theoretical paradoxes of the nineteenth century. A third example is the theory of the "latitude of forms" of which the best-known result is the so-called "Merton theorem", i.e. that the distance traversed by a uniformly accelerated body is equal to the distance it would have traversed if it had had the constant speed at the mid-point of the time interval. The translations of mathematical works from Arabic into Latin were not only responsible for the considerable rise in the level of Western mathematics in the High and Late Middle Ages, but they may also be considered to have been the basis of the mathematics of the seventeenth century and to have a greater influence than Renaissance mathematics.

Bibliography

A. Allard, *Muḥammad ibn Mūsā al-Khwārizmī. Le Calcul Indien (Algorismus). Histoire des textes, édition critique, traduction et commentaire des plus anciennes versions latines remaniées du XIIe siècle*, Paris / Namur 1992.

A. A. Björnbo, "Studien über Menelaos' Sphärik. Beiträge zur Geschichte der Sphärik und Trigonometrie der Griechen", in *Abhandlungen zur Geschichte der mathematischen Wissenschaften* 14, Leipzig 1902, pp. 1–154.

A. Björnbo, *Thabits Werk über den Transversalensatz (liber de figura sectore). Mit Bemerkungen von H. Suter. Herausgegeben und ergänzt durch Untersuchungen über die Entwicklung der muslimischen sphärischen Trigonometrie von Dr. H. Bürger und Dr. K. Kohl*, Erlangen 1924 (= Abhandlungen zur Geschichte der Naturwissenschaften und der Medizin, Heft 7).

B. Boncompagni, *Trattati d'aritmetica. I. Algoritmi de numero Indorum*, Rome 1857.

Ch. Burnett, "*Algorismi vel helcep decentior est diligentia*: the Arithmetic of Adelard of Bath and his Circle", in M. Folkerts (ed.), *Mathematische Probleme im Mittelalter. Der lateinische und arabische Sprachbereich*, Wiesbaden 1996, pp. 221–331.

H. L. L. Busard, "L'Algèbre au Moyen Age. Le 'Liber mensurationum' d'Abû Bekr", *Journal des Savants* (1968), 65–124.

H. L. L. Busard, "Die Vermessungstraktate *Liber Saydi Abuothmi* und *Liber Aderameti*", *Janus* 56 (1969), 161–174.

H. L. L. Busard, "Die Traktate *De Proportionibus* von Jordanus Nemorarius und Campanus", *Centaurus* 15 (1971), 193–227.

H. L. L. Busard / P. S. van Koningsveld, "Der *Liber de arcubus similibus* des Ahmed ibn Jusuf", *Annals of Science* 30 (1973), 381–406.

H. L. L. Busard (ed.), "A Latin Translation of an Arabic Commentary on Book X of Euclid's *Elements*", *Mediaeval Studies* 59 (1997), 19–110.

M. Clagett, "A Medieval Latin Translation of a Short Arabic Tract on the Hyperbola", *Osiris* 11 (1954), 359–385.

M. Clagett, *Archimedes in the Middle Ages*, vol. 1, Madison 1964; vols. 2–5, Philadelphia 1976–1984.

M. Curtze (ed.), *Anaritii in decem libros priores Elementorum Euclidis commentarii ex interpretatione Gherardi Cremonensis in codice Cracoviensi*

569 servata, Leipzig 1899 (= I. L. Heiberg, H. Menge (ed.), *Euclidis Opera Omnia, Supplementum*).

M. Curtze, "Urkunden zur Geschichte der Trigonometrie im christlichen Mittelalter", *Bibliotheca Mathematica*, 3. Folge, 1 (1900), 321–416.

M. Curtze, "Der 'Liber embadorum' des Savasorda in der Übersetzung des Plato von Tivoli", in *Urkunden zur Geschichte der Mathematik im Mittelalter und der Renaissance, 1. Teil*, Leipzig 1902, pp. 1–183.

M. Folkerts, *Euclid in Medieval Europe*, this volume, item III.

M. Folkerts, "Die älteste lateinische Schrift über das indische Rechnen nach al-Ḫwārizmī. Edition, Übersetzung und Kommentar", *Bayerische Akademie der Wissenschaften, Philosophisch-historische Klasse. Abhandlungen. Neue Folge*, Heft 113, Munich 1997.

Ch. H. Haskins, *Studies in the History of Mediaeval Science*, 2nd ed., Cambridge (Mass.) 1927.

J. L. Heiberg, *Apollonii Pergaei quae Graece exstant*, vol. 2, Leipzig 1893.

B. B. Hughes, "The Medieval Latin Translations of al-Khwarizmi's *al-Jabr*", *Manuscripta* 26 (1982), 31–37.

B. B. Hughes (ed.), "Gerard of Cremona's Translation of al-Khwārizmī's *al-Jabr*: A Critical Edition", *Mediaeval Studies* 48 (1986), 211–263.

B. B. Hughes (ed.), *Robert of Chester's Latin Translation of al-Khwārizmī's* al-Jabr. *A New Critical Edition*, Stuttgart 1989.

Istanbul, Beyazıt Library, Kara Mustafa Paşa 379: *The Book of Algebra. Kitāb al-Jabr wa l-muqābala by Abū Kāmil Shujāʿ ibn Aslam (Second half ninth century A.D.)*, Frankfurt 1986 (*Publications of the Institute for the History of Arabic-Islamic Science*, Series C, Volume 24).

S. Ito, *The Medieval Latin Translation of the Data of Euclid*, Tokyo and Boston / Basel / Stuttgart 1980.

G. Junge, "Das Fragment der lateinischen Übersetzung des Pappus-Kommentars zum 10. Buche Euklids. (Nr. 7377A, Fol. 68–70 der Bibliothèque Nationale zu Paris.)", *Quellen und Studien zur Geschichte der Mathematik, Astronomie und Physik*, Abt. B, 3, Studien (1934), 1–17.

A. P. Juschkewitsch, *Geschichte der Mathematik im Mittelalter*, Leipzig 1964.

A. P. Juschkewitsch, "Über ein Werk des Abū 'Abdallah Muḥammad ibn Mūsā al-Ḫuwārizmī al Mağusī zur Arithmetik der Inder", in I. Strube, H. Wussing (ed.), *Beiheft zur Schriftenreihe für Geschichte der Naturwissenschaften, Technik und Medizin. Beiträge zur Geschichte der Naturwissenschaften, Technik und Medizin, hg. zum 60. Geburtstag Gerhard Harigs*, Leipzig 1964, pp. 21–63.

L. Ch. Karpinski, "The algebra of Abu Kamil Shojaʿ ben Aslam", *Bibliotheca Mathematica*, 3. Folge, 12 (1911–12), 40–55.

L. Ch. Karpinski, *Contributions to the History of Science. Part I: Robert of Chester's Latin Translation of the Algebra of al-Khowarizmi*, Ann Arbor 1930.

W. Kaunzner, "Über eine frühe lateinische Bearbeitung der Algebra al-Khwārizmīs in MS Lyell 52 der Bodleian Library Oxford", *Archive for History of Exact Sciences* 32 (1985), 1–16.

W. Kaunzner, "Die lateinische Algebra in MS Lyell 52 der Bodleian Library, Oxford, früher MS Admont 612", in G. Hamann (ed.), *Aufsätze zur Geschichte der Naturwissenschaften und Geographie*, Vienna 1986, pp. 47–89.

S. Koelblen, "Une pratique de la composition des raisons dans un exercice de combinatoire", *Revue d'Histoire des Sciences* 47 (1994), 209–247.

M. Krause, *Die Sphärik von Menelaos aus Alexandrien in der Verbesserung von Abū Naṣr Manṣūr b. 'Alī b. 'Irāq*, Berlin 1936.

P. Kunitzsch, "Die Überlieferung des Almagest, griechisch–arabisch–lateinisch", in R. Schulz, M. Görg (ed.), *Lingua restituta orientalis. Festgabe für Julius Assfalg*, Wiesbaden 1990, pp. 203–210.

M. Levey, *The Algebra of Abū Kāmil. Kitāb fī al-jābr wa'l-muqābala in a Commentary by Mordecai Finzi. Hebrew Text, Translation, and Commentary with Special Reference to the Arabic Text*, Madison / Milwaukee / London 1966.

G. Libri, *Histoire des sciences mathématiques en Italie, depuis la renaissance des lettres jusqu'à la fin du dix-septième siècle*, vol. 1, Paris 1838.

R. Lorch, "Abū Kāmil on the Pentagon and Decagon", in M. Folkerts, J. P. Hogendijk (ed.), *Vestigia Mathematica. Studies in medieval and early modern mathematics in honour of H. L. L. Busard*, Amsterdam / Atlanta 1993, pp. 215–252.

R. Lorch, "The Transmission of Theodosius' *Sphaerica*", in M. Folkerts (ed.), *Mathematische Probleme im Mittelalter. Der lateinische und arabische Sprachbereich*, Wiesbaden 1996, pp. 159–183.

R. Lorch, *Thābit ibn Qurra, On the Sector-Figure and Related Texts. Edited with Translation and Commentary*, Frankfurt/Main 2001.

J. Mogenet, "La traduction latine par Gérard de Crémone du Traité de la Sphère en Mouvement d'Autolycus", *Archives internationales d'histoire des sciences* 5 (1948), 139–164.

J. Mogenet, *Autolycus de Pitane. Histoire du texte suivie de l'édition critique des "Traités de la sphère en mouvement et des levers et couchers"*, Louvain 1950.

J. North, *Richard of Wallingford. An edition of his writings with introductions, English translation and commentary*, 3 vols., Oxford 1976.

G. Sacerdote, "Il trattato del pentagono e del decagono di Abū Kāmil Shogiaʿ ben Aslam ben Mohammad", in *Festschrift zum achtzigsten Geburtstage Moritz Steinschneider's*, Leipzig 1896, pp. 169–194.

W. R. Schrader O.P., *The* Epistola de proportione et proportionalitate *of Ametus filius Iosephi*, Diss. Univ. of Wisconsin 1961.

P. Schub / M. Levey, "Indeterminate Problems of Abū Kāmil (850–930)", *Atti della Accademia Nazionale dei Lincei*, Memorie, Classe di scienze fisiche, matematiche e naturali, Serie VIII, 10 (1970), 23–96.

J. Sesiano, "Les Méthodes d'analyse indéterminée chez Abū Kāmil", *Centaurus* 21 (1977), 89–105.

J. Sesiano, "Survivance médiévale en Hispanie d'un problème né en Mésopotamie", *Centaurus* 30 (1987), 18–61.

J. Sesiano, "La version latine médiévale de l'Algèbre d'Abū Kāmil", in M. Folkerts, J. P. Hogendijk (ed.), *Vestigia Mathematica. Studies in medieval and early modern mathematics in honour of H. L. L. Busard*, Amsterdam / Atlanta 1993, pp. 315–452.

F. Sezgin, *Geschichte des arabischen Schrifttums. Band V: Mathematik*, Leiden 1974.

F. Sezgin, *Geschichte des arabischen Schrifttums. Band VI: Astronomie*, Leiden 1978.

H. Suter, "Die Abhandlung des Abū Kāmil Shoǵāʿ b. Aslam 'über das Fünfeck und Zehneck'", *Bibliotheca Mathematica*, 3. Folge, 10 (1909–1910), 15–42.

P. M. J. E. Tummers, *Albertus (Magnus)' commentaar op Euclides' Elementen der geometrie*, Nijmegen 1984.

K. Vogel, *Mohammed ibn Musa Alchwarizmi's Algorismus. Das früheste Lehrbuch zum Rechnen mit indischen Ziffern. Nach der einzigen (lateinischen) Handschrift (Cambridge Un. Lib. Ms. Ii.6.5.) in Faksimile mit Transkription und Kommentar herausgegeben*, Aalen 1963.

J. Weinberg, *Die Algebra des Abū Kāmil Šoǵāʿ ben Aslam*, Munich 1935.

M. Yadegari / M. Levey, "Abū Kāmil's 'On the Pentagon and Decagon'", *Japanese Studies in the History of Science*, Supplement 2 (1971), 1–54.

II

Early Texts on Hindu-Arabic Calculation

Argument

This article describes how the decimal place value system was transmitted from India via the Arabs to the West up to the end of the fifteenth century. The arithmetical work of al-Khwārizmī's, ca. 825, is the oldest Arabic work on Indian arithmetic of which we have detailed knowledge. There is no known Arabic manuscript of this work; our knowledge of it is based on an early reworking of a Latin translation. Until some years ago, only one fragmentary manuscript of this twelfth-century reworking was known (Cambridge, UL, Ii.6.5). Another manuscript that transmits the complete text (New York, Hispanic Society of America, HC 397/726) has made possible a more exact study of al-Khwārizmī's work. This article gives an outline of this manuscript's contents and discusses some characteristics of its presentation.

1. Indian, Arabic, and Western Arithmetic

The decimal system, with nine digits and the zero, comes from India.[1] At first there were individual symbols for the units and the tens, so that the system differed little from the alphabetical systems used by many peoples, for example the Greeks. The decisive step came when somebody used the same digits for the tens, hundreds, etc. as for the units. So began the "place-value" system for base 10. The oldest example of numbers so written are those of the Gurjara Inscription (595 A.D.), but it is generally assumed that the system goes back at least as far as the first centuries A.D. From the eighth century inscriptions and manuscripts with these numbers become relatively numerous. The oldest known occurrence of the zero in India is in 870, in the Gwalior inscriptions, and was in circular form. It was called *śūnya* (empty). In other documents the zero is written as a dot.[2] There are examples of the "Indian" decimal place-value system from seventh-century Cambodia, Sumatra, and Java.

The decimal place-value system was also known to mathematicians outside India at least as early as the seventh century: the Syrian scholar Severus Sebokht mentioned

[1] For the history of the decimal system see, for instance, Bose, Sen and Subbarayappa 1971, 175–179, Ifrah 1986, Kap. 31, Smith 1925, 65–72, Juschkewitsch 1964a, 102–109, Tropfke 1980, 43–45, Lemay 1982.

[2] A useful survey of the development and the forms of the Indian numerals in India and the Arabic-speaking countries is to be found in Tropfke 1980, 66.

the Indian numerals about 662, saying that their way of writing numbers with 9 digits is beyond all praise. This shows that the new numbers were known in the Near East before the translations into Arabic.

Originally the Indian numbers were used in written documents simply to record numbers; arithmetic operations were made on a calculation board. It appears that at first numbers in each denomination were represented by cowrie shells (or, presumably, by other counters), but numerals were later written with a sharpened stick on dust spread on the board, intermediate results being rubbed out.

In Indian mathematical writings arithmetic was presented in a series of rules. Fractions, particularly unit fractions, were treated in great detail. In contrast to Greek writers, the Indians used the sexagesimal system very little. Procedures to find square and cube roots, either exactly or approximately, were known, and it was recognized that the procedures could be made more exact by multiplying the number whose root was required by a power of ten. There were procedures to check arithmetic operations, for instance by the casting out of nines.[3]

In their many conquests beginning in the seventh century the Arabs took over the forms of the numbers found in the conquered territories – in the eastern Mediterranean, the Greek alphabetic system. Indian numerals, with the zero, were known to educated Muslims at least by the year 760. The oldest known Arabic example is "260" at the end of a papyrus document and may be the Hijra date, i.e. 873–74. Here the zero is represented by a dot.[4] In the following centuries the Indian numerals became known throughout the Arabian empire from the eastern provinces to Spain. In the *Fihrist* (ca. 987) there is a section devoted to the writing of Indian numerals: the nine numerals are mentioned, with an example of each; the tens, hundreds, and thousands being distinguished by one, two, and three dots placed beneath the digits.[5] The form of the numerals was not uniform; in particular, two conventions were developed, one in the East (including Egypt) and the other in North Africa and Spain. They differ in the way some of the numerals were written.[6] Perhaps the oldest known eastern example of all nine digits and the zero is a manuscript which might have been written in Iran about 970,[7] and there are many other examples from later times.[8] In contrast, the western Arabic numerals are known only from late specimens.[9]

[3] The fundamental work on Indian mathematics is still Datta and Singh 1935–1938. A good summary is Juschkewitsch 1964a, 89–174.

[4] See Smith and Karpinski 1911, 56. It is not certain that the number in question is a date, as generally assumed. Gandz (1931, 394) erroneously refers to these numerals as "earliest Arabic documents containing the *ghubâr* numerals."

[5] Flügel 1871-72, 1, 18-19; Dodge 1970, 34-35. See also Karpinski 1910-1911.

[6] The various forms are given, for example, in Tropfke 1980, 66.

[7] Paris, BN, ar. 2457, f. 81r-86r; some scholars believe that the manuscript was not written in the tenth but in the twelfth century. The forms of the numerals in this manuscript are reproduced in line 3 of the table in Smith and Karpinski 1911, 69.

[8] Irani (1955-1956, 4-10, col. A) reproduces the numerals of a manuscript written in 1082. In the same article there are drawn other numerals from more recent eastern Arabic manuscripts.

[9] The oldest examples given by Labarta and Barceló 1988 are from the fourteenth century.

In the year 773 an embassy from India went to the court of Caliph al-Manṣūr in Baghdad. Among the officials was a scholar with astronomical and mathematical expertise. He introduced Indian astronomy to the caliph's court, where an Indian astronomical treatise was translated into Arabic.[10] Under Caliph al-Ma'mūn (reigned 813–833) translation activity reached its climax; many mathematical and astronomical works were translated from Greek. Subsequently the foreign sciences were assimilated into Arabic-Islamic culture.

Among the first who wrote mathematical works in Arabic was Muḥammad ibn Mūsā al-Khwārizmī (ca. 780–ca. 850; fl. 830). He worked in the very center of a group of mathematicians and astronomers (and others) who were active in the "House of Wisdom" (*Bayt al-ḥikma*) in the time of Caliph al-Ma'mūn. He wrote on arithmetic, algebra, astronomy, geography, and the calendar. His treatises on arithmetic and algebra marked the beginning of Arabic writings on these subjects.[11]

Al-Khwārizmī's work on algebra must have been written before his *Arithmetic*, where it is cited. It is still not known which sources he used for the *Algebra*: we may assume that the questions and results of the Babylonians and Indians were available to him, but we do not know in what form. There are indications that there was also an oral tradition, and it may well be that he drew upon this as well as upon written works. He provided geometrical proofs for algebraic rules for solving quadratic equations, but these proofs were not taken over from Greek writings, for example, Euclid's *Elements* and *Data*.

Al-Khwārizmī's *Algebra* is divided into three parts:

- An algebraic section, followed by a short chapter on mercantile calculations illustrating the rule of three after the Indian fashion.

- A relatively short chapter on mensuration by using algebra.

- A section on inheritance problems according to Islamic law. It comprises more than half the book.[12]

There are several Arabic manuscripts of the *Algebra*.[13] The first section was translated twice into Latin in the twelfth century, once by Robert of Chester (Segovia, 1145) and again by Gerard of Cremona (Toledo, second half of the twelfth century).[14]

The arithmetical treatise of al-Khwārizmī is the oldest known Arabic work in which the Indian decimal place-value system and operations with it are described.

[10] See Juschkewitsch 1964a, 179, and Saidan 1978, 6.

[11] For the life and work of al-Khwārizmī, see particularly Toomer 1973 and Sezgin 1974, 228-241.

[12] Analyzed in Gandz 1938.

[13] See Sezgin 1974, 240, 401.

[14] Edited by B. Hughes in Hughes 1986 and Hughes 1989.

Unfortunately, there is no known Arabic manuscript. All that we have are Latin texts based upon a lost translation. Details will be given in part 2.

The oldest extant Arabic texts on Indian arithmetic are from the middle of the tenth century or later, and so more than a hundred years later than al-Khwārizmī himself. There are modern editions of the following works:

- Abū 'l-Ḥasan Aḥmad ibn Ibrāhīm al-Uqlīdisī's *al-Fuṣūl fī 'l-ḥisāb al-hindī* (Chapters on Indian calculation). The treatise was written in Damascus in 952/953 and is the oldest extant work in Arabic on the subject.[15]

- Abū 'l-Ḥasan Kūshyār ibn Labbān's *Fī uṣūl ḥisāb al-hind* (On the elements of calculation of the Indians) from the second half of the tenth century.[16]

- Abū Manṣūr ʿAbd al-Qāhir ibn Ṭāhir al-Baghdādī's *Kitāb al-takmila fī 'l-ḥisāb* (Book on the completion of calculation). Al-Baghdādī died in Isfarāyīn in 1037.[17]

Besides these, there are numerous Arabic texts on calculation. In 1978 A. S. Saidan compared al-Uqlīdisī's text with all Arabic arithmetic texts of which details were known at the time and also with Indian treatises available to him (Saidan 1978).

The Indian numerals were known in the West before the translation of Arabic works into Latin. The oldest known examples are in two Latin manuscripts which were written in Spain towards the end of the tenth century: the *Codex Vigilanus* (976) and the *Codex Emilianus* (992), neither of them a mathematical text. The numerals from 1 to 9 appear here in the western Arabic form.[18]

This form of the numerals became well-known in monastic schools, because they were used on the calculation board (Latin: *abacus*). From the end of the tenth to the twelfth centuries there are numerous descriptions of calculating with the abacus, the oldest by Gerbert (ca. 940–1003). He probably became acquainted with these numerals during his visit to the Spanish Marches shortly before 970. Gerbert used the numerals, however, not for written calculations, but to mark counters for use on the abacus.[19] From the eleventh century there were special names for the numerals so used; some of these were Latinized Arabic words.[20] With the disappearance of Gerbert's abacus in the twelfth century, these names disappeared, too.

The west-Arabic forms of the numerals 1 to 9 are also known by the term *ghubar* numerals (*ghubār* = dust). These forms appeared in the West in treatises on the abacus as well as, somewhat later, in accounts of the new method of written calculation (see

[15] Edited in Saidan 1973; English translation and commentary in Saidan 1978.

[16] Edited in facsimile with English translation in Levey and Petruck 1965.

[17] Edited in Saidan 1985.

[18] See Smith and Karpinski 1911, 138. The relevant parts of both manuscripts are reproduced in van der Waerden and Folkerts 1976, 54, 55.

[19] The standard edition of Gerbert's mathematical writings is still Bubnov 1899. The most important recent studies on Gerbert's arithmetic are Lindgren 1976 and Bergmann 1985.

[20] For the names and their meanings see Ruska 1917, 82-92.

below). There are considerable differences in the manuscripts in the forms of some of the numerals. These may be explained from their being drawn on the top of counters, which can be rotated.[21]

A new phase begins with the translations from Arabic into Latin in the twelfth century. Among the numerous works translated was al-Khwārizmī's *Arithmetic.* Unfortunately, no manuscript of the original Latin text is known. All that we have is a reworking, which begins with *Dixit Algorizmi* and is here referred to as DA (for details see below). The content of DA to a large extent prescribed the content of subsequent treatises on arithmetic in the West: the forms of the numerals and the principles of the place-value system; operations with integers (addition, subtraction, halving, doubling, multiplication, division); operations with fractions (multiplication, division, notation, addition, subtraction, doubling, halving); extracting square roots of integers and fractions. Not treated in DA are: arithmetic and geometric progressions, extraction of cube roots.

Of greater influence than DA were three redactions, all made in the twelfth century: the *Liber Ysagogarum Alchorismi* (LY), the *Liber Alchorismi* (LA) and the *Liber pulveris* (LP).[22]

The LY is a work apparently intended to cover the whole quadrivium. Of the five books, the first three are on arithmetic, the fourth on geometry and the fifth on astronomy. In the title of book 4, music is mentioned as well as geometry. The manuscripts of this text differ markedly. Three versions may be distinguished: LY I, II and III. LY I is usually assumed to be written before 1143, but this is by no means certain. LY II is an expanded version of LY I. In one of its manuscripts a scribe has written *a Magistro A compositus.* It has been supposed that *Magister A* is Adelard of Bath. LY III, written in France, is another redaction of the original *Liber Ysagogarum.*

LA, the second of the texts derived from al-Khwārizmī, shows Spanish influence. The author is named as *magister Iohannes.* It is not clear whether this "Iohannes" was John of Seville (Johannes Hispalensis), who translated several astronomical and astrological texts from Arabic between 1133 and 1142.

LP is partly identical to LA, but also contains original parts. Apparently it is older than LA. We refer here to the common parts as LA/LP.

The three redactions – LY, LA, LP – treat by and large the same material in the same order as in DA, but the presentation is often clearer and more detailed, especially in LA/LP. The authors' efforts to present the material systematically and completely are easily recognizable.

Another treatise on Indian arithmetic was written in the twelfth century, though it was not widely known. One of the two manuscripts gives the author as H. Ocreatus and the title as *Helceph Sarracenicum.*[23] The word "Helceph" comes from the Arabic *al-ḥisāb* (calculation). The author was acquainted with the new methods, but used

[21] See Beaujouan 1948.
[22] For details see the edition in Allard 1992 and the summary in Allard 1991.
[23] Edited with English translation in Burnett 1996, 261-297.

Roman and not Arabic numerals and also terms from the old Western tradition – for example, from Boethius and the abacus treatises in the Gerbert tradition.

On the translation of al-Khwārizmī's *Arithmetic* and the three redactions (and not on the "Ocreatus" text) depend a series of texts on calculating with Hindu-Arabic numerals. They were called *algorismus* treatises, a name that goes back to "al-Khwārizmī," though the historical connection was not understood, and the name was applied to any work of this kind. They were at first written in Latin and described the representation of numbers by Hindu-Arabic numerals and calculation with them (including sexagesimal and common fractions). Between the twelfth and fifteenth centuries a large number of such texts were written. In contrast to the earlier texts they generally treated either integers (*Algorismus de integris*) or fractions (*Algorismus de minuciis*). Among the writings on integers there are at least four texts that may with great probability be ascribed to the twelfth century: two *algorismi* in manuscripts now in London; the "Salem algorismus" and a text from Frankenthal. Though they differ in length, the content of these four treatises is essentially the same.[24]

In the first half of the thirteenth century two well-known authors wrote *algorismus* treatises which were of great influence in the following centuries: Alexander de Villa Dei (d. ca. 1240), *Carmen de algorismo*,[25] and Johannes de Sacrobosco (d. 1236?), *Algorismus vulgaris*.[26] Alexander de Villa Dei's work, written in hexameters, is somewhat earlier than Sacrobosco's. Both treatises are limited to calculations with integers. The success of these textbooks undoubtedly lies in the short and easy expression of each step of the various calculations and the lack of examples. These texts treat the subjects usually occurring in this kind of literature.

There appeared in the thirteenth century a number of treatises of a different character. They seem to have been written partly by Jordanus de Nemore, partly by members of his circle. Besides the usual arithmetical operations, they also treat more general mathematical and philosophical questions.[27]

The best-known textbook on fractions, the *Algorismus de minutiis*, was written by Johannes de Lineriis about 1320.[28] It shows little originality and is characterized by having no proofs, only rules and examples. In general style it is similar to the *Algorismus vulgaris* of Johannes de Sacrobosco and was a text that could easily be studied.

The arithmetical works of Johannes de Sacrobosco, Alexander de Villa Dei and Johannes de Lineriis were soon adopted by universities and became standard textbooks in the faculty of the *artes liberales*. With the help of these books arithmetic with Hindu-Arabic numerals was taught both in universities and grammar-schools.

[24] Edited in Karpinski 1921, Cantor 1865, Allard 1978.
[25] Edited in Steele 1922, 72-80.
[26] Last edited by Pedersen 1983, 174-201.
[27] The various treatises are listed in Thomson 1976, 107-112, with reference to the manuscripts and the secondary literature.
[28] Edited in Busard 1968.

The *maestri d'abbaco* in Italy, the *Rechenmeister* in Germany and similar teachers in other countries, who taught in the vernacular, were ultimately dependent upon these standard Latin textbooks.

2. Al-Khwārizmī's Treatise on Arithmetic

Al-Khwārizmī's arithmetical work is the oldest Arabic work on Indian arithmetic of which we have detailed knowledge. Knowledge of arithmetic procedures undoubtedly existed among the Arabic-speaking peoples before his time; and it is also possible that written texts on the subject earlier than his were available. But unfortunately we have no certain knowledge of either the procedures or the texts. The treatise was written ca. 825, about 130 years earlier than al-Uqlīdisī's *Fuṣūl*, the oldest extant work in Arabic on the subject.

Since no Arabic manuscript of al-Khwārizmī's work is known, the work must be studied in the Latin texts which were based on the translation, now lost, which was made in the twelfth century. As mentioned above, there was an early reworking of the translation which is referred to here as DA. Of this text there are two manuscripts. One of them, Cambridge, University Library, Ii. 6.5, (hereafter called C), has been known for some time. It has been edited three times,[29] published in facsimile three times[30] and translated into Russian,[31] English[32] and French.[33] Manuscript C is apparently of the thirteenth century and was at one time in the monastery of Bury St. Edmunds. It contains on ff. 104r–111v the text of DA, but only a little more than half of the text is present, because the last folios have been lost.

Until recently the existence of a second manuscript has remained unnoticed. In fact, there is a copy of DA in New York, Hispanic Society of America, HC 397/726, ff. 17r–24v (hereafter called N).[34] It was written in Spain, probably in the thirteenth century. Up to the second half of the eighteenth century it belonged to the library of the Universidad Complutense at Alcalá de Henares. Unlike the copy in C, the manuscript is complete. In my book the text of manuscript N is edited for the first time, with parallel German translation.[35] In order to allow a comparison between N and C, the text of C is also edited parallel to N; here a few mistakes in the earlier editions have been corrected. Manuscript N is reproduced in facsimile at the end of the book (Folkerts 1997).

[29] Boncompagni 1857, Vogel 1963, Allard 1992, 1-22.
[30] Vogel 1963, Juschkewitsch 1964b, Juschkewitsch 1983, 185-202.
[31] Kopelevič and Juškevič 1983.
[32] Crossley and Henry 1990.
[33] Allard 1992, 1-22.
[34] For a description of the content, see Folkerts 1997, 19-23. The description in Faulhaber 1983 is misleading for this item in the codex.
[35] Folkerts 1997, 28-107.

With the discovery of manuscript N a more exact study of al-Khwārizmī's work has become possible. Formerly only guesses could be made about a large part of the text on the basis of the redactions LY, LA, and LP. Before discussing this further, we should consider how MS N and C are related to each other. In the part of the text common to the two manuscripts the wording is sometimes different. In the first chapters the differences are small, but they become greater later on. There are striking differences in chapters 3 and 10.1; it is almost as if the two manuscripts followed two different translations in some passages. But on the whole the differences lie in small points of style and do not affect the content. In most cases the wording in N is preferable to that in C: it is more detailed and more accurate. C contains several mistakes not present in N; some of the superior N readings could not reasonably be supposed to be the result of scribal correction. Examples of such mistakes are the error in C in the rule for the resultant place in the multiplication of sexagesimal fractions (chapter 8.1) and the confusion of the terms *minuta* and *secunda* (chapters 10.3 and 10.4). In N there are chapter headings and diagrams with numerical examples, none of which appear in C. But this does not mean that N has the original wording of the translation. In fact, both N and C appear to represent different stages in the reworking of the Latin translation. Nonetheless, N seems to be nearer the original translation than C. This fits well with the observation that N was written in Spain, whereas C was written in England.

Despite all differences, therefore, it seems likely that N and C both represent a common source (DA). Sometimes this common source can be reconstructed. But the differences between the two manuscripts prevent this in many places. The date of DA can be estimated from the dates of the two manuscripts (thirteenth century) and of the translation from the Arabic (twelfth century; see below).

A natural question is: what relation has DA to the original translation? Any answer must be based on the readings of N and C and on equivalent passages in the twelfth-century derivative texts LY and LA/LP.

In the parts common to N and C there are sections in which essentially western ideas are present and in which there are no Arabisms. For example, the discussion of unity (chapter 1.3) is probably to be ascribed to a later redactor. In some passages the text has clearly been "latinized." On the other hand there are many constructions and technical terms which unambiguously show their Arabic origin, no doubt al-Khwārizmī's text. If we compare the complete text DA – as given in N – with the derivative texts LA/LP and LY, we find that the content is essentially the same. It is almost certain that this was the material of the translation. Style and terminology indicate that N – despite some reworking and additions – to a large extent reflects the wording of the Latin translation.

With the help of N we are for the first time in a position to reconstruct the part lost in C – the part dealing with division and root extraction – and thus to establish the whole of al-Khwārizmī's original text. In the part missing in C the derivative texts in LY and LA/LP are unreliable, being greatly expanded and otherwise reworked.

As for the identity of the translator, there are only the slightest indications. Nonetheless, several translators have been suggested by historians of mathematics, always without reason: Adelard of Bath, Hermann of Carinthia, and Gerard of Cremona. No names are mentioned in MS N and C, and any suggestion must be the result of deduction or speculation. Hermann of Carinthia is certainly very unlikely: whereas the al-Khwārizmī text, in the parts that we may be fairly sure represent the words of the author, indicates a fairly literal rendering from the Arabic – with Arabisms and generally stiff Latinity –, Hermann's translating style is notoriously free. The case for Gerard is distinctly better, but by no means free of objection. For the derivative text LY has been dated to before 1143, and this has generally been considered as too early for Gerard, whose activity as translator in Toledo falls rather into the second half of the twelfth century. But the date 1143 is by no means certain. Further, the wording of al-Khwārizmī's quotation of his own *Algebra* in the text of DA (chapter 4.1) is remarkably similar to Gerard's wording in his translation of the *Algebra*. If Gerard was not the translator, then the names of Adelard of Bath and Robert of Chester come to mind. It certainly appears that the translation was made in Spain. For the scribe of MS N (or of an "ancestor" of N) on numerous occasions mistook the Roman numeral *XL* for *X* – a strong indication that its exemplar was written in Visigothic script. Furthermore, the form of the Hindu-Arabic numerals in N is similar to the form of those in the *Codex Vigilanus* and *Codex Emilianus*. But this does not bring us to any conclusions as to the translator.

The contents of al-Khwārizmī's *Arithmetic* may be reconstructed from MSS N and C. It is clear that this work was not a translation of some Indian text, but a compilation of material of Indian origin. The main topics are: form of the numerals and how numbers were written in the decimal place-value system (chapter 1); the six fundamental operations with integers (addition, subtraction, halving, doubling, multiplying, dividing; chapters 2–7); operations with fractions (multiplication, division, notation, addition, subtraction, doubling, halving; chapters 8–12); extraction of square roots of integers and of fractions (chapters 13–19).

The title of the Arabic original cannot be deduced from the Latin reworkings. In MS N the heading runs: *Incipit arismethica Alchoarismi*; in C there is no heading. From Arabic commentaries of al-Khwārizmī's *Arithmetic* mentioned in the *Fihrist* it seems that the title was: *Kitāb fī 'l-jamʿ wa 'l-tafrīq* (Book on gathering [*collectio* = addition] and separation [*dispersio* = subtraction]).

3. Contents of DA[36]

Chapter 1: How numbers are written in the decimal place-value system

< 1.1 > Praise to God. The treatise intends to explain the Indian method of calculation (*numerus*, Arabic *ḥisāb*) with nine digits (*littere*) with which any number

[36] The numbering of the chapters and sections (in angle brackets) corresponds with that of Folkerts 1997.

can be easily represented. With these digits arithmetic, i.e. multiplication, division, addition and subtraction, may be simplified. The praise to God and the introductory passages are typically Arabic.

<1.2> The Indian procedure with numerals is to be demonstrated so that the writing of numbers and operations with them become easier.

<1.3> How the forms of the 9 digits are to be described. Here and in the following text in C only the digits 1, 2, 3, 5 and 0 appear, but in N all nine and the zero are present. They are clearly in the West-Arabic tradition. The forms of 1, 2 and 3 in N are very similar to those in the Codex Vigilanus; digits 4 to 9 are reminiscent of the form on the abacus in the Gerbert tradition. A somewhat changed form of the 4 is found in the 16-place number in section 1.8. The zero is a circle, a little smaller than the other numerals.

There are differences in the forms of the fifth, sixth, seventh, and eighth symbols. If this passage was in al-Khwārizmī's Arabic text, it referred to differences in form within the East-Arabic tradition, to which al-Baghdādī also refers in the case of 2, 3, and 8. It is also possible that the "fifth" to the "eighth" symbols mean the numerals 5, 4, 3, and 2, since in the Latin text the nine numerals are written in order from right to left (copying the Arabic). It all depends on what the translator understood and in which direction he counted.

There follows a section on the nature of "one." At the beginning of his extended discussion the author says that he has shown in his own *Algebra* (*liber algebr et almucabalah*) that any number is composed of "ones." This and two later citations (in chapters 4.1 and 15), which indeed find confirmation in the transmitted text, show that the *Algebra* was written before the *Arithmetic*. This discussion of "one" corresponds to the Greek concept of number, according to which "one" is not a number, but the origin of numbers. The remainder of this chapter has the appearance of an addition by some Latin redactor: "another work on arithmetic" (which might be Boethius' *Arithmetic*) is mentioned, and chapter 1.3 ends with the words: *Redeamus ad librum*.

<1.4> The decimal structure of the integers: by the doubling, tripling . . . , multiplying by 9, of "one" the units are obtained; by the doubling etc. of "ten" the tens are obtained; and so on *ad infinitum*. All numbers that may be expressed in words can be expressed by means of the nine digits. The "zero" is not mentioned here. We note that the author mentions units, tens, hundreds, and thousands and that in Arabic there are words for these, and not for higher, categories; to express other – composed or larger – numbers, these words must be used in combination or derivation.

<1.5> After these preparatory remarks the author explains the place-value principle which the Indians had worked out (*operati sunt*). First the central ideas of decimal place and of the zero are introduced. The term for (decimal) place is *differentia*, which is consistently used in N; in C both *differentia* and *mansio* (remaining, stay, dwelling) are used. It is not clear of which Arabic word *differentia* is a translation; in early Arabic texts either *martaba* (pl. *marātib*) or *rutba* is found. On the other hand, *mansio* is certainly a translation of the Arabic *manzila* (pl. *manāzil*).

It is shown in detail what units, tens, hundreds, thousands and ten-thousands are. According to the place-value system, "one," when put in the next higher place, has the value "ten"; a digit meaning "ten" in the tens means "one" in the units; and so on. This principle is differently expressed in N and C, Arabic practice of writing from right to left being mirrored in N and Latin direction of writing being accommodated in C. The counting of the places begins on the right.

When a "one" is in the second place, i.e. means "ten," a sign in the first place (from the right) is needed to show that it means ten: a sign is put in the first place to show that it is empty. It is in the form of a small *o*. In the entire text the zero is represented by the word *circulus* (Arabic: *dāʾira*). The Indians called the zero *śūnya* (i.e. empty); the Arabic translation of this word is *aṣ-ṣifr*, from which the Latins had the term *cifra* from the twelfth century on. The word *cifra* is also found in a reworking of a translation of al-Khwārizmī's astronomical tables. Whether al-Khwārizmī used the term *aṣ-ṣifr* in either the *Arithmetic* or in his tables, we cannot tell from the Latin texts available to us.

<1.6> There follows a detailed description of how hundreds and thousands are written: two or three (resp.) zeros must be written before (i.e. on the right of) the digits, if these places are not occupied by digits (i.e. 1 to 9). In other cases the appropriate digit must be put in its appropriate place (*in ea differentia que sibi debetur*). The examples 11 and 325 are described in detail.

<1.7> In this section the carrying procedure is described in detail: what is done when in any place ten or more units are assembled (*collecti fuerint*). Various special cases are treated: when just 10 are assembled, the digit in that place must be substituted by a zero; when a "large" number (typically 9) is in some place and 1 is carried from the previous, then two carrying procedures must be done. The description is, by modern standards, lengthy and repetitive.

<1.8> The chapter on how numbers are written ends with a section on writing large numbers and pronouncing them in words. The author gives the names for the places from the first to the thirteenth (i.e. 10^0 to 10^{12}). Since the Arabs have words for numbers only up to a thousand, the higher numbers must be expressed in terms of these; e.g. 10^{12} is expressed as "a thousandfold thousand thousand thousand." From a thousand million on adverbs such as *milies* or phrases such as *milia . . . tribus vicibus* are used.

1,180,703,051,492,863 is the example used for writing large numbers. This number is written incorrectly in N and does not appear in C at all. To facilitate reading, the number is divided into groups of three digits. It is stated that numbers are to be pronounced while being written. This practice is mentioned in numerous texts extant in Arabic.

Chapter 2: Addition and subtraction of integers

<2.1> The general procedure of addition is described. On what object calculations are performed is not stated. But we know from an aside in the chapter

on multiplication – and from extant Arabic texts – that a board (*tabula*, *takht*) was normally used. It was covered with dust, sand, wax, or white powder. On this the numbers were written and in the process of calculation digits were replaced by others or totally erased (Arabic: *maḥā*). In addition, subtraction and multiplication one of the numbers in the calculation was written above the other and its digits were successively replaced by those of the resultant number: at the end of the operation the number written at the top of the calculation was completely replaced by the result. In the addition of two numbers one was placed above the other so that the various decimal places corresponded. The digits in the individual places are to be added and carrying is performed, if necessary. The description is short, and no example is given. In this chapter it is not explicitly stated if one begins at the right or the left side.

<2.2> In subtraction the minuend is written above the subtrahend so that the various places match. In each place the lower number is taken from the upper. The case in which the upper digit is smaller than the lower and the case in which the upper digit is zero are treated in great detail: in such cases a unit must be taken from the next higher place in the upper number and changed into ten units of the current place. The operation begins with the highest place.

<2.3> To explain the general rules of subtraction, N announces two examples and C three. The first example is: 6422 − 3211 = 3211. The individual steps of the calculation are expressed only in words.

<2.4> The second example of subtraction is given to illustrate what happens when "nothing" (*nichil*, C) or "little" (*parum*, N) is left in the (minuend's) places. In C in the example 1144 − 144 only the problem is given, expressed in words. No scheme (i.e. a direct representation of the numbers as they are to be written) is given, the subtraction is not described, and the result is not given. In N there is a slightly different example, 1444 – 144. The scheme is given, the individual steps are described in great detail, and the result is correct. Here and often elsewhere in N the number 40 is incorrectly given as "ten" (*X*). This is a strong indication that in a chain of copies of which N was the end-product there was a manuscript written in Visigothic script: only in this script does the number *XL* resemble the *X* (it is an *X* with a hook on the top right corner).

Chapter 3: Doubling and halving

Halving is treated before doubling. One begins with the lowest decimal place. If it is odd, one halves the even number (i.e. the greatest even number less than the odd number in question) and the half of the remaining 1 is written as 30 (i.e. $\frac{30}{60}$, which is a half in sexagesimal notation) underneath the half of the even number. In all other places an even digit is likewise halved, but in the case of an odd digit the odd 1 is taken to the next lower place as a 5. Doubling is described in only one sentence: one starts with the highest place. If more than 10 result in any one place, ten is carried as a 1 to the next higher place. Examples are not given for either operation.

Chapter 4: Multiplication of integers

< 4.1 > The author starts with a self-quotation from the *Algebra* for a general meaning of multiplication: when a number is multiplied by another, one of them is "repeated" (*duplicari*) according to the number of units in the other. To multiply with Indian numerals the calculator must know the multiplication table (1×1 till 9×9) by heart. In some Arabic and Latin texts a multiplication table is given, but there is none in DA.

Rules follow for writing down the numbers to be multiplied: they are written in two lines, the highest place of the upper (which is the multiplier) being placed above the lowest place of the lower (the multiplicand).

< 4.2 > To explain the individual steps of multiplication, the example 2326×214 is presented. The 4 of the multiplicand 214 is written under the first 2 of the multiplier 2326. The digit which is in the highest place of the multiplier multiplies successively the individual digits of the multiplicand, beginning with its highest place, the result being put directly above the multiplicand (the first digit of the multiplier is replaced in this process by the last digit of the partial result). After the highest place of the multiplier has been dealt with, the multiplicand is shifted one place to the right, and the procedure is repeated with the next digit of the multiplier (3), the partial results being added. Then the multiplicand is shifted again and the process repeated, and so on, until the digits of the multiplier are exhausted.

< 4.3 > In this section special cases are treated, in particular cases in multiplication by zero. In the text there is the equivalent of $0 \times n = n \times 0 = 0$. Then the remainder of the above example is given. At the end the result is 497764.

Chapter 5: Checking multiplication and doubling by casting out nines

The casting out of nines is a check to see whether the multiplication or doubling has been correctly carried out. It is applied in DA, but is not given a special name (only *probatio*). The excess of nines is not found by adding the digits, but by finding the remainder on division by 9. The Latin expression *divide numerum per novem novem* is evidently taken from the Arabic. In the case of multiplication the procedure is applied to the multiplier and multiplicand; the two excesses are multiplied; if the result is the same as the excess of nines of the product, the product is considered correct. The author did not know that the test is a necessary, but not sufficient condition.

It is noteworthy that in DA the casting out of nines is mentioned only in this place. For division in chapters 10.4 and 10.5 another kind of test is given: the dividend is compared with the product of divisor and result. For the other operations no test is given.

Chapter 6: Division of integers. General remarks

< 6 > Chapter 6 treats the general procedures of division of integers; examples follow in chapter 7. The highest place of the divisor is placed under the highest place of the dividend if the digit of the divisor is smaller; if not, the divisor is shifted one

place to the right. The two digits are divided, the result (without fractions) being placed over the dividend, vertically aligned with the last place of the divisor. By this result the successive digits of the divisor are multiplied (starting from the left), the partial products being subtracted from the corresponding digits of the dividend. The dividend has now been changed in some of its digits, and the new number takes the part of the dividend in the next part of the process. This begins by shifting the divisor one place to the right, and so on. In this way the digits of the quotient gradually appear over the (altered) dividend. The result of the division is described in DA what "will be that which corresponds to one" (*ipsum erit, quod debetur uni*). The idea evidently comes from proportions: the division $a \div b = q$ is equivalent to the proportion $a:b = q:1$. If there is a remainder, it is a part which comes from the dividend (*erunt partes unius ex numero quem dividis*). It must be less than the divisor, for otherwise a mistake has been made.

Chapter 7: Division of integers. Examples

<7.1> After saying that division is the inverse of multiplication, the author gives an example which is described in great detail: $46468 \div 324 = 143\frac{136}{324}$. Only the first and last configurations are illustrated by schemes, but all steps are given in words. This procedure can be followed, but with some difficulty.

<7.2> In a second example a number with several places, some of them zeros, is divided by a single-digit number: $1800 \div 9$. The author says that the work of division is shortened when there are zeros in the dividend or produced by the division procedure.

Chapter 8: Multiplication of fractions

<8.1> In chapters 8 to 12 calculation with fractions and mixed numbers (integers mixed with fractions) is treated: first multiplication (chapters 8–9); then division (chapter 10); only then notation, addition, subtraction, halving and doubling (chapter 11); and finally multiplication and division of common fractions (chapter 12). In DA, calculation with common fractions and with sexagesimal fractions are not properly distinguished, but the attempt is made to show that calculation with all kinds of fractions follows the same rules.

There are innumerable names for the fractions, e.g. "half," "third." Some unit fractions and sexagesimal fractions are mentioned, but the author does not know a general word for unit fraction or for sexagesimal fraction. Sexagesimal fractions were used by the Indians. In sexagesimal subdivision there are "places" (*differentie*) as in the decimal representation of integers: integers or "degrees" are in the first place, "minutes" in the second, "seconds" in the third, etc. until the ninth and tenth places. The introductory sentences in DA on fractions make clear that common fractions, particularly unit fractions, were known in the daily life of the Arabs as in other cultures, whereas sexagesimal fractions belonged to another, "scientific," world.

After the general remarks about fractions it is explained how the place of the result of a multiplication of two sexagesimal fractions can be determined. When an integer and a fraction is multiplied, then the kind (*genus*) of fraction is not changed. For instance: $2° \times 2' = 4'$; $3° \times 6''' = 18'''$. When fractions are multiplied, their places are "collected" (*colligis utrasque differentias*), i.e. "added," and the sum is the place of the product. The author applies this principle to multiplying minutes by minutes, seconds by seconds, etc. Two examples are given: $6' \times 7' = 42''$ and $7'' \times 9' = 63''' = 1''3'''$.

<8.2> This section gives an example of multiplying mixed numbers (integers mixed with common fractions). The common fractions are converted into sexagesimal fractions, and the calculation is given as above. The example is: $1\frac{1}{2} \times 1\frac{1}{2}$. The individual steps are: $1\frac{1}{2} \times 1\frac{1}{2} = 90' \times 90' = 8100'' = 135' = 2°15' = 2\frac{1}{4}$. The fractions in each number are written as sexagesimal fractions; then the integral part is expressed in the same unit and added to the fractional part; then they are multiplied decimally. The result is a sexagesimal fraction whose place is determined by the rules of chapter 8.1. It is successively divided by 60 each time reducing the rank of the sexagesimal fraction. Finally the sexagesimal fraction is reconverted to a common fraction. Thus common fractions are here multiplied only after conversion to sexagesimal fractions. This demonstrates the dominating position of sexagesimal fractions over common fractions. Since every factor is expressed in terms of the lowest denomination present in the number after conversion into sexagesimals, only integers must be multiplied, and for this the rules have already been given.

Chapter 9: Multiplication of fractions when they comprise minutes and seconds

In this chapter an example is given to explain how relatively complicated (sexagesimal) fractions, comprising minutes and seconds, are multiplied. The procedure has already been explained in the previous chapter: one converts each of the two numbers to be multiplied into units of the lowest denomination that it contains, the two resulting integers are multiplied, and the product is reconverted to sexagesimal-fraction form by successive divisions by 60. The example is: $2°45' \times 3°10'0'' = 8°43'52''30'''$.

At the end there is a remark that there is a shorter method. Presumably the direct multiplication of the constituent sexagesimal places, the application of the rule determining the place of the result and the addition of the partial results is meant. This procedure would require a multiplication table of 1×1 up to 59×59. Such tables exist in extant Arabic texts, but not in DA.

Chapter 10: Division of fractions of any kind: general remarks; examples of divison of sexagesimal fractions; check

<10.1> Chapter 10 treats the division of fractions. Since no distinction is made between sexagesimal fractions and common fractions, it could be assumed that the

following procedures are valid for all fractions. What actually follows are rules for division of sexagesimal fractions (but see chapter 10.2); the division of common fractions is left until chapter 12.

First the general rule is given for dividing fractions or mixed numbers (i.e. numbers mixed of integers and fractions): each of the numbers is reduced to units of the lowest denomination present in either number; the division between the resulting integers is made; the result is in *gradus*, i.e. whole numbers, because the numbers that were divided are of the same rank (*unius generis*).

<10.2> To explain the rule, three examples are given. The first is: $\frac{15}{3} \div \frac{6}{3} = 2\frac{1}{2}$. It has been supposed that the original text ran 15 thirds divided by 6 thirds. This would maintain a sexagesimal treatment in this part of the text.

<10.3> The second example illustrates the case in which the dividend is smaller than the divisor: $10'' \div 5' = 10'' \div 300'' = 600''' \div 300'' = 2'$.

<10.4> To verify the division, a check can be performed: the quotient is multiplied by the divisor, and the result should be the dividend. Two examples of this are given: for the divisions $50 \div 10 = 5$ and $10'' \div 5' = 2'$.

<10.5> The third example is: $10' \div 5'''$. The procedure is: $10' \div 5''' = 36000''' \div 5''' = 7200°$. The check is $7200° \times 5''' = 36000''' = 600'' = 10'$.

Chapter 11: How numbers made up of integers and sexagesimal fractions are written; arithmetical operations with them: addition, subtraction, halving and doubling

<11.1> The chapter begins with the way of writing (*constituere*) mixed numbers: the integers and their sexagesimal parts are written one under the other. If there is no entry in a sexagesimal place, two zeros are written. An example (in modern notation) is: $12°30'45''0'''50''''$.

<11.2> In this section addition is treated, but the text is short and unclear. It is not stated how the two numbers are to be placed on the board, and a word for "addition" is not explicitly mentioned. It is mentioned, however, that when more than 60 units are accumulated in one place, 60 of them are carried as 1 to the next place.

<11.3> When one sexagesimal fraction is subtracted from another, the procedure begins with the highest place. From each place of the minuend the corresponding place of the subtrahend is subtracted. When some place of the minuend is less than the corresponding place of the subtrahend, 1 is borrowed from the next-highest place and becomes 60 in the place of the subtrahend, so that the subtraction can be performed. No example is given; and because of the clumsy verbal expression, the text is hard to understand.

<11.4> In doubling sexagesimals the rule is to begin with the highest place. When more than 60 is thereby accumulated in any place, the excess over 60 is written down, and 1 is added to the next-highest place. Halving sexagesimals, by contrast, begins with the lowest place. There are no examples. We note that doubling is here treated before halving, though in chapter 3 it was the other way round.

Chapter 12: Multiplication and division of common fractions

< 12.1 > Until now the only fractions with which calculations were made were sexagesimal fractions. Now multiplication and division with common fractions are explained. The procedure is analogous to the multiplication with sexagesimal fractions: the fractions are reduced to units of the "lowest denomination" (*ex genere inferioris differentie*). Then these units are multiplied. Finally the result is divided out to obtain integers.

< 12.2 > To explain the procedure, two examples are given. The first is: $\frac{3}{7} \times \frac{4}{9}$. According to DA, the seventh and the ninth should be considered "as minutes" and their product "as seconds." To find out whether there are integers and, if so, how many, the result of the multiplication must be divided by the product (in this case: seventh times ninths). The author gives the result of this example as $\frac{12}{63}$, i.e. "12 parts of 63 parts of one." In N, but not in C, there is a long explanation of the individual steps in the example. The procedure involves several unnecessary operations.

< 12.3 > The second example is $3\frac{1}{2} \times 8\frac{3}{11}$. In contrast to the previous problem, the presentation is clear. The two mixed numbers are written vertically one beside the other, so that the integers are written at the top, then the numerators and then the denominators. The integers and the fractions are converted into units of the corresponding denominator. Now the numerators are multiplied together, the denominators are multiplied together, and one product is divided by the other. Thus: $7 \times 91 = 637$; $2 \times 11 = 22 \Rightarrow 637 \div 22 = 28\frac{21}{22}$.

In the middle of the calculation the text in C breaks off, so that the subsequent text until the end of the whole work is transmitted only in N.

< 12.4 > To divide common fractions the rule is: convert both fractions to fractions of the same sort (i.e. with the same denominator) and divide "one by the other" (i.e. the new numerators). The result is an integer. The procedure is explained by the example $20\frac{2}{13} \div 3\frac{1}{3}$. Since neither of the terms "numerator" and "denominator" are mentioned, the presentation is not easy.

Chapter 13: General statements about roots

< 13.1 > With chapter 13 the last part of the book begins. It deals exclusively with the extraction of roots. Chapter 13 treats the general properties of (square) roots (*radix*; Arabic *jidhr*).

Without defining the term "root," the author states that, when two numbers – be they integers or fractions – are equal, their product "has a root," i.e. is a square number. This simple result is justified by observing that in this case the multiplication of one number by another is equal to the multiplication of a number by itself. Then comes the theorem: if a and b are square numbers, then so is $a \times b$. There are no examples in this section.

< 13.2 > The next theorem runs: for any numbers a and b, when $a \div b$ is a square number, then $a \times b$ is also a square number. The example with $a = 18$ and $b = 8$ is given. Then $18 \div 8 = 2\frac{1}{4} = (1\frac{1}{2})^2$, a square, and $18 \times 8 = 144 = 12^2$, also a square. It is

stated further that $b \div a$ is also a square number, for $8 \div 18 = \frac{4}{9} = (\frac{2}{3})^2$. Another example, with $a = 8$ and $b = 2$ is mentioned, but not given in detail.

< 13.3 > In this section the author treats a question of some importance for the extraction of roots: from which decimal places can roots be extracted? Expressed in modern terms, he reaches the result that 10^n has a square root if, and only if, n is even. The principles of the Indian decimal place system are first repeated (cf. chapter 1). Then it is asked whether the units, tens, hundreds and thousands have roots or not: one has a root, i.e. 1; 10 has no root; 100 has a root, i.e. 10; 1000 has no root, because it is equal to 100×10, i.e. a square multiplied by a non-square. Finally the author states that 1, 10, 100 . . . are alternately squares and non-squares, i.e. the places with odd rank have square roots – and those with even rank have not.

< 13.4 > Next comes the question: from which fractions can roots be extracted? It is shown that sexagesimal fractions of odd rank (i.e. minutes, thirds, fifths . . .) have no square roots, but those of even rank (i.e. seconds, fourths, sixths . . .) do have. The author treats common fractions analogously: he considers a class of "first fractions" (*prime fractiones*) which have no root (i.e. fractions whose denominators have no root); "second fractions" (*secunde fractiones*, i.e. first fractions multiplied by themselves) do have roots; "third fractions" (fractions of the *tertia differentia*, i.e. first fractions multiplied by themselves twice) have no roots, etc. We should note here that the author tries again to force common fractions into the rules valid for sexagesimal fractions.

Chapter 14: General procedure for extracting square roots

This chapter gives in detail the procedure for finding the square root of any integer; an example is not given here, but only in chapter 16. First the decimal place-value system is yet again described, this time with a reference to the beginning of the work. Then comes a repetition of the statement of the previous chapter, that the decimal places of odd rank have roots and those of even rank have none. The procedure begins by counting the decimal places in the number whose root is required: if the highest place is of odd rank, then you "work in it" (*operaberis in ea*); if it is of even rank, then you "do not work in it," but go to the next place (to the right). This means, of course, that the places are counted from right to left and marked in pairs, as is done today. The greatest square number less than or equal to the number (or pair of numbers) in the left-most group must now be found and subtracted from it; the difference is written in the place of the group being operated on. The root of the square number is the highest place of the square root sought. Then this partial result is doubled and written one place to the right. The greatest single digit, the next place of the result, is now sought such that when it is written alongside the doubled partial result to form one number, and when this number is multiplied by the digit, the product is less than the number formed by the first two groups on the left side. This procedure is continued, until there is no remainder or the remainder is less than the double of the result so far calculated. If there is no remainder, the result is the exact square root (*radix integra*). If there is a remainder, an approximate root is obtained by adding to the partial result a fraction comprising the remainder divided by twice the partial result. This corresponds to the modern formula $\sqrt{N}=\sqrt{a^2+r}\approx a+\frac{r}{2a}$. The very abstract presentation is made easier by the example in chapter 16.

Chapter 15: Theoretical questions about the extraction of square roots; the reason why the partial results need to be doubled

This chapter complements the previous chapter and is apparently intended to justify the doubling of the partial results in the procedure for extracting roots. The author refers to "another book by himself" (*patefeci in alio libro meo*: evidently his *Algebra*) for the general rule that when two numbers both with *nodi* and *unitates* (i.e. numbers with two decimal places) are to be multiplied, four separate multiplications must be made. In modern notation: $(10a+b)\times(10c+d)=10a\times 10c+10bc+10ad+bd$. A similar rule for squaring is evidently implied, so that the reason for doubling of the partial result becomes clear. The text in DA is sometimes not very clear and perhaps has suffered at the hands of copyists.

Chapter 16: Examples of extracting the square root of integers

<16.1> According to the procedure abstractly described in chapter 14, $\sqrt{5625}=75$ is now calculated in detail.

<16.2> This section treats the special case in which a pair of places are both zero. No example is given.

<16.3> Another, "shortened," procedure (*adbreviatio*) is explained for finding the roots of numbers ending in a series of zeros: half the number of zeros is put at the end of the root of the number shorn of its zeros. In this case one does not need to go through the procedure of doubling etc. as in the general case. The example of $\sqrt{10000}$ is given, the result being 100.

Chapter 17: Calculation of the roots of fractions

<17.1> To find the root of a fraction or of a mixed number, the first step is to reduce the number to the lowest place of the fractional part. When this place is such that "a root exists" (i.e. in the case of sexagesimal fractions: seconds, fourths, . . .), then the procedure of chapter 16 is applied, and the result is reconverted to larger units. When the lowest place "has no root" (i.e. in the case of sexagesimal fractions: minutes, thirds, fifths, . . .), the number must be reduced to units of a place which has a root; and again the root is found and then reconverted to larger units. The procedure is made the more exact (*propius veritati*) the lower the place ("having a root") to which it is reduced.

<17.2> To explain this procedure, two examples are given. The first is: $\sqrt{2}$. The 2 is converted into seconds; if the 2 were converted into fourths or sixths, the result would be still more exact. Thus $\sqrt{2}$ is found as $\sqrt{7200}'=84'$ + remainder = 1 integer 24′ + remainder. (A better approximation for $\sqrt{2}$ is found in chapter 19.2.)

<17.3> The second example is of a root of a mixed number comprising integers and common fractions: $\sqrt{2+\frac{1}{3}+\frac{1}{13}}$. The number must be reduced to units "of one species" (*ex uno genere*); in modern terms, on the product of the denominators as denominator. Then numerator and denominator of the fraction must be multiplied by a number to make the denominator a square. Implicit here is the formula $\sqrt{\frac{a}{b}}=\frac{\sqrt{ab}}{b}$.

The individual steps are:

$$\sqrt{2+\frac{1}{3}+\frac{1}{13}}=\sqrt{\frac{78+13+3}{39}}=\sqrt{\frac{94}{39}}=\sqrt{\frac{94\times 39}{39^2}}=\frac{\sqrt{3666}}{39}$$

$$=\frac{60}{39}+\text{remainder}=1\frac{21}{39}+\text{remainder}.$$

To obtain a still more exact result, the fraction would have to be converted to "fourths," i.e. its numerator and denominator would be multiplied twice more by 39 (the denominator). Then the steps would be (in N there are some incorrect numbers):

$$\frac{\sqrt{3666}}{39}=\frac{\sqrt{3666\times 39^2}}{39^2}=\frac{\sqrt{5575986}}{39^2}=\frac{2361+\text{remainder}}{39^2}$$

Chapter 18: On extracting square roots depending on what the Indians have established

There follows a chapter which, according to its title, did not belong directly to the Indian procedure (*opus*), but was somehow dependent on it; it is supposed to be easier and more exact (*levius opere Indorum atque subtilius*). The words *sed ego extraxi illud secundum quod constituerunt Indi* show unambiguously that this section is due to the author, al-Khwārizmī, himself, but depends on Indian procedures.

Al-Khwārizmī here demonstrates a method of improving the accuracy of root extraction: he adds an even number of zeros to the number whose root is to be extracted. After extraction the root is reduced in the rank of its decimal places by the appropriate number – half the number of zeros which were added to the original number. The following relation is implicit:

$$\sqrt{\mathrm{N}}=\frac{1}{10^k}\sqrt{\mathrm{N}\times 10^{2k}}$$

The fractional part is re-expressed in sexagesimal fractions. The result is more exact the more zeros are added. The presentation is understandable, but clumsy.

Chapter 19: Extraction of roots of integers or of mixed numbers (by adding zeros)

< 19.1 > In this chapter the extraction of the roots of integers or mixed numbers (the fractional part being either a vulgar or sexagesimal fraction) is again treated. It is surprising that this topic comes up yet again, since it is amply treated in chapters 14–16 for integers and in chapter 17 for fractions. Indeed, nothing is essentially new in this chapter: in 19.1 the procedure of the previous chapter is repeated, and in 19.2 an example of it is presented.

The general procedure is first indicated: the elements of the mixed number are reduced to units of its lowest place; if this place is not one that "has a root," the number must be further reduced to the next-lower place. Now an even number of zeros can be placed at the end of the number (cf. chapter 18). After extracting the root the remainder is converted to sexagesimal fractions by repeated multiplication by 60. The whole section could be greatly shortened or omitted entirely, since the general procedure has already been described in chapter 18.

< 19.2 > The procedure is demonstrated by the example $\sqrt{2}$. In this case, 6 zeros are added. The steps are:

$$\sqrt{2} = \sqrt{\frac{2.000.000}{10^6}} = \frac{1}{1000} \times \sqrt{2.000.000} = \frac{1}{1000} \times [1414 + \text{remainder}]$$

$$\approx \frac{1}{1000} \times 1414 = 1(\textit{unit}) + \frac{414}{1000} = 1 + \left(\frac{414 \times 60}{1000}\right)' = 1 + \frac{24840}{1000}$$

$$= 1 + 24' + \left(\frac{840 \times 60}{1000}\right)'' = 1 + 24' + \left(\frac{50400}{1000}\right)'' = 1 + 24' + 50'' + \left(\frac{400 \times 60}{1000}\right)'''$$

$$= 1 + 24' + 50'' + 24'''.$$

Thus by the use of the extra zeros a more exact result has been obtained for the same example than with the simpler procedure of chapter 17.2, which only produces minutes.

With this example the text ends. There follows only the *explicit* with praise and thanks to God. Of this whole formula the original Arabic is easily understood: "It is finished (*tamma, Finit*) the book (*al-kitāb, liber*) with the help of God (*bi-ʿaun Allāh, cum dei adiutorio*) and praise to Him (*wa-ḥamdihi, et eius laude*)".

If DA – as it may be deduced from manuscripts N and C – is compared with later Arabic texts on arithmetic, the following remarks may be made:

- *Very detailed, sometimes long-winded presentation*

This applies, for instance, to the statements in chapter 1 on the decimal system, the representation of numbers, the idea of place (*differentia*) and the use of zeros in writing multiples of the powers of 10. Many things mentioned in this chapter are repeated in the chapter introducing fractions (13.3). – The zero receives special treatment not only in the first chapter, but also in one of the sections on extraction of roots (16.2). – The method of increasing the accuracy of root extraction by adding zeros to the number comes in chapter 18 and again immediately afterwards in chapter 19.1. – It is perhaps understandable that the properties of the zero should be described in more detail and perhaps more often than properties of the "common" digits, since the zero has the strange property of being at once "nothing" and at the same time having the power of increasing the value of a digit ten times.

- *Missing technical terms*

For several important terms in arithmetic there are no technical terms in DA; e.g. there are no proper terms for "minuend," "subtrahend," "multiplier," "multiplicand," "divisor," "dividend," "numerator," "denominator," "square number." Even for the zero there is no special expression (such as: *ciffra* found in other texts), only the description *circulus*. Because of the lack of technical terms the presentation of arithmetic operations becomes hard to understand – a good example is the multiplication and division of fractions in chapter 12.

- *Inappropriate order of the material*

In chapter 1.7 things connected with addition are described, although addition is formally only described later. – In one place (chapter 3) doubling is treated after halving and in another (chapter 11.4) before halving. – Multiplication and division of fractions are treated in chapters 8–10, i.e. before the notation of fractions (chapter 11).

- *Unclear or awkward presentation*

Often in the text there is a verbal description of a complex process without the "schemes" that would clarify the steps of the operation. Examples are: multiplication (chapter 4), division (chapter 7.1), multiplication of common fractions (chapter 12.1). – The discussion of fractions is often obscured by the forced analogy of common fractions and sexagesimal fractions: clearly the author had the idea that the rules for these two classes of fractions were the same; in chapter 8 he even tries to prove this. – An example of division of fractions (chapter 10.2) appears to be spoiled by the confusion of the common fraction "thirds" (*tercie*) with the sexagesimal "thirds" (*tercia*). – The description of the addition of sexagesimal fractions is very unclear; it is even not stated in the text that addition is being treated (chapter 11.2). – The pairing off of the digits in the initial stages of root extraction (chapter 14) is not clearly described. – Also obscure is the explanation of why doubling is necessary in the procedure for extraction of roots (chapter 15). – The example of $\sqrt{2}$ is given a second time (chapter 19.2; cf. chapter 17.2) without reference to the former, less accurate, calculation.

- *Unclear understanding by the author or redactor*

Some passages may be unclear because the author or redactor did not understand the matter. An example is the lack of justification of the different treatment of the odd 1 in the operation of halving (chapter 3) – sometimes 30 is added to the next-lower place and sometimes 5. – Further he tries, without success, to show that common fractions and sexagesimal fractions follow the same rules in extraction of square roots (chapter 13.4). – The notion of *prime fractiones* etc. remains unclear (chapter 13.4).

- *Old-fashioned presentation*

Some of the descriptions of arithmetic processes give the impression of having been written when these processes were new. An example is the presentation of the casting out of nines (chapter 5): the excess of nines is not found by adding the digits but by finding the remainder of the original number on division by 9. (The passage is obscured by the apparently mixed term of *divide per novem novem.*) This check is applied to doubling and multiplication, but not to division – for which the lengthy check by multiplication is suggested (chapter 10.4). – The multiplication of common fractions is made unnecessarily complicated by the multiplication of the numerators with the corresponding denominators (chapter 12.2; only in manuscript N).

Even if it is not certain whether all of these curiosities were in the translation from the Arabic or whether they crept in at a later stage, they all point in the same direction: that the text of DA represents a relatively early stage of the knowledge of Indian arithmetic. At such a time it was difficult to understand the content in detail, and many technical terms had not been coined. The presentation of the material in DA is not uniform: sometimes it is long-winded and sometimes so short as to be scarcely comprehensible. All this accords well with the fact that al-Khwārizmī wrote the text in a time in which the 'new arithmetic' was first introduced into the Arabic-speaking world. The suggestion of A. P. Youshkevitch, that al-Khwārizmī expressed himself clearly and that the oddities and inconsistencies of the Latin text were entirely due to a Latin redactor of the twelfth or thirteenth centuries, appears quite unjustified.

Not surprisingly, the texts on Indian arithmetic written after al-Khwārizmī's time present the material more clearly. This is true for both Arabic and Western authors. The writings of al-Uqlīdisī, Kūshyār ibn Labbān and al-Baghdādī – all of them more than 100 years later than al-Khwārizmī – show that the knowledge taken over from the Indians was not only understood, but also developed. *Inter alia*, technical terms were introduced, procedures were found which simplified the calculations and the methods were more clearly explained. In the West, also, there were reworkings which show that the problems of understanding the material, found in DA, had been solved as early as the twelfth century: the derivative texts of LA/LP (which in one place explicitly cites *Alchorismus*, i.e. the text of the translation effectively present in DA) and LY have the same structure as DA, but have a developed terminology and present the new arithmetic in a clearer form – whether or not this is done in extended form (as in LA/LP) or in a short presentation (as in LY). It is easy to understand that these later texts or other simple compilations such as those of Johannes de Sacrobosco or Alexander de Villa Dei were much more popular than DA, which was closer to the original translation, but not so easy to understand. An indication of this is the existence of only two manuscripts of DA.

References

Allard, André. 1978. "A propos d'un algorisme latin de Frankenthal: une méthode de recherche." *Janus* 65:119–141.

Allard, André. 1991. "The Arabic origins and development of Latin algorisms in the twelfth century." *Arabic Sciences and Philosophy* 1:233–283.

Allard, André. 1992. *Muḥammad ibn Mūsā al-Khwārizmī: Le Calcul Indien (Algorismus). Histoire des textes, édition critique, traduction et commentaire des plus anciennes versions latines remaniées du XIIe siècle.* Paris: Albert Blanchard/Namur: Société des Etudes Classiques.

Beaujouan, Guy. 1948. "Étude paléographique sur la 'rotation' des chiffres et l'employ des apices du Xe au XIIe siècle." *Revue d'histoire des sciences* 1:301–313.

Bergmann, Werner. 1985. *Innovationen im Quadrivium des 10. und 11. Jahrhunderts. Studien zur Einführung von Astrolab und Abakus im lateinischen Mittelalter.* Stuttgart: Franz Steiner.

Boncompagni, Baldassarre. 1857. *Trattati d'aritmetica. I. Algoritmi de numero Indorum.* Rome: Tipografia delle scienze fisiche e matematiche.

Bose, D. M., S. N. Sen, B.V. Subbarayappa. 1971. *A Concise History of Science in India.* New Delhi: Indian National Science Academy.

Bubnov, Nicolaus. 1899. *Gerberti postea Silvestri II papae Opera Mathematica (972–1003).* Berlin: R. Friedländer u. Sohn.

Burnett, Charles. 1996. "*Algorismi vel helcep decentior est diligentia:* The Arithmetic of Adelard of Bath and his Circle." In *Mathematische Probleme im Mittelalter – der lateinische und arabische Sprachbereich*, edited by Menso Folkerts, 221–332. Wiesbaden: Harrassowitz.

Busard, Hubert L. L. 1968. "Het rekenen met breuken in de middeleeuwen, in het bijzonder bij Johannes de Lineriis." In *Mededelingen van de Koninklijke Vlaamse Academie voor Wetenschappen, Letteren en Schone Kunsten van België.* Klasse der Wetenschappen. Jaargang 30, nr.7. Brussels.

Cantor, Moritz. 1865. "Ueber einen Codex des Klosters Salem." *Zeitschrift für Mathematik und Physik* 10:1–16.

Crossley, John N., Alan S. Henry. 1990. "Thus Spoke al-Khwarizmi: A Translation of the Text of Cambridge University Library Ms. Ii. vi.5". *Historia Mathematica* 17:103–131.

Datta, B., A. N. Singh. 1935–38. *History of Hindu Mathematics. A Source Book.* Lahore: Motilai Banarsi Das.

Dodge, Bayard. 1970. *The Fihrist of al-Nadim. A Tenth-Century Survey of Muslim Culture.* English translation. 2 vols. London/New York: Columbia University Press.

Faulhaber, Charles B. 1983. *Medieval Manuscripts in the Library of the Hispanic Society of America.* New York: Hispanic Society of America.

Flügel, Gustav. 1871–72. *Kitāb al-Fihrist. Mit Anmerkungen herausgegeben von Gustav Flügel*, 2 vols. Leipzig: Vogel.

Folkerts, Menso. 1997. *Die älteste lateinische Schrift über das indische Rechnen nach al-Ḫwārizmī. Edition, Übersetzung und Kommentar* von M. Folkerts unter Mitarbeit von Paul Kunitzsch. München: Bayerische Akademie der Wissenschaften, Philosophisch-historische Klasse, Abhandlungen, Neue Folge, Heft 113.

Gandz, Solomon. 1931. "The Origin of the Ghubār Numerals, or the Arabian Abacus and the Articuli." *Isis* 16:393–424.

Gandz, S. 1938. "The Algebra of Inheritance." *Osiris* 5:319–391.

Hughes, Barnabas B. 1986. "Gerard of Cremona's Translation of al-Khwārizmī's *al-Jabr*: A Critical Edition." *Mediaeval Studies* 48:211–263.

Hughes, Barnabas B. 1989. *Robert of Chester's Latin translation of al-Khwārizmī's* al-Jabr. *A new critical edition.* Stuttgart: Franz Steiner.

Ifrah, Georges. 1986. *Universalgeschichte der Zahlen.* Frankfurt, New York: Campus Verlag.

Irani, Rida A. K. 1955–56. "Arabic numeral forms." *Centaurus* 4:1–12.

Juschkewitsch, A. P. 1964a. *Geschichte der Mathematik im Mittelalter.* Leipzig: B. G. Teubner Verlagsgesellschaft.

Juschkewitsch, A. P. 1964b. "Über ein Werk des Abū ʾAbdallah Muḥammad ibn Mūsā al-Ḫuwārizmī al-Maǧūsī zur Arithmetik der Inder." In *Beiheft zur Schriftenreihe für Geschichte der Naturwissenschaften, Technik und Medizin. Beiträge zur Geschichte der Naturwissenschaften, Technik und Medizin,* hg. zum 60. Geburtstag Gerhard Harigs von I. Strube und H. Wußing, 21–63. Leipzig: B. G. Teubner Verlagsgesellschaft.

Juškevič, A. P. 1983. "O trude po arifmetike Muchammada ibn Musy al-Chorezmi." In *Muchammad ibn Musa al-Chorezmi. Matematiceskie traktaty,* 150–202. Tashkent: Izdatelstvo "FAN."

Karpinski, Louis Charles. 1910–1911. "Hindu numerals in the Fihrist." *Bibliotheca Mathematica* 3. Folge. 11:121–124.

Karpinski, L. C. 1921. "Two Twelfth Century Algorisms." *Isis* 3:396–413.

Kopelevič, Ju. Ch., A. P. Juškevič. 1983. "Muchammad ibn Musa al-Chorezmi, Kniga ob indijskom sčete. Perevod s latinskogo Ju. Ch. Kopelevič i A. P. Juškeviča." In *Muchammad ibn Musa al-Chorezmi. Matematičeskie traktaty,* 5–19. Tashkent: Izdatelstvo "FAN."

Labarta, Ana, Carmen Barceló. 1988. *Números y cifras en los documentos arábigohispanos.* Córdoba: Área de Estudios Árabes e Islámicos.

Lemay, Richard. 1982. "Arabic numerals." In *Dictionary of the Middle Ages,* edited by Joseph R. Strayer, vol. 1, 382–398. New York: Scribner.

Levey, Martin, Marvin Petruck. 1965. *Kūshyār ibn Labbān. Principles of Hindu Reckoning. A translation with introduction and notes by Martin Levey and Marvin Petruck of the Kitāb fī usūl hisāb al-hind.* Madison and Milwaukee: University of Wisconsin Press.

Lindgren, Uta. 1976. *Gerbert von Aurillac und das Quadrivium.* Wiesbaden: Franz Steiner.

Pedersen, Fritz S. 1983. *Petri Philomenae de Dacia et Petri de S. Audomaro opera quadrivialia.* Pars I: *Opera Petri Philomenae.* Copenhagen: Societas linguae et litterarum Danicarum.

Ruska, Julius. 1917. *Zur ältesten arabischen Algebra und Rechenkunst.* Heidelberg (Sitzungsberichte der Heidelberger Akademie der Wissenschaften, Philosophisch-historische Klasse, Jahrgang 1917, 2. Abhandlung): Carl Winter's Universitätsbuchhandlung.

Saidan, Aḥmad Salīm. 1973. *Abū l-Ḥasan Aḥmad ibn Ibrāhīm al-Uqlīdisī. Al-Fuṣūl fī l-ḥisāb al-hindī.* Edited by A. Saidan. Amman: Mansurat al-Lagna al-Urdunniya.

Saidan, A. S. 1978. *The Arithmetic of al-Uqlīdisī. The Story of Hindu-Arabic Arithmetic as told in* Kitāb al-Fuṣūl fī al-Ḥisāb al-Hindī *by Abū al-Ḥasan Aḥmad ibn Ibrāhīm al-Uqlīdisī, written in Damascus in the year 341 (A.D. 952/3). Translated and Annotated, from the unique copy, ms. 802 at Yeni Cami Library, Istanbul, written in the year 582 (A.D. 1186).* Dordrecht/Boston: Reidel.

Saidan, A. S. ed. 1985. *ʿAbd al-Qāhir ibn Ṭāhir al-Baġdādī: at-Takmila fī ʾl-ḥisāb.* Kuwait: Mansūrāt mahad al-maṭūṭāt al-arabīya.

Sezgin, Fuat. 1974. *Geschichte des arabischen Schrifttums.* Band V: *Mathematik.* Leiden: E. J. Brill.

Smith, David Eugene. 1925. *History of Mathematics.* Vol. 2. Boston: Ginn.

Smith, D. E., and L. Ch. Karpinski. 1911. *The Hindu-Arabic numerals.* Boston/London: Ginn.

Steele, Robert. 1922. *The Earliest Arithmetics in English.* London: Milford.

Thomson, Ron B. 1976. "Jordanus de Nemore: Opera." *Mediaeval Studies* 38:97–144.

Toomer, Gerald J. 1973. "Al-Khwārizmī." *Dictionary of Scientific Biography.* vol. 7, 358–365. New York: Charles Scribner's Sons.

Tropfke, Johannes. 1980. *Geschichte der Elementarmathematik*, 4. Auflage, Berlin/New York: Walter de Gruyter.

Vogel, Kurt. 1963. *Mohammed ibn Musa Alchwarizmi's Algorismus. Das früheste Lehrbuch zum Rechnen mit indischen Ziffern. Nach der einzigen (lateinischen) Handschrift (Cambridge Un. Lib. Ms. Ii. 6.5.) in Faksimile mit Transkription und Kommentar herausgegeben.* Aalen: Otto Zeller.

van der Waerden, B. L., Menso Folkerts.1976. *History of Mathematics. Counting, Numerals and Calculation.* 3: *Written Numbers.* Milton Keynes: The Open University.

III

Euclid in Medieval Europe*

Between 1964 and 1984 Marshall Clagett published the five volumes of his *Archimedes in the Middle Ages*. These volumes, the fruit of many years of research in printed and manuscript material, contain editions (with English translations) of all relevant texts and give detailed information about the textual history of Archimedes' writings in the Latin West. Clagett's work has become a model for the history of mathematics in the Middle Ages.

It is impossible to publish a similar series on *Euclid in the Middle Ages*: even if this were restricted to Euclid's main work, the *Elements*, its length and the complexity of its transmission in the Middle Ages would prevent it. But on the other hand, in the last decades almost all relevant Latin texts of this time have been edited, most of them for the first time and almost all by H. L. L. Busard. This scholarly activity began with Clagett's fundamental article from 1953[1], which gave a new impetus to study the transmission of the *Elements*, especially since the time of the translations from the Arabic. Thus it is now possible to give a summary of the development which takes into account the numerous new results that were obtained in the last half century.

Euclid's *Elements* was the most important mathematical textbook, at least until modern times. Even in Greek antiquity, it was much copied, commented and reworked; later it was translated into many languages. In the Middle Ages, too, it was the object of much scholarly activity, both in the Arabic-Islamic area and in the Latin West. This article is limited to the Western Middle Ages. Its aim is to present the various translations, reworkings and commentaries, beginning with Boethius' translation of about 500 AD and ending with the Greek *editio princeps* of 1533 and the first translations into the vernacular in the fifteenth and early sixteenth centuries; but some information on the translations into Arabic is included, upon which the Latin translations of the twelfth century were based. The first part of this contribution is on the transmission of the *Ele-*

*A booklet with the same title was published 1989 in Winnipeg (Canada). This contribution is a totally revised version of it, taking into account the numerous new results found since that time.

[1] Clagett, 'Medieval Latin Translations'.

ments in the medieval West and the variety of texts which were written in that time, though the specific characteristics of the different versions are not given here[2]. The second part of this article gives, for all versions, a list of manuscripts which transmit them and contains information about modern editions and translations.

The situation can be summarized as follows: before the twelfth century, the *Elements* was in general only known by the Latin translation which Boethius had made about 500 AD, but only parts of it were available from the eighth and ninth century on. In the twelfth century the complete text of the *Elements* became known: it was three times translated, independently, from the Arabic and once from the Greek. But much more important than these translations were two reworkings: that of Robert of Chester (ca. 1140), which was based upon the Arabic translations and the Boethius text, and that of Campanus (ca. 1255), which itself used Robert's text and enriched it by contemporary material. From the end of the thirteenth century, Campanus' version superseded Robert's text and was dominant until the beginning of the sixteenth century. Greek manuscripts were not used in the West to establish the Euclid text before the second half of the fifteenth century. The first complete printed edition of the *Elements* was that of Campanus' version (1482). In the first half of the sixteenth century this text competed with the Greek-Latin translation of Zamberti (first published 1505). The first edition of the Greek text was published in 1533. From the fifteenth century there were also translations into Italian; between 1543 and 1570 the first Italian, German, French and English translations appeared in print.

* * *

I. The translations, reworkings and commentaries

1. Translations from the Greek before the humanistic movement

1.1 Boethius

The most important direct translation of the *Elements* from the Greek was that of Boethius about the year 500[3]. From two testimonies by Cassiodorus we know that Boethius made a Latin translation of Euclid; and

[2]For this, see the article 'Probleme der Euklidinterpretation' in this volume, item IV.

[3]The following is a summary of what we know about this translation and its transmission in the Middle Ages. For details, see Folkerts, 'Importance', pp. 187–191, and Folkerts, 'Boethius Geometrie II', pp. 69–82.

it seems very likely that he translated all of the 13 (or 15) books of the *Elements*. But only parts of this translation have survived. They are extant in four different texts. These four groups of excerpts from Boethius' translation of Euclid's *Elements* are usually called **Ma**, **Mb**, **Mc** and **Md**, respectively. The first three texts seem to have originated in the eighth century in Corbie, which at that time was the capital of the medieval world for geometrical texts; the fourth text was compiled in Lorraine in the first half of the eleventh century. I add some remarks on these excerpts and on the treatises or compilations to which they belong.

The excerpts **Ma** belong to the third recension of Cassiodorus' *Institutiones*. This recension probably originated in the eighth century – certainly before Hrabanus Maurus' *De Institutione Clericorum* (819). In **Ma**, the definitions of books 1 and 5 are present[4], but the propositions are missing.

Mb is a group of excerpts from Boethius' translation which was included in a new redaction of the *Corpus Agrimensorum* that very probably originated in Corbie and was brought together to serve as geometrical material for the *quadrivium*. The two oldest manuscripts of this redaction are now in the Vatican and in Wolfenbüttel. They contain the definitions, postulates and axioms of book 1 and the first three propositions with proofs[5].

The third group of excerpts, **Mc**, belongs to the so-called *Geometria I*, a compilation in five books which in the manuscripts is ascribed to Boethius, but was in reality put together from various sources, presumably in the eighth century and also in Corbie. **Mc** contains the definitions, axioms and postulates and most of the enunciations of Euclid's *Elements*, books 1–4[6].

The excerpts **Md** form an essential part of the so-called *Geometria II*, a treatise in two books which in the manuscripts is also generally attributed to Boethius, but was compiled in Lorraine in the first half of the eleventh century. The excerpts **Md** very much resemble **Mc** and **Mb**: **Md** has the same interpolations and transpositions as **Mc**, but since **Md** contains certain propositions which are absent in **Mc**, we may presume that the author of the *Geometria II* used a collection of excerpts that was more complete than the Pseudo-Boethian *Geometria I*. Similar to **Mc**, **Md** contains the definitions, axioms and postulates and most enun-

[4] For details about the extant definitions and propositions, see p. 32.

[5] For details about the extant definitions and propositions, see p. 33.

[6] For details about the extant definitions and propositions, see p. 34.

ciations of books 1–4 of the *Elements*. In addition, like **Mb**, it contains the first three propositions of book 1 with proofs[7].

It is possible to reconstruct from these fragments the text of Boethius' original translation of books 1–4 and of the definitions of book V[8].

One should also mention that the twelfth-century *Liber Ysagogarum Alchorismi* has been found to contain axioms and enunciations which may belong to the Boethius tradition but are not among the extant Boethius fragments[9].

1.2 Other early texts based upon the Greek

There are two other anonymous Latin Euclid fragments from the fifth and ninth centuries respectively. The older of them is a palimpsest now in the Biblioteca Capitolare of Verona, Ms. XL (38). From the original text there are three double folios with parts of Euclid's *Elements*, books 11–13[10]. Geymonat, who edited the text[11], considered that the writing could be dated to the last years of the fifth century and we have here part of the original Boethius' translation.

The second fragment was written at the beginning of the ninth century in a north-east French scriptorium, perhaps Corbie; it may be connected with the Palatine library of Charlemagne[12]. This fragment, now in the University Library at Munich[13], contains Euclid, I.37–38 and II.8–9. The translator obviously did not understand the mathematical content nor had he mastered Latin grammar; he took Greek letters attached to figures or used to denote the endpoints of segments as numerical symbols. Therefore it seems very unlikely that this fragment has anything to do with the Boethian tradition.

1.3 Greek-Latin translation from the twelfth century

Another direct translation was made in South Italy or Sicily in the twelfth century. It was discovered by J. E. Murdoch in 1961. According to Mur-

[7] For details about the extant definitions and propositions, see pp. 35f.

[8] This has been done in Folkerts, 'Boethius Geometrie II', pp. 173–217.

[9] For the *Liber Ysagogarum*, see p. 18.

[10] There are parts of the propositions XI.24–25; XII.2–3. 8; XIII.2–3. 7. In the manuscript the books 12 and 13 are marked as XIIII and XV.

[11] Geymonat, 'Fragmenta Veronensia'.

[12] See Bischoff, 'Studien', p. 14 (*Panorama der Handschriftenüberlieferung aus der Zeit Karls des Großen*) and p. 158 (*Die Hofbibliothek Karls des Großen*).

[13] Ms. 2° 757. Besides of older editions and comments by M. Curtze and A. A. Björnbo, there is a modern edition by Geymonat, 'Nuovi frammenti'.

doch this translation was made a little before 1175[14]. The Norman kingdom of Southern Italy and Sicily was in the twelfth century a meeting-point of Greek and Latin culture in which Greek, Latin, and Arabic civilizations lived side by side in peace and toleration. The production of translations was encouraged by the Sicilian kings, from Roger to Frederick II and Manfred. Thus it is no surprise that at the same time in this region (South Italy and Sicily) not only were the *Elements* translated from the Greek, but also the *Almagest*, some minor writings of Euclid (*Data*, *Optics*, *Catoptrics*), the *Elementatio physica* of Proclus and Zenodorus' *On Isoperimetric Figures*. Whether all of these translations were done by the same person or the same group is an unsolved question[15].

J. E. Murdoch made the first analysis of this translation of the *Elements*[16], and H. L. L. Busard has edited the text[17]. It should be noted that we do not possess book 14 of the South Italian Euclid translation in its original form. What we do have is an abbreviated version of books 14–15 (and also the original version of book 15). The abbreviation is found in only one manuscript[18]. In book 15 it follows the Arabic-Latin tradition. It is possible that Leonardo Fibonacci was the author of this compendium. He was acquainted with the integral version of books 1–13 and 15 of the Greek-Latin translation as well as with the epitome of books 14 and 15 of the *Elements*[19].

Unfortunately, we cannot say exactly how the manuscript used by the translator is related to the extant Greek manuscripts. It should perhaps be added that, despite Heiberg's analysis of the manuscripts available to him (see his 'Prolegomena'), many details of the transmission of the Greek text of the *Elements* are still unclear.

The Greek-Latin translation of the *Elements* was not very influential. We know of only two mathematicians who used it, Leonardo Fibonacci and Campanus in his redaction of the *Elements*[20].

* * *

[14]Murdoch, 'Euclides Graeco-Latinus', p. 270.

[15]For the different opinions, see Busard, 'Campanus', p. 31. Richard Lemay believed that Hermann of Carinthia at an advanged age learnt Greek and that he was the translator of the *Almagest* from Greek into Latin (Lemay, 'Itinéraire').

[16]Murdoch, 'Euclides Graeco-Latinus'.

[17]Busard, 'Translation ... from the Greek'.

[18]Paris, BnF 7373, f. 167v–172v. Edition in Busard, 'Translation ... from the Greek', pp. 399–411.

[19]See Folkerts, 'Leonardo Fibonacci's Knowledge', where a detailed argument is given.

[20]For Campanus, see p. 23.

2. The Arabic Euclid

Questions about the Arabic-Latin translations of Euclid cannot be separated from questions about the state of the Arabic text. Which Arabic versions could the translators have had before them in the twelfth century?

For a long time it was known that the *Elements* had been translated, or reworked, in the ninth century by al-Ḥajjāj, Isḥāq b. Ḥunayn and Thābit b. Qurra. The serious study of the Arabic tradition of the *Elements* started with an article by M. Klamroth written in 1881[21]. He was the first who tried to classify the Arabic manuscripts known at that time. Not much happened up to 1980, when in Harvard two Ph.D. theses by John Engroff and Gregg De Young were presented, editions of books 5 and 7–9, respectively, of the "Isḥāq–Thābit version"[22]. Since that time many new results have been obtained, especially by De Young[23] and Brentjes[24]. The following outline is based upon these works[25].

According to the *Fihrist* of Ibn al-Nadīm, Euclid's *Elements* were translated by al-Ḥajjāj, and this translation was later revised by him. The second version, "for al-Ma'mūn", he says is better than the first, "for Hārūn". Another translation was made by Isḥāq b. Ḥunayn and this translation was revised by Thābit ibn Qurra. Thus there are at least two traditions: Ḥajjāj and Isḥāq–Thābit. Books 14–15 seem to have been translated by Qusṭā ibn Lūqā (d. early tenth c.), for many Arabic manuscripts explicitly assert this[26]. Two manuscripts (Copenhagen 81 and Istanbul, Fatih 3439) say at the beginning of book 11 that books 11–13 were translated by al-Ḥajjāj.

There has been much discussion to which version(s) the extant Arabic manuscripts belong. For a long time it seemed undisputed that the original version ("Ḥajjāj I") is lost, but that we have the first six books and a few definitions of book 7 of "Ḥajjāj II", together with Nayrīzī's commentary, in the manuscript Leiden 399,1[27]. S. Brentjes found another manuscript, now in Qum (Iran), which belongs to the same tradition; it is incomplete, too, but can fill some lacunae in the Leiden

[21]Klamroth, 'Euklid'.

[22]Engroff, 'Arabic Tradition'; De Young, 'Arithmetic Books'. These dissertations have not been printed.

[23]De Young, 'Textual Traditions'; 'New traces'; 'Latin Translation'.

[24]Brentjes, 'Textzeugen'; 'Observations'; 'Relevance'.

[25]It relies especially upon the article of De Young, 'Latin Translation'.

[26]See Sezgin, 'GAS 5', p. 96.

[27]The Arabic text of this manuscript was edited in Besthorn, 'Codex Leidensis'.

manuscript[28]. Brentjes showed that al-Nayrīzī's commentary is a heavily edited text connected with the Ḥajjāj tradition but cannot serve as pure document of this text[29]. There are traces of the Ḥajjāj text in other Arabic manuscripts, too, but no pure text of this version is known.

All Arabic manuscripts of the *Elements* which are currently known fall within the Isḥāq–Thābit tradition, though there are textual differences between them. By using patterns of macro and micro variations, De Young was able to classify them into two groups, A and B[30]. The larger number of manuscripts fall into Group A. One of the manuscripts of Group B is Tehran, Malik 3586, by far the oldest manuscript: it was written in A.H. 343 = AD 954/55[31]; it is noteworthy that even this very early manuscript has a remarkable number of macro-variants, especially in terminology, which distinguishes it from the other Arabic manuscripts[32].

The *Elements* have been richly commented upon in the Arabic-Islamic culture[33]. In the thirteenth century Naṣīr al-Dīn al-Ṭūsī established an Arabic redaction of the *Elements*, a so-called *Taḥrīr*. He used material from the Ḥajjāj and from the Isḥāq–Thābit traditions and made several remarks about the differences between them. His *Taḥrīr* was very influential, and numerous manuscripts have come down to us. There is another *taḥrīr*, not by Ṭūsī, but attributed to him. It was printed in Rome in 1594 and was the first Arabic text of the *Elements* which became known in the West in print. But in the Middle Ages Ṭūsī's redaction had no influence in the West, because it was not translated into Latin.

* * *

3. Translations from the Arabic

We come now to the twelfth century Latin texts which are either translations from the Arabic or reworkings of these translations.

---

[28] Brentjes, 'Relevance', p. 203.

[29] Brentjes, 'Relevance', p. 206.

[30] De Young, 'Latin Translation'. "Macro variations" are large-scale divergences, e.g. major interpolations, re-orderings of propositions or substitutions of variant demonstrations; "micro variations" are the minor textual variants which can be accommodated within a critical appatus.

[31] The missing part at the end of book 7 of this manuscript is now Tehran, Dānishgāh 2120.

[32] In books 11–13 it shares these macro-variants with the manuscript Uppsala, O. Vet. 20.

[33] A list of 60 comments, reworkings etc. is given in Sezgin, 'GAS 5', pp. 104–115.

The systematic examination of the Arabic-Latin tradition of Euclid's *Elements* began in 1953 with Marshall Clagett's fundamental article[34]. For the first time three different texts were distinguished: version I, "a translation"; version II, "an abridgement"; version III, "an editio", all of which were ascribed by Clagett to Adelard of Bath. From an examination of the most important manuscripts Clagett was able to characterize these three versions. He also tried to find the relations between them. He further gave a general description of the other Euclid translations from Arabic, namely those of Gerard of Cremona and Hermann of Carinthia, and also of medieval reworkings of the texts. After this article was written it was the accepted theory that there are three Adelard versions of Euclid and a reworking by Campanus. Later, J. E. Murdoch showed that this gives too simple a picture and pointed to a number of "mélanges" and reworkings[35]. Of all these versions and redactions only the Campanus reworking was in print before 1967 – and that in editions of the fifteenth and sixteenth centuries. Since this time H. L. L. Busard has published critical editions of nearly all Latin translations and reworkings. In the introductions to his editions Busard has collected the available information about the authors and their texts, and in an article[36] he gave information on further, hitherto unknown, reworkings of the text. Yet other results about the connections between the Arabic and Latin versions were published in several articles by P. Kunitzsch[37] and by S. Brentjes[38]. Some questions were also treated by R. Lorch and M. Folkerts in the Adelard Colloquium held in 1984 and again by Ch. Burnett, M. Folkerts and R. Lorch in an *Addendum* published in the proceedings of this conference[39], but it should be noted that some results presented there must be reconsidered in the light of studies published since that time.

We now know that Euclid's *Elements* were translated three times from the Arabic in the twelfth century: by Adelard of Bath, Hermann of Carinthia and Gerard of Cremona. In addition, at least three commentaries of the *Elements* were translated from Arabic. In this chapter some information about these translations will be given. Latin reworkings and

[34] Clagett, 'Medieval Latin Translations'.

[35] Murdoch, 'Medieval Euclid'.

[36] Busard, 'Early Adaptations'.

[37] Kunitzsch, 'Findings'.

[38] Especially Brentjes, 'Remarks', and 'Observations'.

[39] Lorch, 'Remarks'; Folkerts, 'Adelard's Versions'; Burnett, Folkerts, Lorch, 'Addendum'.

commentaries of the twelfth, thirteenth and fourteenth centuries will be mentioned in Section 4.

3.1 Adelard of Bath (formerly called: "Adelard I")

First the text that since Clagett's article has been called "Adelard I"[40]. No single manuscript is known which presents the complete text of the *Elements*. What we have are four manuscripts which contain books I–VIII, one for books X.36–XV.2, one for book X.36–49, but none at all for books IX to X.35. From the style it is clear that the different parts belong to the same work. It is quite possible that Adelard himself was its author: in the oldest manuscript, Oxford, TC 47, this text is attributed to Adelard[41]. But there are other witnesses to Adelard's authorship: the spelling *el* for the Arabic article (normally, Gerard and others used *al*); and the inflecting of an Arabic word according to Latin inflexion – both being known features of Adelard's work[42]. "Adelard I" is clearly a translation, not a commentary. It must have been translated from the Arabic, since many of its technical terms are transcriptions from Arabic, some of which are not found in other versions. Curiously, there are yet more Arabic terms in the margins of one – just one – manuscript, Bruges 529. Technical terms are not always consistently used. Thus we sometimes find *inconiunctivus* for *equidistans*, *quadratura* for *ductu*, *differens* for *residuum*. Busard supposes that the translator was constantly seeking the right Latin expressions for the various Arabic terms[43]. Certainly there is no Greek influence in this version – so *hypotenusa*, *gnomo*, *parallelogrammum*, *ysosceles*, *ortogonaliter*, and so on do not appear.

We cannot with certainty give the date of Adelard's translation. But it must have preceded his *De opere astrolapsus*, since in that treatise he says that he had translated the *Elements*[44]. According to Haskins[45] the work on the astrolabe was composed between 1142 and 1146, according to Dickey[46] in 1149/50. Anyway, Adelard's translation of the *Elements* may be dated to the second quarter of the twelfth century.

Which Arabic translation or mixed version Adelard used in the translation of the *Elements*, is not clear. It seems that the text which we have

---

[40]For general remarks on this translation, see Busard, 'Campanus', pp. 2–5.

[41]"per Adelardum Batoniensem ex arabico in latinum sermonem translata".

[42]See Kunitzsch, 'Findings', p. 124.

[43]Busard, 'Adelard of Bath', p. 17.

[44]"apud Euclidem a quindecim libris artis geometrice quos ex arabico in latinum convertimus sermonem esse conniciendam" (Haskins, 'Studies', p. 25).

[45]Haskins, 'Studies', pp. 28f.

[46]Dickey, 'Adelard of Bath', pp. 7–11.

is not the original translation of Adelard: some sections show a complete literalness against the Arabic, while other parts are latinized[47].

3.2 Hermann of Carinthia

There is one Latin translation of Euclid's *Elements* which is extant in only one manuscript, and this contains only books 1–12 (Paris, BnF, lat. 16646). Another witness to this translation, a few citations – though not by name – is in MS Vat. Reg. lat. 1268; they reflect a more reliable text than the Paris manuscript[48]. The manuscript now in Paris was left by Gerard d'Abbeville to the Sorbonne in his legacy of 1271, and it seems that it is identical with an item of the *Biblionomia*, which was composed by Richard de Fournival some time close to 1246. Richard de Fournival described the Euclid text in this manuscript as "commentarius Hermanni secundi"[49]. This "Hermannus secundus" was the famous translator Hermann of Carinthia. There are other indications which point out that this work was indeed written by Hermann of Carinthia: the style, some phrases and the penchant for using words of Greek origin are typical for Hermann and can also be found in his *De essentiis* (written in 1143)[50]. On the other hand, the author used a number of Arabic transliterations which are not present in Adelard's translation[51].

Until now, no Arabic manuscript has been found that could have been the one which Hermann translated. After having checked book 1 of the *Elements*, S. Brentjes thinks that "Hermann's text represents in Book I a mixed text based on an Arabic edition which was derived from al-Ḥajjāj's tradition, but also contains contaminations from the Isḥāq/Thābit version, interpolations from Arabic translations of additional Greek material or Arabic commentaries transmitting parts of them, and variations introduced by Arabic authors, scribes, or students."[52] But there are good reasons to assume that some proofs in MS Vat. Reg. lat. 1268 are closer to the original Hermann proofs than those in the Paris manuscript and that in general the text in the Paris manuscript is a reworking. Anyway, Hermann's translation was not directly influential: apart from one of the

[47]Kunitzsch, 'Findings', p. 120; Burnett, 'Latin and Arabic influences', pp. 120–122.

[48]Busard, 'Early Adaptations', pp. 133f.; Busard, 'Campanus', pp. 24f.

[49]"Euclidis geometria, arithmetica et stereometria ex commentario Hermanni Secundi. In uno volumine cuius signum est littera D." (Birkenmajer, 'Richard de Fournival', p. 162.)

[50]See Burnett, 'Arabic-Latin translators', p. 63, and Busard, 'Campanus', p. 24.

[51]See Busard, 'Campanus', pp. 23f.

[52]Brentjes, 'Observations', pp. 51f.

reworkings of the *Elements* found in MS Vat. Reg. lat. 1268 (see p. 54) the only example of the use of it is the compilation of Robert of Chester[53].

3.3 Gerard of Cremona

Gerard of Cremona was the most famous translator from Arabic into Latin in the twelfth century. We know that he went to Toledo "for love of Ptolemy's *Almagest*" (*amore almagesti*), which he could not find in the Latin world. In Toledo he learned Arabic and succeeded in translating not only the *Almagest*, but also many other texts in different fields[54]. He died in Toledo in 1187 at the age of seventy-three.

Gerard's translation of Euclid's *Elements* was found by A. A. Björnbo in 1901 and edited by H. L. L. Busard in 1983[55]. Some years before, Busard had edited some *scholia* which seem to originate with Gerard and which in the main are transmitted in manuscripts other than those containing the translation of the *Elements* itself[56]. Most of the *scholia* belong to books 13 and 14 of the *Elements*. It is noteworthy that no old manuscript of Gerard's translation exists: all manuscripts are from the fourteenth and fifteenth centuries.

Gerard is well known for his extreme literalness in translating Arabic texts. As it stands, however, the "Gerard" translation of the *Elements* is not always word for word: it seems, indeed, that a literal translation – no doubt by Gerard – has been reworked giving it a more standardized and uniform wording and a better Latin style (not so severely "Arabicized" as Gerard's translations normally are)[57]. This would explain the fact that the earliest manuscripts are from the fourteenth century. In the transmitted Latin text there are some alternative proofs, which, as we can read, are "from another book". Thus it it possible that Gerard himself used more than one Arabic manuscript.

In spite of these indications of revision, the Gerard text, as we have it, is superior to the other Latin translations and is the one closest to the Greek tradition. Unfortunately, it was used in the Middle Ages far less than the compilation by Robert of Chester[58]. Traces of Gerard's translation can be found only in some isolated manuscripts of other texts, in

---

[53] For Robert's compilation, see p. 14.

[54] For his translation activities, see, e.g., Kunitzsch, 'Gerard's translations', and Lemay, 'Gerard', pp. 176–188.

[55] Busard, 'Gerard of Cremona'.

[56] Busard, 'Euklid-Scholien'.

[57] See Kunitzsch, 'Findings', pp. 119f.

[58] For this, see p. 14.

which some formulations have been borrowed from Gerard or marginal notes agree with his text[59]. In almost all cases the references to Gerard's translation are given anonymously. Only in one manuscript, which contains Campanus' reworking of the *Elements*[60], do we read that three propositions were taken over from Gerard's translation[61].

3.4 Commentaries to the *Elements* translated from the Arabic

Besides the three translations of the complete text of the *Elements*, at least three commentaries on the *Elements* or on parts of it were translated from Arabic into Latin in the twelfth century.

One of these commentaries was that of al-Nayrīzī (latinized as Anaritius), who lived in the ninth century AD in Bagdad and died there at the beginning of the tenth century AD[62]. His commentary on the *Elements* is especially important because it informs us about the activities of some Greek mathematicians (e.g. Hero, Simplicius) whose writings are lost in Greek. What we have in Arabic are two manuscripts which are incomplete and in which Nayrīzī's commentary is combined with an Arabic version of the *Elements*[63]. Nayrīzī's commentary was translated into Latin by Gerard of Cremona in the twelfth century. The transmitted Latin text ends with book 10. It was edited from only one manuscript by M. Curtze in 1899[64]. Of books 1–4 P. M. J. E. Tummers published a modern critical edition in 1994[65].

Gerard of Cremona also translated another commentary from the Arabic into Latin, this time only of book 10 of the *Elements*. No Arabic manuscript of this commentary has been found. M. Curtze, the first editor of this translation, used only one manuscript (Cracow 569), in which this commentary follows immediately the commentary by al-Nayrīzī, and thus thought that it belonged to the Nayrīzī text. Later, when better

---

[59] For details, see Busard, 'Campanus', pp. 27f.

[60] For Campanus, see p. 23.

[61] "Nota quod in translacione magistri Gerardi Cremonensis sunt tres propositiones que non sunt in translacione Boecij" (Prague, Universitní Knihovna, IV.D.5, at the end of book 8).

[62] For al-Nayrīzī, see Sezgin, 'GAS 5', pp. 283–285.

[63] One of them, Leiden 399/1, ends with the beginning of book 7; it was edited, together with a Latin translation, by Besthorn and others in the years 1893–1932 (Besthorn, 'Codex Leidensis'). The other manuscript, now in Qum, was found only recently. The text of this manuscript has not yet been edited; some information can be found in Brentjes, 'Relevance', p. 203; see p. 6 above.

[64] Curtze, 'Anaritii … commentarii', pp. 1–152.

[65] Tummers, 'Anaritius', pp. 1–124.

manuscripts were found, it became clear that it is a different text. From the best manuscript (Paris, BnF, lat. 9335), where the name of the author is given as "Abbacus", H. Suter concluded that Muḥammad ibn ʿAbd al-Bāqī might have been the author[66]. F. Sezgin doubts this, especially because a treatise of this Muḥammad, who lived in Bagdad and died in 1141 AD, would scarcely have been available in Toledo during Gerard of Cremona's lifetime, and considered Abū Yūsuf Yaʿqūb b. Muḥammad al-Rāzī, who wrote a commentary on book 10 of the *Elements*, as possibly the author[67]. H. L. L. Busard, who published a modern critical edition based upon all six known manuscripts, brought into consideration a third candidate, Yūḥannā al-Qass, because he was an expert on the *Elements* and because there are various propositions in the commentary which can be regarded as lemmas of the Greek Euclid[68]. The content of this commentary was described by Suter[69]. The most characteristic feature of it are the numerical examples which are given to illustrate various proofs.

The third commentary on the *Elements* which became known in the West through the translations from the Arabic is the commentary on book 10 by Pappus of Alexandria (4th c. AD). The Greek text is lost, but we have a translation into Arabic by Abū ʿUthmān al-Dimashqī[70]. The Arabic text was translated into Latin, as it seems, by Gerard of Cremona, but only the beginning (ca. the first half of part 1) of the Latin text has survived, in a single manuscript, from which it was edited by G. Junge[71].

* * *

[66]Suter, 'Autorennamen', pp. 22f., and Suter, 'Kommentar', p. 235.

[67]Sezgin, 'GAS 5', pp. 388f.

[68]Busard, 'Commentary on Book X', p. 24.

[69]Suter, 'Kommentar'.

[70]Edition of the Arabic text with English translation and with German notes (by Gustav Junge) and an English introduction (by William Thomson) in Thomson, 'Commentary of Pappus'.

[71]Junge, 'Pappus-Kommentar'. See also Suter, 'Autorennamen', p. 25, and Sezgin, 'GAS 5', pp. 174f.

4. Reworkings, commentaries and related texts from the twelfth to the fourteenth centuries

4.1 Robert of Chester ("Version II"; formerly called: "Adelard II")

More influential than the three translations of the *Elements* into Latin was a reworking for which M. Clagett in his article of 1953 coined the name "Adelard II". As we shall see, this name is not justified; we shall call it here "Version II"[72]. Until the second half of the thirteenth century, when Campanus compiled his text of the *Elements* – which then became the standard medieval "Euclid" (see p. 23) –, Version II was the most important Euclid text in the Latin West. More than 50 manuscripts are still extant which transmit the complete text or parts of it. Most of the Latin commentaries on the *Elements* which were composed in the thirteenth and fourteenth centuries, including that of Campanus, used the enunciations of the propositions contained in Version II, but provided new proofs.

This version did not come into being at once, but there were several stages of composition, as we can deduce from the different forms of the transmitted text. Version II appears to have originated as a collection of definitions, axioms, postulates, and enunciations; the proofs came later and piecemeal. The earliest stage took place about 1140 – and thus the author could not know Gerard of Cremona's translation from the Arabic, which came later. At this stage no proofs were included; this means that the text consisted only of the definitions, postulates, axioms and enunciations together with the porisms, if any. The earliest known manuscripts (Chartres 498, written in the 1140s, and Munich, Clm 13021, written between 1163 and 1168) belong to this first stage. The second form of Version II is represented by MS Oxford, Trinity College 47. It does not contain Euclid's proofs of the theorems, but replaces them in books 1–6 with "directions for proof", which show, above all, how the solution of one theorem can be arrived at by the use of the axioms, postulates and

[72]The principal characteristics of Version II and its relationship with the other early Latin versions of the *Elements*, including that of Campanus, were first treated in Murdoch, 'Medieval Euclid'. During the preparation of the critical edition (Busard / Folkerts, 'Robert of Chester'), the situation became clearer. More detailed information can be found in the introduction of this edition (pp. 11–31). For additions, see Burnett, 'Latin and Arabic influences'; Burnett pointed out the importance of the manuscript Oxford, TC 47, in the formation of the text, which represents a special stage in the composition of Version II. Our present knowledge about Version II is summarized in Busard, 'Campanus', pp. 5–10.

the previous theorems. These directions are generally very brief, often consisting simply of the numbers of the previous propositions whose results are needed to prove the one at hand. The third stage of Version II is represented by the manuscripts which contain full proofs in books 7–15, as well as the "directions for proof" for books 1–6[73]. But the situation is more complicated, because in some manuscripts there are proofs, especially in books 7–9, which are quite different from those in the other manuscripts. It seems that these special proofs are reworkings of the original ones. This means that there was no canonical text for Version II in the Middle Ages but that it was at an early stage reworked and commented.

For discussion of the question of authorship, the various stages should be treated separately. The first stage, i.e. the collection of enunciations etc., was not a translation from Arabic, but was compiled from Latin sources. In this part of the text there are only four Arabic expressions in transliteration, and all of them can be found in the Adelard or Hermann text[74]. It is noteworthy that several enunciations and definitions in Version II are almost identical in wording with the Boethius excerpts. In some of them the text is so complex that the agreement cannot be a coincidence. But other enunciations have an "Arabic" formulation. Here Adelard, Version II and Hermann sometimes agree; Adelard and Version II sometimes agree and differ from Hermann, or Hermann and Version II agree and differ from Adelard, but Adelard and Hermann never agree and differ from Version II[75]. Thus, the author of the enunciations had at his disposal both translations from the Arabic, Adelard and Hermann, and excerpts from the Boethius tradition.

The question of authorship of the first stage of Version II is connected with the date of its composition. A *terminus ante quem* is given by the earliest known manuscript, Chartres 498. Together with Chartres 497, it contains the *Eptateuchon*, that is, a collection of texts fundamental for teaching the seven liberal arts, selected by Thierry of Chartres and deposited in the cathedral library some time before his death in 1151. The *Eptateuchon* was probably written before 1141, though the date of

---

[73] Unlike all the Greek-Latin and Arabic-Latin translations and the Greek and Arabic traditions, where the porisms are inserted at the end of the proofs, the porisms in Version II have been appended to the enunciations. This confirms that the text was formed as is described in this article.

[74] *elmuhain*, *elmunharifa*, *alkaide* and *mutekefia*.

[75] See Burnett, Folkerts, Lorch 1987, p. 67.

composition may be doubted[76]. We know neither the exact date of the translation of Adelard nor that of the translation of Hermann, but we do know that in 1138 Hermann was already sufficiently familiar with Arabic to produce his translation of Zael. Thus it is possible that Hermann completed his translation of the *Elements* as early as the late thirties.

Considering this date and the first witness of the Version II text in the *Eptateuchon* by Thierry of Chartres, we must assume that Version II was compiled by someone who knew the works of Adelard and Hermann and was probably acquainted with Thierry. There are good reasons to assume that this man was Robert of Chester (Ketton). Robert lived in Spain ca. 1141–1147, was archdeacon of Pamplona in 1143 and lived in London about 1147–1150[77]. Of the literary partnership and close friendship of Hermann with Robert there is abundant evidence. We know that Peter the Venerable, abbot of Cluny, found Robert and Hermann in 1141 in the region of the Ebro where they were engaged in astrological studies. At this time Robert worked with the assistance of Hermann on the first Latin translation of the *Qur'ān*. Robert's translation of the algebra of al-Khwārizmī, dated Segovia 1145, may be said to mark the beginning of European algebra. Other works of Robert were his revision of Adelard's version of the tables of al-Khwārizmī[78] and his translation of the *Iudicia* of al-Kindī. He dedicated this translation to Hermann. In his dedication Robert states that he has occupied himself with Euclid's *Elements*[79]. On the other hand Hermann dedicated to Robert his translation of Abū Ma'shar's *Introductorium maius* (tr. 1140) and the *De essentiis* (written in 1143). Also in 1143 Hermann completed his translation of Ptolemy's *Planisphaerium*, which he dedicated to Thierry of Chartres. Hermann was a pupil of Thierry and there was a close contact between them. It is remarkable that almost all geometrical, arithmetical and astronomical writings which Thierry inserted into his *Eptateuchon* are texts that were known before the translations from Arabic; the only "modern" texts in this collection of the liberal arts are the Version II of the *Elements* and the astronomical tables of al-Khwārizmī. This cannot be a coincidence.

Let us make a first conclusion: the author of the first stage of Version II was acquainted with Hermann's translation of the *Elements*. It seems that this translation was finished not later than about 1140. Only

[76]See Haskins, 'Studies', p. 226.

[77]See Haskins, 'Studies', especially chapter III and pp. 120–123.

[78]The *incipit* of this text in the only known manuscript (Madrid, BN 10016) is very similar to that of the Version II manuscripts.

[79]See Burnett, 'De essentiis', pp. 5f.

a short time later (in any case, before 1151) Version II was incorporated by Thierry of Chartres into his *Eptateuchon*. Hermann of Carinthia was in close contact with his teacher Thierry. In the early 1140s Robert of Chester and Hermann of Carinthia lived and worked together in Northern Spain and were engaged in translating Arabic texts and writing reworkings. Robert himself attests that he studied Euclid. All these arguments give very good reasons for supposing that Robert of Chester was the author of the first stage of Version II and that he or Hermann sent this work to Thierry, who inserted it in his *Eptateuchon*.

The question of the author of the proofs is more problematic. It is remarkable that in most of the Version II manuscripts the proofs are written *before*, not *after*, the enunciations. In some of the oldest manuscripts[80] the proofs are in the margin next to the propositions. This leads to the suggestion that the proofs were first written in the margin and that by later copyists they were placed mostly before the enunciations. This would also explain confusions within some manuscripts, where the proofs are partly before and partly after the enunciations.

The proofs often contain interesting Arabic transcriptions and other indications of Arabic origin. Some of these words are not present in either Adelard or Hermann. Thus it seems that the compiler had access to Arabic sources, not only to the Latin translations of Adelard and Hermann. A *terminus ante quem* for the proofs is provided by the manuscript Munich, Clm 13021. This manuscript contains no proofs. But it does contain diagrams, and these are lettered in a way characteristic for Version II but not found in other Euclid versions of the twelfth and thirteenth centuries. Therefore the proofs must have existed by the date of this manuscript, i.e. 1163–1168. For these and other reasons[81] it is possible that the proofs, too, were compiled by Robert of Chester. If this assumption is correct, then it is likely that he first compiled only stage 1 of Version II and that he continued his work after the completion of his work on the translation of the *Qurʾān* in 1143, for in the letter to Peter the Venerable which accompanies it he says: "[Your wisdom] has compelled me meanwhile to lay aside my main study of astronomy and geometry"[82].

In any case, differences between the proofs and additional proofs given in some manuscripts suggest that from the beginning the work was continually being changed as it passed through the hands of the students of

[80] E.g. Oxford, TC 47 (which represents stage 2 of Version II), and London, BL, Royal 15 A XXVII.

[81] Busard / Folkerts, 'Robert of Chester', pp. 26–29; Busard, 'Campanus', pp. 8f.

[82] See Burnett, 'De essentiis', p. 6.

geometry. In a later stage of development the proofs seem to have been enlarged by more than one person, and this may explain some of the differences in the proofs within the manuscripts. Some obscure names, such as *Ocreatus*, *Eggebericus*, *Lincol' Zeob' Rog' Hel'* and *Reginerus*, which are given in some proofs, may indicate that these persons contributed to the text of the proofs or have given additional proofs.

4.2 *Liber ysagogarum Alchorismi*

In the twelfth century a mathematical compendium with the title *Liber ysagogarum Alchorismi* was composed by a certain "Magister A". The geometrical part of book 4 is in general in the tradition of the Roman land surveyors (*agrimensores*)[83]. But there exist interpolated versions in some manuscripts[84] which contain additional enunciations. Some of them have been borrowed from Robert's redaction[85]; others seem to belong to the Boethius tradition but are not among the extant Boethius fragments[86].

4.3 *Mélanges* of the Robert and the Boethius texts

Already in the twelfth century scholars attempted to combine Robert's compilation of the *Elements* with the available part of Boethius' translation of Euclid. We know of two independent *mélanges* of this kind. One of these contains most of the definitions, postulates, axioms, and enunciations of the first four books of Euclid[87]. Of this text we know only one manuscript, which was written in northern Germany about 1200 and which was presented by a *magister Helmoldus* to the Michaelis monastery in Hildesheim[88]. The compiler of this text was an able mathematician who composed his text carefully from the two sources. For the definitions and postulates he preferred Boethius' formulation, while in the axioms and enunciations he mostly follows Robert. The compiler often improved the defective text of the Boethius tradition; in many places he wrote his own text, which deviates from both Boethius and Robert.

The second *mélange* is known from five manuscripts, all from the twelfth century, which all go back to a common source, but present dif-

[83]For the content, see Curtze, 'Algorismus-Schrift', pp. 14–16.

[84]Milan, Bibl. Ambros., A 3 sup.; Oxford, BL, Lyell 52; Paris, BnF, lat. 16208. See the list of manuscripts on p. 45.

[85]I.42; III.19, 24, 35; VI.4. See Dickey, 'Adelard of Bath', p. 306 and p. 310, and the text in Busard / Folkerts, 'Robert of Chester', pp. 128, 145, 146, 152, 172.

[86]There is a list of these enunciations on p. 46. See Burnett, 'Adelard of Bath', pp. 174, and Dickey, 'Adelard of Bath', pp. 87–93.

[87]Facsimile edition with transcription in Folkerts, 'Ein neuer Text'.

[88]Now Luneburg, Ratsbibliothek, MS. misc. D 4° 48.

ferent levels of reworking[89]. Two of these manuscripts contain only the first three books of the *Elements*[90] and one consists of fragments from books 3 and 4. All 15 books of the *Elements* are transmitted in only one of these five manuscripts, Paris, BnF, lat. 10257 (**P**)[91].

This manuscript **P** is of special importance. It was also written in the twelfth century and was originally in Chartres[92]. It has enunciations (but no proofs) for all fifteen books. A notable feature is the presence of what appear to be direct Latin transliterations of Greek terminology, e.g. for the various irrational quantities in book 10. But all of these terms appeared long before in the well-known work *De nuptiis Mercurii et Philologiae* of Martianus Capella and thus there is no reason to assume that **P** incorporated material direct from a Greek text. The Greek-Latin terminology found in **P** is also to be seen in a commentary from the early thirteenth century[93], and some enunciations of **P** appear in the margin or as interlinear glosses in two manuscripts containing Robert's redaction[94]. Besides the enunciations, **P** contains in book 1 proof sketches for 11 propositions. These proofs, which were later added by a scribe with some mathematical education, are independent from the known Arabic-Latin versions of the *Elements*, but allude to discussions found in Greek and Arabic commentators. Thus S. Brentjes thinks that they go back to another, unknown, Latin translation[95].

4.4 Johannes de Tinemue (formerly called: "Adelard III")

Now for the version which Clagett called "Adelard III". He thought that it was "an edition" of Version II. Later research, especially by Wilbur Knorr and by H. L. L. Busard, who edited the text, led to new results,

---

[89] For details, see Folkerts, 'Euklidbearbeitungen'. This article was written at a time when Version II was thought to be a reworking by Adelard; thus "Adelard II" should be replaced everywhere by "Version II" or "Robert of Chester's version". The manuscript Munich, Clm 23511, does not transmit this *mélange*, but the text of Version II.

[90] Munich, Clm 13021, and Oxford, BL, Digby 98. From book 4 on, the Munich manuscript contains the text of Version II.

[91] The fifth manuscript, which formerly belonged to the Honeyman collection in San Juan Capistrano (California), is a copy of the Paris manuscript.

[92] See Goldat, 'Early Medieval Traditions', who has edited the text of this manuscript.

[93] See Section 4.5 on p. 21.

[94] See Busard, 'Campanus', p. 12.

[95] Brentjes, 'Remarks', p. 135.

which will be presented here[96]. Accordingly, this version should no longer be called "Adelard III". It will be called here "Version III".

Version III was written after Version II: it quotes some words from the Version II proof of VI.28 (in Heiberg's numeration) word by word, calling this version a *commentum*, and it copied some statements after the proofs of XIV.7 and 8 which are found in Robert's text. Version III refers to propositions I.5 and I.47 (the Pythagorean theorem) by the words *elnethea* or *elnefea* and *dulck* (= *dulcarnon)*, respectively; these words appear in the margin of some Version II manuscripts[97]. On the whole, the enunciations of Version III are very close to those of Robert of Chester's Version II. But Version III has additional material, e.g. commentaries on the definitions in book 2 and subsequent books; and there are some additional proofs, which were very likely borrowed from Aristotle. In Robert's text the letters to indicate points are given in the Arabic sequence, i.e. *a b g d h z*; in Version III they are in the Latin order, i.e. *a b c d e f*. In books 11 and 12 Robert sometimes gives the Arabic letters *alif*, *gim*, *del* to indicate points; in Version III they have been replaced by Latin letters.

A special feature of the style of Version III is the use of the word *falsigraphus* as a technical term to mark off the hypothesis assumed within an indirect proof; this means that the *falsigraphus* becomes the advocate of the false hypothesis. In Version II we find *adversarius* instead of *falsigraphus*, but it should be noted that *falsigraphus* in Version III is more frequently used than *adversarius* in Version II. This very special use of the word *falsigraphus* links the Version III text with the treatise *De curvis superficiebus*, a tract presenting an epitome of Archimedes' proofs on the measurement of the surfaces and volumes of cones, cylinders and spheres[98], where the term *falsigraphus* is used in the same way. There are other striking similarities between the style of Version III and *De curvis superficiebus*[99]. This leads to the conclusion that both treatises are by the same author. Since the *De curvis superficiebus* was written by an (otherwise unknown) Johannes de Tinemue, he was responsible for Version III, too.

Three manuscripts of Version III contain an introduction which is derived partly from Arabic and partly from Latin sources. It includes mate-

---

[96] See Busard, 'Johannes de Tinemue', pp. 12–16, and the summary in Busard, 'Campanus', pp. 12–21.

[97] For the meaning of these words, see Kunitzsch, 'Peacock's Tail'.

[98] Edited with English translation in Clagett, 'Archimedes 1', pp. 450–506.

[99] For details, see Knorr, 'Medieval tradition', and Busard, 'Campanus', pp. 14–16.

rial, common in twelfth-century Latin works, on the liberal arts and the classification of the sciences and also material which ultimately goes back to Proclus' commentary on book 1 of the *Elements* and which was known through Arabic sources. In contrast to the Greek text, Version III has the tendency to multiply cases due to possible variations in the constructions and to use particular numerical values in the proofs of propositions dealing with continuous magnitudes. The author also inserted material from Boethius' *Arithmetic.*

As for the date of composition of Version III it would be helpful to know when Johannes de Tinemue wrote his *De curvis superficiebus.* But, unfortunately, this date is unknown[100]. Since Campanus knew Version III[101], this version must have been composed in the first half of the thirteenth century at the latest. According to the medievalist and paleographer R. A. B. Mynors, the earliest manuscript which transmits the text (Oxford, Balliol College 257) is written in a style of chancery-hand familiar from the late twelfth century. Thus it is possible that Version III originated in the late twelfth century. Its influence was very limited. Besides Campanus, Roger Bacon knew the version; he quotes it in his *Geometria speculativa*, giving it the title of *editio specialis Alardi Bathoniensis*, and distinguishing it from what he calls the *commentum*, i.e. the Robert version.

4.5 Commentary from the early thirteenth century

No less important than Version III was another commentary on the *Elements*, which is only slightly younger than Version III, but was unknown to M. Clagett. It was first mentioned by H. L. L. Busard in 1985[102] and edited by him in 1996[103]. This commentary exists in five manuscripts[104]. It also depends on Version II, from which it took over not only the text of the definitions, postulates, axioms and enunciations, but also some characteristics in style and in the wording of the proofs. In this commentary there are additions that can be found, too, in the text or in the margin of some Version II manuscripts. But in contrast to Version II, the proofs, especially in books 1–3, are here complete and detailed. When a

[100]We only know that *De curvis superficiebus* was cited by Robert Grosseteste in a work written in all likelihood before 1232; see Clagett, 'Archimedes 3', pp. 1253f.

[101]See p. 24.

[102]Busard, 'Early Adaptations', pp. 131f.

[103]Busard, 'Thirteenth-Century Adaptation'.

[104]See p. 47. For details, see Busard, 'Thirteenth-Century Adaptation'. pp. 20–31, and the same, 'Early Adaptations', pp. 131f.

problem admits of several cases, the author proved all of them, and sometimes he provides an alternative proof. In some propositions of book 10 the author used terms for the different types of irrationalities that are transliterated Greek words. These terms, which apparently go back to Martianus Capella's encyclopedic work, are also found in the *mélange* in the manuscript Paris, BnF, lat. 10257[105], but this commentary is the only place where they are used not in the enunciations, but in the proofs.

Some proofs in books 6 and 9 resemble the corresponding proofs in the redaction of Campanus, but, as Busard has shown, this does not mean that the compiler knew Campanus' text, because these proofs can be found in versions anterior to Campanus' redaction. If Campanus' text had existed in the time when this version was composed, in all likelihood the compiler would have used it. There are good reasons to assume that this version originated in the first half of the thirteenth century.

It is of special interest that this anonymous commentary was one of the main sources of the commentary which some scholars ascribe to Albertus Magnus[106].

4.6 Another commentary from the thirteenth century (formerly called: "Adelard IIIB")

For another commentary on the *Elements* Clagett has coined in his 1953 article the name "Adelard IIIB". This version is preserved in a single manuscript (Oxford, BL, Savile 19) and contains only books 1–5[107]. Although this manuscript "agrees with the thirteenth-century manuscript Oxford, BL Digby 174 in proposition I.1"[108] and these folios in Digby 174 contain a Version III text, this does not mean that the Savile text depends on Version III. On the contrary, there are many differences between them: the author of the Savile text used the term *adversarius* more frequently than *falsigraphus* and he gives no comment on the definitions of books 2–4; many of the proofs are notably shorter than those of Version III[109]. But it is evident that the anonymous author of this reworking also used the enunciations of Robert's redaction, but reworked the proofs in his own style. Busard has found that there are some stylistic similarities between the Savile text and Jordanus de Nemore's *Liber philotegni*. One

105 For this, see p. 19.

106 For this, see p. 25.

107 From book VI.1 on we have in this manuscript a straight Campanus text.

108 Clagett, 'Medieval Latin Translations', p. 25.

109 For details, see Busard, 'Campanus', p. 21f.

example: both texts refer to the preceding proposition by means of the phrase "per proximam", while in Version III we find "per premissam".

Since in the Savile manuscript the proofs from book 6 on are those of Campanus, it seems plausible that an unknown person wrote the first five books in his own redaction and from book 6 on took over the Campanus text. If this is true, then the Savile text was written after Campanus had completed his redaction. Busard found that both texts used the same strategy in some proofs which differs from that in the other versions[110]. In any case, the version transmitted in the Savile manuscript was written in the thirteenth century.

4.7 Campanus of Novara

The most important medieval Latin text of the *Elements* by far was the reworking which Campanus of Novara compiled in the middle of the thirteenth century. It replaced the Version II text, which had hitherto been the dominant Latin Euclid text, and was to become *the* text of the *Elements* up to the sixteenth century. The importance of Campanus' text of the *Elements* in the late Middle Ages is shown by the large number of manuscripts which have survived[111]. But Campanus' version was also the only complete medieval Latin text of the *Elements* which was printed before the twentieth century: the first edition, which was published in Venice by Erhard Ratdolt in 1482, is a milestone in the history of early printing and marks the beginning of mathematical typesetting; and in the sixteenth century there were several reprints of Campanus' text, mostly together with Zamberti's translation from the Greek[112]. A new critical edition with commentary and notes has been prepared by H. L. L. Busard[113].

The time of composition follows from the earliest dated extant codex of the work, which was written in 1259. There are some arguments that Campanus wrote it between 1255 and 1259[114].

Contrary to what is stated in many, even in modern, bibliographical works and in books on the history of mathematics, Campanus' work is not a translation. Like most of the Latin texts of the *Elements* before him, it is a reworking and is based upon the text written by Robert of

[110] Busard, 'Campanus', p. 22.

[111] See the list on pp. 48ff.

[112] These editions are listed in Steck, 'Bibliographia Euclideana', pp. 47f. For Zamberti, see p. 29.

[113] Busard, 'Campanus'.

[114] See Busard, 'Campanus', p. 32.

Chester in the twelfth century: Campanus adopted most of the enunciations from Robert, but his proofs are largely independent of Robert's. Especially notable are the additions which Campanus made to his basic text of the *Elements*. He had didactic interests and tried to make the text into a self-contained book by discussing and elucidating many points which were obscure and should be debated. By doing this, he succeeded in writing a handbook of mathematics which was very suitable for the mathematical studies in the Faculty of Arts, and thus it became one of the most important books in the university curriculum. Campanus' method of working has been analyzed by Murdoch[115]. Particularly notable is Campanus' habit of including contemporary and near-contemporary material in his reworking. The most important additions came from Jordanus' *Arithmetica*, from Nayrīzī's commentary on the *Elements*[116], from the so-called "Version III" of the *Elements* (which was written by Johannes de Tinemue)[117] and from the direct translation from the Greek[118]. It is not the place here to go into details. Suffice it to remark that especially in books 7–9 Campanus added a considerable number of definitions, postulates and axioms and even enriched the text of the enunciations, especially by material from Jordanus and al-Nayrīzī, and that he also added much material to books 12–15[119]. Some additions to the text seem to be original with Campanus[120]. There are several remarks that became large comments or even small monographs, e.g. on the definitions of book 5 or on the horn-like angle.

Thus it it no suprise that Campanus' Euclid was very influential. It was used by almost all mathematicians of the fourteenth and fifteenth centuries[121], and if Euclid is cited at this time without giving special information about the version, normally Campanus' reworking is meant. Even when Zamberti's translation from the Greek became available in the early sixteenth century, Campanus' redaction did not fall into oblivion, as can be seen from the various parallel editions of the text by Campanus and by Zamberti.

---

[115] Murdoch, 'Medieval Euclid'.

[116] See p. 12.

[117] See p. 19.

[118] See p. 4.

[119] For details, see Busard, 'Campanus', pp. 33–39.

[120] See Busard, 'Campanus', pp. 48–50.

[121] Three case studies are given in this volume: Johannes Regiomontanus (item VII), Piero della Francesca (item X) and Luca Pacioli (item XI).

4.8 Other reworkings and commentaries from the thirteenth and fourteenth centuries

Besides the reworkings mentioned above, other commentaries to Euclid's *Elements* and related material were written in the thirteenth and fourteenth centuries. In the main they did not concern the complete text of the *Elements* and most of them are known from only one manuscript[122]. Of greater interest are three texts which have been written by, or may be attributed to, well-known scholars: Albertus Magnus, Roger Bacon and Nicole Oresme[123].

There is a single manuscript which contains an otherwise unknown commentary on the first four books of Euclid's *Elements*[124]. At the top of the first folio of this text we find in a hand other than that of the main text the note: *Primus Euclidis cum commento Alberti*. From this Bernhard Geyer argued in an article completed in 1944 and published in 1958 that this commentary must be attributed to Albertus Magnus (d. 1280)[125]. If this were true, this text would be the only known mathematical treatise of Albertus Magnus. Geyer first thought that the Vienna manuscript was very probably an autograph of Albertus Magnus, but later he somewhat modified his enthusiasm. P. M. J. E. Tummers, who published in 1980 an article on this commentary[126] and in 1984 edited the first book of it[127], listed similarities between other writings of Albertus Magnus and some passages in this commentary and concluded that the commentary was a work of Albertus Magnus, but for paleographical reasons he thought that it was a copy and not an autograph. Another possibility was raised by Paul Hossfeld, who proposed that the text had been dictated by Albertus Magnus to a scribe[128]. Anthony Lo Bello, who translated the first book of this commentary from Tummers' edition into English, is absolutely convinced that this commentary is not only by Albertus Magnus, but that it also was dictated by himself about the year 1250. The main argument for Lo Bello is the opinion of Michael I. Allen (Chicago), described as a palaeographer, who dates the manuscript to this time; from this Lo Bello concludes: "Roma locuta est, causa finita est"[129]. The matter is not so

[122]See items 4.8.1–4.8.3 and 4.8.6 in the list of manuscripts in part II of this article.

[123]See items 4.8.4, 4.8.5 and 4.8.7 in the list of manuscripts in part II of this article.

[124]Vienna, Dominikanerkloster 80/45, f. 105r–145r.

[125]Geyer, 'Die mathematischen Schriften'. Geyer's arguments were taken over by Hofmann, 'Euklid-Bearbeitung'.

[126]Tummers, 'Commentary of Albert'.

[127]Tummers, 'Albertus (Magnus)' commentaar', deel II, pp. 1–102.

[128]Hossfeld, 'Euklidkommentar'.

[129]Lo Bello, 'Commentary of Albertus Magnus', p. xxii.

clear to me: anyone who is experienced with medieval manuscripts knows that a manuscript of the thirteenth or fourteenth century can scarcely be dated beyond doubt to a short period; and there are those who date the manuscript to the fourteenth century.

For the composition of his commentary the author employed two principal sources: al-Nayrīzī's commentary on the *Elements*, which had been translated by Gerard of Cremona, and the commentary from the early thirteenth century (see p. 21)[130]. Thus this commentary is not very original.

In a manuscript of the fifteenth century there is a commentary on the *Elements* of which the beginning is lost[131]. As Busard has shown, there are good reasons to assume that Roger Bacon was the author[132]. The subjects which are treated in this commentary have to do with different parts of the *Elements*: general remarks about the category to which an angle belongs or about the knowledge of irrationals required to construct the five Platonic polyhedra; the quadrature of the circle[133]; an arithmetical proof of the incommensurability of the diagonal of a square with its side; remarks on some definitions and propositions in books 1, 5 and 6 of the *Elements*; and a theorem which is needed to criticize a proposition in the *Elementa Jordanis super demonstrationem ponderum.*

Nicole Oresme, who has a special place within the development of science in the fourteenth century, wrote *Quaestiones super geometriam Euclidis.* This treatise is not a "normal" commentary on the *Elements*, but discusses in the form of *quaestiones*, which was typical in the scholastic literature of the time, problems that came from special topics and formulations in Euclid's *Elements.* The text was edited by Busard in 1961 from two manuscripts in the Vatican library[134]. With the help of another manuscript, which Guy Beaujouan had found in Seville, some corrections were made[135]. There exists a fourth, hitherto unknown, manuscript in Greifswald which seems to have come from a German university; this shows that Oresme's work was known in the fourteenth century in Germany, too. The Greifswald manuscript has some similarity with

[130]Tummers, 'Albertus (Magnus)' commentaar', vol. I, pp. 56, 59 *et passim.*

[131]Edited in Busard, 'Roger Bacon'. For the manuscript, see p. 54 below.

[132]Busard, 'Roger Bacon', pp. 199–201.

[133]This part of the commentary agrees with the so-called "Abbreviated Version of the Pseudo-Bradwardine text" which goes back to Archimedes' *Quadrature of the circle.*

[134]Busard, 'Oresme'.

[135]See the essay review in Murdoch, 'Oresme'.

that in Seville[136]. Taken broadly, Oresme's *Quaestiones* dealt with four (not especially Euclidean, but typically scholastic) subjects: mathematics and the infinite, in particular infinite series (*Quaestiones* 1–5); the notions of commensurability and incommensurability (*Quaestiones* 6–9); the graphic representation of "intensible" and "remissible" forms, i.e. the geometry of qualities, which was a specialty of Oresme (*Quaestiones* 10–17); the nature and continuity property of angles (*Quaestiones* 18–21)[137].

* * *

5. Humanistic translations from Greek into Latin; early translations into the vernacular

There is little scholarly work on Euclid's *Elements* in the West during the fifteenth century. Some manuscripts of the fifteenth century seem to be students' notes or copies, but no systematical research on them has been done[138]. In general, these texts were based upon Campanus' reworking of the *Elements.*

The humanists tried to reconstruct the original texts of classical Greek works with the help of Greek manuscripts, which became available to a greater extent in the West during the fifteenth century. These activities were supported by Pope Nicholas V (1447–1455), who built up a spectacular humanist library in Rome, hired copyists and translators and sent agents to buy manuscripts abroad. Nicholas was responsible for new translations from the Greek of Archimedes' works (by Jacobus Cremonensis) and of Ptolemy's *Almagest* (by George of Trebizond)[139]. Some years later Johannes Regiomontanus (1436–1476) tried to reconstruct the original text of the *Elements* with the help of several Latin and Greek manuscripts, some of them made available for him from Bessarion's library[140].

[136]See item 4.8.7 in the list of manuscripts in part II of this article.

[137]For details, see Busard, 'Oresme', pp. 71–176, and Murdoch, 'Oresme', pp. 68–78.

[138]The manuscripts known to me are listed in Section 5.1 of the enclosed list of codices, but this list will be incomplete.

[139]On Nicholas' activities and his interest in the mathematical sciences see Rose, 'Renaissance', pp. 36–38.

[140]For the mathematical manuscripts in Bessarion's collection, see Rose, 'Renaissance', pp. 44–46. The activities of Johannes Regiomontanus in this respect are analyzed in item VII of this volume, where special emphasis is given to his autograph in Nuremberg (Stadtbibliothek, Cent. VI.13).

The first text of the *Elements* which was accessible in print was the edition of Campanus' reworking; it was published in Venice in 1482[141]. The so-called "Geometry II" attributed to Boethius, which contains parts of Boethius' translation of the *Elements*[142], was first published as part of the Collected Works of Boethius in Venice in 1492 and reprinted in 1499[143].

Many propositions of the *Elements*, together with proofs and scholia, were taken over by Giorgio Valla (d. 1499) in his gigantic encyclopaedia entitled *De expetendis et fugiendis rebus* (published Venice, 1501); these extracts were scattered among material translated from other Greek scientific texts. In 1498 Valla had already published in Venice a collection of 25 texts which he had translated from Greek into Latin; among them were books 14 and 15 of the *Elements*[144]. The rubrics in the printed book are unclear: what Valla calls "book 14", is in reality book 15 and what he calls "Hypsicles' translation of the same book of Euclid", is book 14, which indeed was later added by Hypsicles to the 13 genuine books of the *Elements*. Besides the printed edition, there are two manuscripts which contain Valla's translation of the two books, but here in the correct order, though the rubric is as misleading as in the printed text[145]. Valla, who taught humanities at Venice, used his position to further the cause of mathematics. In 1492 he gave courses not only on strictly humanistic topics, but also on Euclid and Vitruvius. In the 49 books of his *De expetendis et fugiendis rebus* mathematics, natural philosophy and medicine have a special place. Valla hoped that his translations of Eu-

---

[141]For Campanus' reworking, see p. 23. Facsimiles of the first page and of the *impressum* are in Steck, 'Bibliographia Euclideana', pp. 218f.

[142]See p. 3.

[143]For details of these editions, see Folkerts, 'Boethius Geometrie II', pp. 41–43. There were similar editions in the sixteenth century; see *idem*, pp. 43–45.

[144]The title page runs: "Georgio Valla Placentino Interprete. Hoc in volumine hec continentur: Nicephori logica. Georgij valle libellus de argumentis. Euclidis quartus decimus elementorum. Hypsiclis interpretatio eiusdem libri euclidis ..." Facsimiles of this page and of two other pages are in Steck, 'Bibliographia Euclideana', pp. 222–224.

[145]Piacenza 211 and Vat. Ottob. lat. 2273; see Section 5.1. Both manuscripts have identical rubrics: "Hypsiclis in deputatum Euclidi volumen interprete Georgio Valla Placentino" (= book 14) and "Euclidis elementorum quartus decimus liber Georgio Valla Placentino Interprete" (= book 15). The translated text in both manuscripts and in the printed edition is identical, but the manuscripts have different letters of dedication: in the Piacenza manuscript, on f. 1r–2v, to "Petrus Barocius Patavii episcopus"; in the Vatican manuscript, on f. 1r–4v, to "princeps Albertus pius Carpensis". The latter dedication is also found in the printed book.

clid and Archimedes would rescue mathematics from the "rust and mud" with which they were covered[146].

The first published edition of a complete Latin text of the *Elements* based upon the Greek text appeared in print in Venice in 1505. It was prepared by Bartolomeo Zamberti (b. 1473, d. not before 1539), a lawyer and a humanist who lived in Venice and tried to restore mathematics and to establish it as a bridge between nature and theology[147]. Zamberti was fully aware that his translation, because it was based upon a Greek manuscript, was superior to the older Latin texts, which were "impure" because they were in the Arabic-Latin tradition. He labelled Campanus as *ille interpres barbarissimus* and complained that his reworking was full of "wondrous ghosts, dreams and fantasies" (*miris larvis, somniis et phantasmatibus*); therefore he intended to turn the student to the more humanistic pleasures of his own edition. Zamberti's main objective was the reconstruction of the *Elements* in its original perfect state, but because the Greek manuscript that he used was poor, his success was somewhat limited. He translated not only the *Elements*, but also the commentary of Proclus on book 1[148].

Zamberti's edition provoked a defence of Campanus by Luca Pacioli, who published, also in Venice, a revised version of the Campanus text in 1509[149]. On the title page[150] Pacioli called Campanus *interpres fidissimus*; the errors of his redaction came not from him, but from careless copyists.

The conflict between the texts of Campanus and Zamberti was not solved in the following decades. Instead of deciding, in 1516 the French humanist Jacques Lefèvre d'Etaples published both versions in Paris in a composite volume, and until 1558 several other parallel editions of Cam-

[146] *rubigine et squalore (quibus), quod vehementer doleo, bonae demersae sunt disciplinae*; cited from Rose, 'Renaissance', p. 49. For Valla's activities in this field see *id.*, pp. 46–50. Valla's translations of Euclid were based upon a Greek manuscript now in Modena; see Heiberg, 'Prolegomena', p. CII. The fundamental study on Valla is Heiberg, 'Beiträge'.

[147] On Zamberti see Rose, 'Renaissance', pp. 50f., and especially the detailed monograph by Weißenborn, which on Zamberti is still the definitive study (Weißenborn, 'Übersetzungen'). Facsimiles of the title-page and the *impressum* are in Steck, 'Bibliographia Euclideana', pp. 227f.

[148] Besides the printed text, there are a few manuscript copies; one of them also contains his translation of Proclus' commentary (see p. 57).

[149] For Pacioli's edition, see item XI in this volume.

[150] Facsimile in Steck, 'Bibliographia Euclideana', p. 234.

panus and Zamberti were printed; and also both versions were published separately[151].

To a certain degree, the medieval development came to an end in 1533, when the Greek text of the *Elements* was first published in Basel. This edition, which included the Greek text of Proclus' commentary as well, was brought out by the theologian Simon Grynaeus (d. 1541)[152] and was the only complete edition of the Greek before the eighteenth century.

The history of the text of Euclid's *Elements* in modern times cannot be given here[153]. But we may end with some remarks on the first translations into the vernacular.

The earliest of them were translations into Italian, and some of them were made before 1500[154]. In most cases the translators are not known. An Italian translation of the Campanus version is found in two fifteenth-century manuscripts now in Siena. One manuscript, dated 1485 and now in Pisa, is associated with Raffaello di Giovanni di Guglielmo Canacci, who was a *maestro d'abbaco* in Florence and may have been the translator. From the middle of the sixteenth century we know about Italian translations by an (otherwise unknown) Costanzo Bolognese (who was public lector of astronomy in Milan), by Fabrizio Sanseverini of Ferrara and by the well-known Florentine humanist Benedetto Varchi, who translated the text from the Greek, apparently from the *editio princeps* of 1533[155].

At almost the same time the first vernacular text of the *Elements* was printed: the Italian translation by Niccolò Tartaglia (Venice, 1543)[156]. In the next three decades there followed translations into German, French and English: a German translation of books 7–9 by Johann Scheubel, professor at the University of Tübingen, was printed in Augsburg in 1555[157]; there followed a German translation of books 1–6 by Wilhelm Holtzmann

---

[151] For details, see Rose, 'Renaissance', p. 52, and Steck, 'Bibliographia Euclideana', pp. 46–48.

[152] Facsimiles of the title-page and the first page of Proclus' commentary are in Steck, 'Bibliographia Euclideana', pp. 251f.

[153] A summary can be found in Murdoch, 'Transmission', pp. 448–452.

[154] A list of the oldest Italian manuscripts is given on pp. 57f.

[155] See Pagli, 'Volgarizzazioni', pp. 53–55 (summary), 55–90 (detailed description of every manuscript).

[156] Title-page in Steck, 'Bibliographia Euclideana', p. 257.

[157] Title-page in Steck, 'Bibliographia Euclideana', p. 271. An unpublished Latin commentary by Scheubel to books 11–15 of the *Elements*, dated 1561, is in manuscript Vatican, Palat. lat. 1350; a student's copy of his lectures on Euclid held at the University of Tübingen is in Tübingen, UB, Mb 6.

(Xylander; Basel 1562)[158]. The first printed French translation was that of Pierre Forcadel (Paris, 1564)[159]. An English translation was published by Henry Billingsley, with a preface by John Dee (London, 1570).

* * *

II. Medieval manuscripts containing Euclid's *Elements*

This is a list of all known Latin and vernacular manuscripts up to the beginning of the sixteenth century that contain the text of Euclid's *Elements* or reworkings, commentaries and related material. Included is a list of the known manuscripts of translations from Greek into Arabic, because these Arabic texts were used for the translations into Latin in the twelfth century. In 'Printed editions' usually only the latest edition is given, which mostly refers to older ones. The 'modern' Euclid texts – mostly from the fifteenth century – are listed in Section 5.1 without distinguishing the different texts; there is some information about them, but further research is necessary to obtain certain results.

The parts of Euclid's *Elements* contained in each manuscript are indicated. For example, "I–X" means that the complete text of each book 1 to 10 respectively (including the definitions, postulates and axioms) is found. "I.1" indicates proposition 1 of book 1, i.e. the theorem with its proof. If there is no proof, this is indicated by "enunciation only" or "without proof".

1. Translations from the Greek before the humanistic movement

1.1 Boethius tradition

1.1.1 Cassiodorus, "Institutiones", third recension (**Ma**)

Incipit:

I. def. 1: *Punctum est cuius pars nulla est.*

Manuscripts:

Berne, Burgerbibliothek, Ms. 87, f. 9rv, a. 1004 (excerpts of the *Institutiones* II.5–6 and extracts of Euclid [II. def. 2; V. deff.])

Berne, Burgerbibliothek, Ms. 212, f. 1r–110r, s. 9

Berne, Burgerbibliothek, Ms. 234, f. 1r–53v, s. 9/10

Cambridge, TC, R.15.14 (939), f. 81r–87r, s. 10 (excerpts of the *Institutiones* II.5–6 and extracts of Euclid)

[158] Title-page in Steck, 'Bibliographia Euclideana', p. 279.

[159] Facsimile of one page in Steck, 'Bibliographia Euclideana', p. 284.

Chartres, Bibl. de la Ville, Ms. 90, s. 10 (The manuscript was burnt during World War II)
Chartres, Bibl. de la Ville, Ms. 102, f. 1r–61v, s. 10 (The manuscript was burnt during World War II)
Glasgow, UL, Hunter. 281, s. 10/11
Karlsruhe, Badische LB, Augiensis 106, s. 10
Munich, BSB, Clm 13084, f. 54rv, s. 10 (only II. def. 2; V. def. 1–8. 11. 9. 10. 13. 12. 14–16. 18. 17; does not contain the text of Cassiodorus)
Naples, BN, V A 13, f. 15v–16v, 37r–39r, s. 9 (excerpts of the *Institutiones* II.5–7 and extracts of Euclid, as Munich, Clm 13084)
New York, CU, MS Plimpton 164, f. 16v–17v, s. 16 (excerpts of the *Institutiones* II.5–7 and extracts of Euclid, as Munich, Clm 13084)
Oslo and London, The Schøyen Collection, MS 1800 (formerly Cheltenham, Phillipps MS 16278), s.9[2]
Paris, BnF, lat. 8679, s. 10 (incomplete)
Paris, BnF, lat. 12963, s. 10[1]
Paris, BnF, lat. 13048, f. 59r–82r, s. 10
private possession (formerly Cheltenham, Phillipps MS 7071), f. 40r–42v, s. 12 (excerpts of the *Institutiones* II.5–6 and extracts of Euclid)
Reims, Bibl. Mun., Ms. 975, f. 33ff., s. 10
St. Gallen, Stiftsbibliothek, Ms. 199, p. 115–374, s.10
Valenciennes, Bibl. Mun., MS 195, s. 9

PRINTED EDITIONS OF THE EUCLID EXCERPTS:

Mynors, 'Cassiodori Senatoris Institutiones', pp. 169–172; Folkerts, 'Boethius Geometrie II', pp. 177–185, 189–191 (**Ma**).

COMMENTS:

See p. 3 and Folkerts, 'Boethius Geometrie II', pp. 70–82. Extant are: I. def. 1–12. 14. 13. 15–23, post. 1–5, ax. 1. 3. 2. 7; II. def. 2; V. def. 1–8. 11. 9. 10. 13. 12. 14–16. 18. 17.

1.1.2 EUCLID EXCERPTS IN MANUSCRIPTS OF THE "CORPUS AGRIMENSORUM" (Mb)

INCIPITS:

I. def. 1: *Punctum est cuius pars nulla est.*
I.1, enunciation: *Super datam rectam lineam terminatam triangulum aequilaterum constituere.*
I.1, proof: *Sit data recta linea terminata AB. Oportet igitur super eam, quae est AB, triangulum aequilaterum constituere ...*

MANUSCRIPTS:

Berne, Burgerbibliothek, Ms. 87, f. 16v, a. 1004 (only I.1–3 with proofs)
Brussels, BR, MS 4499–4503, f. 4v–5r, s. 12in. (only I. def. 19–23)

Brussels, BR, MS 10615–10729, f. 47r–v, s. 12 (copy of Vatican, Palat. lat. 1564)

Cambridge, TC, R.15.14 (939), f. 96r–97v, s. 10 (only I.1–3 with proofs)

Chartres, Bibl. de la Ville, Ms. 498, f. 153r, s. 12^1 (only I. def. 19–23. The manuscript was burnt during World War II, but a microfilm has survived)

Munich, BSB, Clm 13084, f. 53v–54r, s. 10 (only I. def. 19–23, post. 1–5, ax. 1. 3. 2. 7)

Naples, BN, V A 13, f. 30v–31r, s. 9 (only I. 1–3 with proofs)

Paris, BnF, lat. 8679A, f. 19r–21r, s. 16 (copy of Wolfenbüttel, HAB, Cod. Guelf. 105 Gud. lat.)

Vatican, Ottob. lat. 1862, f. 26r, s. 12^2 (only I. def. 19–23)

Vatican, Palat. lat. 1564, f. 79r–82r, s. 9

Wolfenbüttel, HAB, Cod. Guelf. 105 Gud. lat., f. 43r–45v, s. 9

PRINTED EDITION:

Folkerts, 'Boethius Geometrie II', pp. 177–185. 215–217 (**Mb**). As to older editions, see p. 48 and p. 174.

COMMENTS:

See p. 3 and Folkerts, 'Boethius Geometrie II', pp. 70–82. The main manuscripts (Vatican, Palat. lat. 1564, and Wolfenbüttel, HAB, Cod. Guelf. 105 Gud. lat.) contain: I. def. 1–12. 14. 13. 15–23, post. 1–5, ax. 1. 3. 2. 7, prop. 1–3 with proofs.

1.1.3 PSEUDO-BOETHIUS, "GEOMETRY I" (**Mc**)

INCIPITS:

Text: *Geometria est disciplina magnitudinis immobilis ...*

I. def. 1: *Principium mensurae punctum vocatur, cum medium tenet figurae. Punctum est cuius pars nulla est.*

I.2, enunciation: *Ad datum punctum datae rectae lineae aequalem rectam lineam collocare.*

MANUSCRIPTS:

Bamberg, Staatliche Bibliothek, Ms. class. 55 (formerly H.J. IV 22), f. 1v–16r, s. 10in.

Berne, Burgerbibliothek, Ms. 87, f. 1v–8v, a. 1004

Berne, Burgerbibliothek, Ms. 299, f. 1r–14r, s. 11

Cambridge, TC, R.15.14 (939), f. 4r–43v, s. 10

Cesena, Bibl. Malat., Plut. sin. XXVI.1, f. 221r–249r, ca. 1450–1465

Chartres, Bibl. de la Ville, Ms. 498, f. 141v–143r, s. 12^1 (excerpts; no Euclid texts. The manuscript was burnt during World War II, but a microfilm has survived)

Copenhagen, KB, Gl. kgl. S. 277 fol., f. 28v–29r, s. 13 (fragment)

Einsiedeln, Stiftsbibliothek, Cod. 298, p. 1–22, s. 10/11

Einsiedeln, Stiftsbibliothek, Cod. 358, p. 1–20, 23–37, s. 10/11

Florence, BLaur., Plut. 29.19, f. 1r–27r, ca. 1250
London, BL, Addit. 47679, f. 77rv, 97r–124v, 130v, s. 12
Madrid, BN, Ms. 9088 (formerly Aa 53), f. 130r–142r, s. 11
Munich, BSB, Clm 560, f. 122r–149r, s. 9
Munich, BSB, Clm 4024a, f. 3r–9v, s. 13 (fragment)
Naples, BN, V A 13, f. 1v–15r, s. 9
New York, CU, MS Plimpton 164, f. 1r–15v, s. 16 (copy of Naples V A 13)
Oxford, BL, Douce 125, p. 1–67, s. 10
Paris, BnF, lat. 8732B, f. 99rv, 118r–119v, s. 17 (excerpts)
Paris, BnF, lat. 13020, f. 59v–83r, s. 9
Paris, BnF, lat. 13955, f. 107r–123v, s. 9
Paris, BnF, lat. 14080, f. 65r–72v, 80r–87v, 79r–v, 73r–77v, s. 9
Prague, UK, IX.C.6 (= 1717), f. 47r–63v, s. 11
private possession (formerly Cheltenham, Phillipps MS 7017), f. 1r–22v, s. 12
Rostock, UB, Ms. phil. 18 (formerly IV 111,4), p. 1–17, s. 12
St. Gallen, Stiftsbibliothek, Ms. 830, p. 283–309, s. 11
Vatican, lat. 3353, f. 379r–383r, s. 16 (incomplete)
Vatican, lat. 3906, f. 189r, s. 16 (fragment)
Vatican, lat. 6017, f. 72r–90v, s. 12
Vatican, Barb. lat. 92, f. 19v–22v, 38v–44r, ca. 1100
Vatican, Ottob. lat. 1862, f. 37r–44v, s. 12
Vatican, Regin. lat. 1023, f. 16r, s. 9 (fragment)
Vienna, ÖNB, Ms. 55, f. 1r–13v, 22r–v, 14r–21v, s. 10
Vienna, ÖNB, Ms. 2269, f. 220r–223r, s. 11
Wrocław, BU, Rehdig. 55, f. 2r–26r, s. 12 (copied from Rostock, Ms. phil. 18)

PRINTED EDITION:

Edition of the Euclid excerpts: Folkerts, 'Boethius Geometrie II', pp. 176–214 (**Mc**). For older editions of the Euclid excerpts, see p. 48 and p. 174. There is no complete edition of the Geometry I text; the partial editions are listed in Folkerts, 'Altercatio', pp. 88f.

COMMENTS:

See p. 3 and Folkerts, 'Boethius Geometrie II', pp. 70–82. Extant are the following Euclid parts: I. def. 1–12. 14. 13. 15–23, post. 1–5, ax. 1. 3. 2. 7; II. def. 1. 2, prop. 1; III. def. 1–6. 8–11; IV. def. 1. 2, prop. 1; III. def. 6. 8; I.2–4. 6–8. (9.) 10–18. 21. 23. 26–28. 31–37. 39–41. 43. 42. 44–48; II.1. 3–6. 9–12. 14; III.3. 7 beginning. 22 end. 27. 30–33; IV.1–4. 6. 8. 12. 11; III.7 end. 9. 12. 10. 13. 14. 16. 18. 19. 24. 22 beginning (all propositions without proofs). – The folio numbers given here refer to the beginning and end of the Geometry I text, not of the Euclid excerpts.

1.1.4 Pseudo-Boethius, "Geometry II" (Md)

Incipits:

Text: *Quia vero, mi Patrici, geometrum exercitissime ...*

I. def. 1: *Principium autem mensurae punctum vocatur. Punctum est cuius pars nulla est.*

I.1, enunciation: *Supra datam rectam lineam terminatam triangulum aequilaterum constituere.*

I.1, proof: *Sit data recta linea terminata AB. Oportet igitur super eam, quae est AB, triangulum aequilaterum constituere ...*

Manuscripts:

Berlin, SB, Ms. lat. oct. 162, f. 69r–70r, 71v–76r, s. 12^1 (excerpts; no Euclid text)

Brussels, BR, MS 4499–4503, f. 28r–v, s. 11 (excerpts; no Euclid text)

Cesena, Bibl. Malat., Plut. sin. XXVI.1, f. 201r–220r, ca. 1450–1465

Chartres, Bibl. de la Ville, Ms. 498, f. 155r–166r, s. 12^1 (The manuscript was burnt during World War II, but a microfilm has survived)

Erlangen, UB, MS 379, f. 1v–76r, s. 11

Groningen, Bibl. RU, MS 103, f. 207v–216r, s. 16 in. (excerpts)

London, BL, Addit. 47679, f. 170r–171r, s. 12 (excerpts; no Euclid text)

London, BL, Arundel 339, f. 49r–59r, s. 13

London, BL, Harley 3595, f. 57r–68r, s. 11

London, BL, Lansdowne 842B, f. 51r–65v, s. 15

Munich, BSB, Clm 4024^a, f. 1v–2v, s. 13 (fragment)

Munich, BSB, Clm 13021, f. 194r–202r, s. 12^2

Munich, BSB, Clm 23511, f. 30r–51v, s. 12/13

New York, CU, MS Plimpton 250, f. 98r–129r, ca. 1500 (copy of Vat. lat. 3123)

Paris, BnF, lat. 7185, f. 62v–80r, s. 12ex.

Paris, BnF, lat. 7377C, f. 18v–24r, 25v–34v, s. 11ex.

Rostock, UB, Ms. phil. 18 (formerly IV 111,4), p. 29–30, s. 12 (fragment)

Vatican, lat. 3123, f. 65v–84r, s. 12

Vatican, Barb. lat. 92, f. 3r–16v, ca. 1100

Vatican, Ottob. lat. 1862, f. 26v–33v, s. 12^2

Vatican, Regin. lat. 1071, f. 56r–58v, s. 12 (incomplete)

Wrocław, BU, Rehdig. 55, f. 38r–39r, s. 12 (fragment; copied from Rostock, Ms. phil. 18)

Printed edition:

Folkerts, 'Boethius Geometrie II', pp. 113–135 and 177–211, 215–217 (**Md**). As to older editions, see pp. 41–49 and p. 174.

Comments:

See p. 3 and Folkerts, 'Boethius Geometrie II', pp. 70–82. Extant are the following Euclid parts: I. def. 1–12. 14. 13. 15–23, post. 1–5, ax. 1. 3. 2. 7; II. def. 1.

2; III. def. 1–6. 8–11; IV. def. 1. 2; I.1–8. (9.) 10–41. 43. 42. 44–48; II.1. 3–6. 9–12. 14; III.3. 7 beginning. 22. 27. 30–33; IV.1–4. 6. 8. 12. 11 (all propositions without proofs); further, I.1–3 with proofs. – The folio numbers given here refer to the beginning and end of the Geometry II text, not of the Euclid excerpts.

1.2 Other early texts based upon the Greek

1.2.1 Verona palimpsest

Incipit:
XI.24, enunciation: *Si solidum parallelis planis circumdatum est ...*

Manuscript:
Verona, Bibl. Capit., MS XL (38), f. 331v–r, 326v–r, 341r–v, 338r–v, 336r–v, 343r–v, ca. 500

Printed edition:
Geymonat, 'Fragmenta Veronensia'.

Comments:
See p. 4. Fragments of the following propositions are present: XI.24–25; XII.2–3. 8; XIII.2–3. 7.

1.2.2 Munich fragment

Incipit:
I.38, enunciation: *Quae triangula quo aequalis grados sunt et autem in ipsis utraque aequales alterutrum sunt ...*

Manuscript:
Munich, UB, MS 2° 757, 2ff., s. 9in.

Printed edition:
Geymonat, 'Nuovi frammenti', with remarks on the older edition by Curtze.

Comments:
See p. 4. Extant are the propositions I.37–38 and II.8–9 (beginning).

1.3 Greek-Latin translation from the twelfth century

Incipits:
I def. 1: *Punctus est cuius pars nulla.*
I.1, enunciation: *Super datam rectam terminatam trigonum isopleurum constituere.*
I.1, proof: *Esto data recta terminata* ab. *Oportet ergo super* ab *rectam trigonum isopleurum constituere ...*

MANUSCRIPTS:

Paris, BnF, lat. 7373, f. 2r–167v, 173r–175v (I–XIII, XV); f. 167v–172v (abbreviated version of books 14 and 15), s. 13. (f. 2r–v has definitions, postulates, and common notions of book 1 in the Campanus version, s. 14)

Florence, BNC, Conv. soppr. C I 448, f. 1r–104v, s. 14 (incomplete; ends within X.48)

Venice, BNM, fondo antico 271 (= 1642), f. 1r–v, s. 15 (I. def. 1–20)

PRINTED EDITION:

Busard, 'Translation ... from the Greek'.

COMMENTS:

See p. 4 and Busard, 'Translation ... from the Greek'. The abbreviated version of books 14–15, which is transmitted only in the Paris manuscript, was known to Leonardo Fibonacci; see Busard, 'Translation ... from the Greek', pp. 17–20.

* * *

2. The Arabic Euclid

2.1 Isḥāq–Thābit text

MANUSCRIPTS:

Cambridge, UL, Addit. 1075, 233ff., 13th c. AD (I. def. 19 – XV)

Copenhagen, KB, MS 81, 214ff. (V–XV)

Dublin, Chester Beatty Library, MS 3035, AH 669 / 1270–71 AD, 126ff. (I.39–XV)

*Dunedin, Otago University Library, De Beer 8, AH 873 / 1468–69 AD (I. def. 10–III)

El Escorial, Real Biblioteca de San Lorenzo, MS ar. 907, 184ff., about 13th c. AD (I–XV), Maghrebi script

Istanbul, Süleymaniye Kütüphanesi, Fatih 3439/1, f. 2r–65v, salkh Dhī 'l-Ḥijja 586 / 28 January 1191 (III.1–7, IV.14–XV)

*Kabul, Kitābkhāne-i Wizārat-i Maʿārif, 241ff., 16th c. AD or later

Kastamonu, Il Halk Kütüphanesi, Ms. 73, f. 1v–69v, 14th c. AD (I–XV, enunciations and diagrams only)

Oxford, BL, Huntington 435, 221ff. (I.13–XV)

Oxford, BL, Thurston 11, 212ff., finished 13 Jumādā I 635 / 1 January 1238 (I–XV)

*Paris, BnF, hébr. 1381 (Arabic text written in Hebrew characters)

Rabat, Al-Khizāna al-ʿĀmma (Bibliothèque Générale), MS 53, pp. 46–178, Dhū 'l-Qaʿda 1015 / March 1607 (I–V), Maghrebi script

Rabat, Al-Khizāna al-ʿĀmma (Bibliothèque Générale), MS 1101, 125 ff., 23 Muḥarram 683 / 8 April 1284 (I–XV), Maghrebi script

*Rampur, Raza Library, MS 3656 ('Arshī 200), 214ff., 15th c. AD (I–XV)

St. Petersburg (formerly Leningrad), Akademija Nauk, Institut Vostokovedenija, C 2145, 279ff., Jumādā II 584 / August 1188 (I–XV)

Tehran, Dānishgāh 2120, pp. 1–12 (the missing part, VII.20–39, from MS Tehran, Malik 3586)

Tehran, Majlis-i Shūrā, MS 200, 226ff., 15th c. AD (I–XV) (almost certainly copied from the Dublin manuscript)

Tehran, Malik 3586, 240ff., AH 343 / 954–55 AD (I–XV) (the missing part, VII.20–39, is MS Tehran, Dānishgāh 2120)

Uppsala, UB, O. Vet. 20, 202ff., Rabīʿ I 434 / Oct.–Nov. 1042 (I–XV)

EDITIONS:

There are only (unpublished) editions of books 5 (Engroff, 'Arabic Tradition') and 7–9 (De Young, 'Arithmetic Books').

COMMENTS:

All the known manuscripts seem to represent the Isḥāq–Thābit text, more or less often interspersed with quotations or portions obviously from al-Ḥajjāj. An exact identification of all parts of text in all manuscripts needs further detailed studies. The data of the Arabic manuscripts described here were collected by Paul Kunitzsch from the microfilms of manuscripts kept in the "Lehrstuhl für Geschichte der Naturwissenschaften" (LGN), Munich. Manuscripts of which no microfilms exist in the LGN, Munich, and which therefore could not be inspected, are, here and in section 2.3, marked *.

2.2 al-Nayrīzī's commentary

MANUSCRIPTS:

Leiden, Bibl. RU, Or. 399/1, 81ff., 8 Rabīʿ I 539 / 8 September 1144 (I – beginning of VII; lacuna at the beginning of I)

Qum, The Public Library of Ayatallah Al-Uzma Al-Marashi Al-Najafi, MS 6526, 112ff. (I. def. 4–V.25)

EDITIONS:

Besthorn, 'Codex Leidensis' (with Latin translation of the Arabic). An English translation of book 1 is in Lo Bello, 'Commentary of al-Nayrizi', pp. 86–203.

COMMENTS:

See p. 6. The commentary was translated into Latin by Gerard of Cremona; see p. 12.

2.3 Related manuscripts

*Mumbai (Bombay), Mullā Fīrūz R I.6, 81 ff., 12th c. AD (according to Brentjes, 'Al-Karābīsī's Commentary': "an anonymous edition of the 'Elements'"; according to Sezgin, 'GAS 5', p. 400: "an anonymous commentary (or a *Kitāb Aghrāḍ al-Uṣūl*), Book of the Intentions of the Elements")

Paris, BnF, ar. 2500, f. 95v–109r, 17 Rajab 806 / 30 January 1404 (according to Kunitzsch, 'Peacock's Tail', an anonymous abbreviation of I)
*Paris, BnF, pers. 169 (Arabic, extracts from II; according to Brentjes, 'Textzeugen', not a direct copy of an Arabic translation)

EDITIONS:
Paris, BnF, pers. 169 (in Brentjes, 'Textzeugen').

COMMENTS:
According to Brentjes, Mumbai and Paris pers. 169 may contain portions from the Ḥajjāj II version.

* * *

3. Translations from the Arabic

3.1 Adelard of Bath (formerly called: "Adelard I")

INCIPITS:
I. def. 1: *Punctus est illud cui pars non est.*
I.1, enunciation: *Nunc demonstrandum est quomodo superficiem triangulam equalium laterum super lineam rectam assignate quantitatis faciamus.*
I.1, proof: *Sit linea assignata* ab. *Ponaturque centrum supra* a *occupando spacium quod est inter* a *et* b *circulo, supra quem* gdb ...

MANUSCRIPTS:
Bruges, Stadsbibliotheek, MS 529, f. 1r–48v, s. 13 (I–VIII.25)
Florence, Bibl. Ricc., MS 2968/2, f. 23r, s. 14 (enunciations of X.24 and X.17)
Leipzig, UB (formerly Stadtbibliothek), Rep. I, 68c, f. 59r, s. 13 (enunciations of X.24 and X.17)
London, BL, Burney 275, f. 302r–308r, s. 14 (VII.3–VIII.25)
Oxford, BL, Auct. F.3.13, f. 223r–224r, s. 13^2 (IV. def.–IV.7)
Oxford, BL, D'Orville 70, f. 39r–61v, 63r–71v, s. 13/14 (X.36–XV.2)
Oxford, TC, MS 47, f. 171r–180v, 163r–170v, 155r–162v, 147r–154v, 139r–146v, s. 12 (I–VIII.22)
Paris, BnF, lat. 16201, f. 35r–82r, s. 12 (I–VIII.22)
Paris, BnF, lat. 16648, f. 58r, s. 13 (enunciations of X.24 and X.17)
Vatican, Regin. lat. 1137, f. 73v–74r, s. 13^2 (enunciations of X.24 and X.17)
Vienna, ÖNB, Ms. 83, f. 64v–65v, s. 14 (X.36–49)

EDITIONS:
Busard, 'Adelard of Bath'.

COMMENTS:
See p. 9 and Busard, 'Adelard of Bath'. Books IX and X.1–35 are missing, but the enunciations (not the proofs) of X.17 and X.24 are in MSS Florence, Bibl. Ricc., MS 2968/2; Leipzig, UB, Rep. I 68c; Paris, BnF, lat. 16648; Vatican, Regin. lat. 1137. These enunciations (together with the proofs, which are from

Version II) are edited in Busard, 'Adelard of Bath', pp. 9f., and Busard / Folkerts, 'Robert of Chester', p. 88.

3.2 Hermann of Carinthia

INCIPITS:

Introduction: *Septem sunt omnis discipline fundamenta ...*

I. def. 1: *Punctus est cui pars non est.*

I.1, enunciation: *Primum igitur equilaterum triangulum supra rectam et definite quantitatis lineam collocamus.*

I.1, proof: *Data siquidem linea recta inter* a *et* b *puncta, acceptaque punctorum distancia, id est linee spacio fixo circino supra* a *centrum fiat circulus* bgd ...

MANUSCRIPTS:

Paris, BnF, lat. 16646, f. 2r–108r, s. 13 (I–XII)

Oxford, BL, Digby 174, f. 160r–v, s. 13 (XI–XI.4, going back to a Hermann version)

Vatican, Regin. lat. 1268, f. 113r–v, s. 14in. (XI–XI.4, in agreement with Digby 174)

EDITIONS:

Busard, 'Hermann of Carinthia'.

COMMENTS:

See p. 10 and Busard, 'Hermann of Carinthia'.

3.3 Gerard of Gremona

INCIPITS:

Introduction: *Ea a quibus procedit scientia ...*

I. def. 1: *Punctum est cui pars non est.*

I.1, enunciation: *Super rectam lineam definite quantitatis triangulum equilaterum constituere.*

I.1, proof: *Verbi gratia: Ponatur linea recta* ab *definite quantitatis, et super centrum* a *secundum quantitatem spatii quod est inter* a *et* b, *circumducatur circulus, super quem sunt* g, d, b ...

Scholia: *In scriptis que transtulit Ysaac hoc quod sequitur invenitur ...*

MANUSCRIPTS:

Boulogne-sur-Mer, Bibliothèque de la Ville, MS 196, f. 1r–144r (I–XV), 144r–147v (scholia I–XIX), s. 14

Bruges, Stadsbibliotheek, MS 521, f. 1r–109v (I–XV), 109v–113v (scholia I–XVII), s. 14

Florence, BNC, Conv. soppr. J V 30, f. 55r–57v, s. 14 (scholia IV–XVII, XX–XXII)

Madrid, BU, MS 102 (117-z-6), f. 137r–196v, s. 14 (I–X.101)

Milan, Bibl. Ambr., D 186 inf., f. 1r–107v (I–X.109, XIII.11–XV.5), 107v–112r (scholia I–XVII), s. 15
Oxford, BL, Canon. misc. 145, f. 137r, s. 15 (III. def., III.1 porism, III.2–7 enunciations)
Oxford, BL, Digby 174, f. 160v–173v, s. 13/14 (XI.5–XIV.1)
Palermo, Biblioteca Comunale, 2 Qq E 98, f. 1r–83v, s. 14 (I–X.17)
Paris, BnF, lat. 7216, f. 1r–107v (I–XV), 107v–108r (scholia XVIII–XIX), s. 15
Paris, BnF, lat. 7377B, f. 1r–9r, s. 14/15 (scholia I–XVII; scholium XVII twice: at the end and after scholium III)
Paris, BnF, lat. 11247, f. 78r–91v, s. 16 (scholia IV–XVII, XX–XXII)
Rome, Bibl. Vallic., MS F.86, f. 49r–54r, s. 14 (I–I.5)
Vatican, lat. 7299, f. 1r–136r (I–XV), 136r–141r (scholia I–XVII), s. 14
Vatican, Regin. lat. 1268, f. 113v–142v (XI.5–XV), 142v–143r (scholia I–III), s. 14
Vatican, Regin. lat. 1904, f. 86r–90r, s. 15/16 (XV.1–5, XIII.16)
Vatican, Rossiano 579, f. 1r–126v (I–XV), 126v–130v (scholia I–XVII), s. 14

EDITIONS:
The main text is edited in Busard, 'Gerard of Cremona'. The *scholia* are edited in Busard, 'Euklid-Scholien'.

COMMENTS:
See p. 11; Busard, 'Gerard of Cremona' and Busard, 'Euklid-Scholien'. The *scholia*, most of which belonging to books 13 and 14, differ within the manuscripts.

3.4 Commentaries to the *Elements* translated from the Arabic

3.4.1 AL-NAYRĪZĪ, COMMENTARY ON BOOKS 1–10

INCIPITS:
I. def. 1: *Dixit Euclides: Punctum est quod partem non habet ...*
I.1: *In primo theoremate sunt quinque figure, una Euclidis et 4 Irini.*

MANUSCRIPTS:
Cracow, BJ, 569, f. 1r–23r (p. 7–51), s. 14
Madrid, BN, Ms. 10010, f. 13v–36v, 49v–50r, s. 13/14
Oxford, BL, Digby 168, f. 124r–125r, s. 14 (fragments)
Vatican, Regin. lat. 1268, f. 144r–183v, 206r–207v, s. 14in.

EDITIONS:
Curtze, 'Anaritii ... commentarii', pp. 1–252. Edition of book 1 in Tummers, 'Albertus (Magnus)' commentaar', deel II, pp. 121–190; of books 1–4 in Tummers, 'Anaritius', pp. 1–124. English translation of book 1, from Curtze's edition, in Lo Bello, 'Gerard of Cremona's Translation', pp. 14–106.

COMMENTS:

Commentary by al-Nayrīzī on Euclid I–X in the translation of Gerard of Cremona. See p. 12.

3.4.2 MUḤAMMAD IBN 'ABD AL-BĀQĪ (?), COMMENTARY ON BOOK 10

INCIPIT:

Cum quantitates ad invicem comparantur, alie earum sunt communicantes, alie incommunicantes ...

MANUSCRIPTS:

Cambridge, UL, Mm. 2.18, f. 49v–65r, s. 14
Cracow, BJ, 569, f. 23r–37v (p. 51–80), s. 14
Madrid, BN, Ms. 10010, f. 36v–49v, s. 13/14
Paris, BnF, lat. 7377A, f. 1r–33v, s. 14
Paris, BnF, lat. 9335, f. 92v–110v, s. 13
Vatican, Regin. lat. 1268, f. 183v–205v, s. 14in.

EDITIONS:

Curtze, 'Anaritii ... commentarii', pp. 252–384. Modern edition: Busard, 'Commentary on Book X', pp. 33–110.

COMMENTS:

The author of this Arabic commentary on book 10 of the *Elements* could have been Muḥammad ibn 'Abd al-Bāqī, Abū Yūsuf Ya'qūb b. Muḥammad al-Rāzī or Yūḥannā al-Qass. It was translated into Latin by Gerard of Cremona. See pp. 12f. and Busard, 'Commentary on Book X'.

3.4.3 PAPPUS, COMMENTARY ON BOOK 10

INCIPIT:

Intentio in tractatu decimo libri Euclidis in radicibus est inquisitio de magnitudinibus communicantibus et seiunctis et rationalibus et surdis ...

MANUSCRIPT:

Paris, BnF, lat. 7377A, f. 68r–70v, s. 13 (fragment)

EDITION:

Junge, 'Pappus-Kommentar' (edition on pp. 7–17).

COMMENTS:

The text is the beginning of Pappus' commentary on Euclid, book 10. The Pappus text was translated into Arabic by Abū 'Uthmān al-Dimashqī and from Arabic into Latin, as it seems, by Gerard of Cremona. See p. 13.

* * *

4. Reworkings, commentaries and related texts from the twelfth to the fourteenth centuries

4.1 Robert of Chester ("Version II"; formerly called: "Adelard II")

INCIPITS:

I. def. 1: *Punctus est illud cui pars non est.*

I.1, enunciation: *Triangulum equilaterum super datam lineam rectam collocare.*

I.1, proof: *A duobus terminis date linee ipsam lineam occupando cum circino duos circulos sese invicem secantes describe ...*

MANUSCRIPTS:

Berlin, SB, Ms. lat. qu. 510, f. 1r–59v, s. 13 (I–XV)

Budapest, private library of Walter Endrei, 1 folio, s. 13in. (I.44 end – II.9 beginning)

Cambridge, Fitzwilliam Museum, McClean 168, f. 1r–6v, s. 13[2] (X.84–XI. def. 8, XI.14–31)

Cambridge, Gonville & Caius College 504/271, f. 30v–86r, s. 13 (I–XV)

Cambridge, UL, Dd. XII.61 (= 778), f. 2v–124v, s. 12ex. (I–XV)

Chartres, Bibl. de la Ville, Ms. 498, f. 122v–124v, 141r–v, s. 12 (VII. def. 12–IX, XIV.6–XV. Before f. 141 many folios are missing. The manuscript was burnt during World War II, but a microfilm has survived)

Copenhagen, KB, Gl. kgl. S. 277 fol., f. 102r–113v, s. 13 (I–XV, enunciations only)

Darmstadt, Hessische LHB, MS 739, f. 1r–39v, s. 13 (I–VI.32)

Dresden, Sächsische LB, Db 86, f. 1r–48v, s. 13 (I–XV)

Erfurt, UB, Amplon. qu. 23, f. 1r–70v, s. 13 (I–XV)

Erfurt, UB, Amplon. qu. 352, f. 23v–62v, s. 13in. (I–VIII.16)

Exeter, Cathedral Library, MS 3503, f. 1r–96v, s. 13[2] (I–XV)

Florence, BNC, Conv. soppr. J I 18, f. 136r–159v, s. 13 (I–VI.23)

Florence, BNC, Conv. soppr. J I 32, f. 49r–103r, s. 13 (I–XV)

Florence, Bibl. Ricc., MS 2968/2, f. 1r–22v (I–XV; proofs only up to I.5); f. 23r (proofs of X.24 and X.17); f. 23r (introduction to X), s. 14

Hildesheim, Dombibliothek, MS 744, f. 2r–108r, s. 13[1] (I–IX.38, X.26–XV.5)

Leipzig, UB (formerly Stadtbibliothek), Rep. I, 68c, f. 1r–59r (I–XV); f. 59r (proofs of X.24 and X.17); f. 59r–60r (introduction to X), s. 13

Lincoln, Cathedral Library, MS 238, f. 1r–8r, s. 13[1] (VI. def. 2; VI.1–32, enunciations only)

London, BL, Addit. 33381, f. 186r–v, 187r–188v, s. 14in. (III.35–IV.4, V.17–VI.3)

London, BL, Addit. 34018, f. 1r–78v, s. 12[2] / 13[1] (I–XV)

London, BL, Burney 275, f. 293r–302r, ca. 1300 (I–VII.2)

London, BL, Royal 10 A VII, f. 213r, s. 13 (I. def., ax., post.; I.1 without proof)
London, BL, Royal 15 A XXVII, f. 1r–47r, s. 12ex. (I–XV; proofs only in I.1–32)
London, BL, Royal 15 B IV, f. 154r–157v, s. 13^1 (X.92–XII.3; proofs only)
London, BL, Sloane 285, f. 14v–65v, s. 13 (VI.5–XV.5)
London, BL, Sloane 1044, f. 80r–v, s. 13^2 (XI.31–36)
London, Royal Society, MS 28, f. 1r–74v, s. 13 (I–XV.2)
Munich, BSB, Clm 3523, f. 3r–21v, s. 13 (I–VI.21)
Munich, BSB, Clm 4377, f. 142r–v, s. 14ex. (I. def., post., ax.)
Munich, BSB, Clm 13021, f. 170r–186v, s. 12^2 (IV–XV; enunciations only. On f. 186v–187r there follow commentaries to some propositions of the *Elements*; see Munich, BSB, Clm 23511)
Munich, BSB, Clm 14111, f. 1r–4r, s. 15 (introduction to book 10: XI.6–XV; enunciations only)
Munich, BSB, Clm 14353, f. 78r–93v, s. 13/14 (I–XI.5; proofs only up to IV.1)
Munich, BSB, Clm 23511, f. 1r–27r, ca. 1200 (IV–XV; enunciations only). On f. 28v–29r there follow commentaries to *Elements* I.18, I.5, III.1–4.
Munich, BSB, Clm 26889, f. 187r–v, s. 14^2 (I. def., post., ax.)
Naples, BN, VIII C 22, f. 1r–44r, s. 13^2 (I–XV)
Oxford, Balliol College, MS 257, f. 2r–v, s. 12ex. (I. def.–I.1)
Oxford, BL, Ashmole 357, f. 71r–75v, s. 14 (XI. def., XI.1–35)
Oxford, BL, Auct. F.3.13, f. 1r–48v, s. 13^2 (I–XV)
Oxford, BL, Auct. F.5.28, f. ii r–xli v, 1r–15r (I–XV); f. 74r (addenda to VII.1 and X.105), s. 13^1
Oxford, BL, Auct. F.5.29, f. 132r–158r, s. 13ex. (I–VIII.10)
Oxford, BL, Digby 174, f. 154r–159v, s. 13 (VI.11–X end; proofs only in book 6)
Oxford, BL, D'Orville 70, f. 38r–v, s. 13 (introduction to book 10)
Oxford, CCC 224, f. 114r–138r, s. 13 (I–XV; enunciations only)
Oxford, CCC 251, f. 22r–83r, s. 13 (I–XV)
Oxford, CCC 283, f. 50r–65v, ca. 1300 (X–XI.3)
Oxford, TC, MS 47, f. 104v–138r, s. 12^2 (I–XV; proofs only up to the end of book 6)
Paris, BnF, lat. 7215, f. 4r–105r, s. 13ex. (I–XV)
Paris, BnF, lat. 7292, f. 247r–267v, s. 15 (VII–X.24)
Paris, BnF, lat. 7374, f. 1r–13v, 109v–111r, s. 13^1 (I–I.46; introduction to book 10)
Paris, BnF, lat. 7374A, f. 1r–57r, s. 13ex. (I–IX.13)
Paris, BnF, lat. 7420A, f. 80r–85r, s. 13^2 (I–III.35)
Paris, BnF, lat. 11245, f. 1r–57v, s. 13^1 (I–VII.36, X.1–20, X.85–90)
Paris, BnF, lat. 16647, f. 3r–92r, s. 13^1 (X. def.–XV end; f. 2r: second proof of X.42 ascribed to Eggebericus)

Paris, BnF, lat. 16648, f. 58r, s. 13 (proofs of X.24 and X.17)
Prague, UK, III.H.19 (= 572), f. 1r–127v, s. 12^2 (I–XII.4)
Vatican, Ottob. lat. 1862, f. 1r–19r, ca. 1200 (I–XV; proofs only up to VI.32)
Vatican, Regin. lat. 1137, f. 1r–73v (I–XV); f. 73v–74r (proofs of X.24 and X.17), s. 13^2
Venice, BNM, fondo antico 332 (= 1647), f. 87r–233r, s. 13^2 (I–XV, together with Version III)
Vienna, ÖNB, Ms. 83, f. 63v–64r, s. 14 (introduction to book 10)
Wolfenbüttel, HAB, Cod. Guelf. 51.9 Aug. 4to, f. 53r–60v, 61r–68r, s. 13 (I–III.7, V–VIII.5)
Worcester, Chapter Library, Q 89, f. 1r–3r, s. 13 (I–I.20)
Wrocław, BU, Mil. II.10, f. 1r–115v, s. 13 (I–XV)

EDITIONS:
Busard / Folkerts, 'Robert of Chester'.

COMMENTS:
See p. 14 and Busard / Folkerts, 'Robert of Chester'.

4.2 *Liber ysagogarum Alchorismi*

INCIPITS:
Preface: *Quoniam de quarta introducendis matheseos nos fari disciplinarum presens tempus ammonuit ...*
Text: *Instans pars temporis est cuius nulla pars est ...*

MANUSCRIPTS:
Admont, Stiftsbibliothek, fragment 4, 1f., s. 14
Genova, BU, E.III.28, f. 231r–232v, s. 15 (incomplete)
Milan, Bibl. Ambr., A 3 sup., f. 1r–20r, s. 12 (interpolated version)
Munich, BSB, Clm 13021, f. 27r–31v, s. 12^2
Munich, BSB, Clm 18927, f. 31r–33r, s. 13
Oxford, BL, Lyell 52, f. 21r–34r, s. 14 (interpolated version)
Paris, BnF, lat. 16208, f. 67r–71r, s. 12^2 (interpolated version)
Vienna, ÖNB, Ms. 275, f. 27r, a. 1143 (excerpts)

EDITIONS:
Of the arithmetical part there are editions by Nagl, 'Algorismus-Schrift' (from MS Vienna 275), by Curtze, 'Algorismus-Schrift' (from Munich, Clm 13021), and by Allard, 'Calcul Indien', pp. 23–61. The complete text has been edited by Dickey, 'Adelard of Bath', pp. 251–328.

COMMENTS:
See p. 18 and Burnett, 'Adelard of Bath', pp. 173f. For the manuscripts, see Allard, 'Calcul Indien', pp. XXXVI–XXXVIII. – This is a work in five books on the four parts of the quadrivium. It contains *inter alia* parts of Euclid's *Elements* in the Boethian tradition with the following passages which are not

found in the Boethian excerpts (see 1.1 – 1.4): I. ax. 5, I.9, II.13, III.1, 3, 20, 25, 35, 36; IV.15; VI.2, 4, 9. The text differs within the manuscripts. – The folio numbers given here concern the beginning and end of the *Liber ysagogarum* text, not of the Euclid excerpts.

4.3 *Mélanges* of the Robert and the Boethius texts

INCIPITS:

Text: *Quia, mi Patricii, geometrum exercitatissime ...* (Luneburg)

I. def. 1: *Principium mensure punctum vocatur cum medium tenet figure. Punctum vero est cuius pars nulla est* (Munich, Oxford, Paris); *Principium autem mensure punctus. Punctus est cuius nulla pars est* (Luneburg)

I.1, enunciation: *Supra datam rectam lineam terminatam triangulum equilaterum constituere* (Munich, Oxford, Paris); *Supra datam rectam lineam terminatam equilaterum constituere triangulum* (Luneburg)

MANUSCRIPTS:

Leiden, Bibl. RU, Voss. lat. qu. 92, f. 2rv, 1rv, s. 12ex. (fragment: III.34–36 Heiberg = 33–35 Version II; IV.14–16 Heiberg = 14–19 Version II; with additions from Ps.-Boethius, *Geometry II*)

Luneburg, Ratsbibliothek, Ms. misc. D 4° 48, f. 13r–17v, ca. 1200 (I–IV)

Munich, BSB, Clm 13021, f. 164r–169v, s. 12^2 (I–III)

Oxford, BL, Digby 98, f. 78r–85v, s. 12 (I–III. def. 4)

Paris, BnF, lat. 10257, f. 1r–88r, s. 12 (I–XV)

(formerly) San Juan Capistrano, Honeyman MS 50 (MS Math. 1), f. 1r–25r, s. 12^2 (I–XV; copied from Paris, BnF, lat. 10257)

EDITIONS:

The Paris text was edited by Goldat, 'Early Medieval Traditions', pp. 189–401. The Luneburg text was edited by Folkerts, 'Euklidbearbeitungen', pp. 22–40; there is a facsimile edition (in Folkerts, 'Ein neuer Text'). There are no editions of the other texts.

COMMENTS:

See p. 18. No manuscript contains proofs. There are striking differences within the manuscript texts; see Folkerts, 'Euklidbearbeitungen'.

4.4 Johannes de Tinemue (formerly called: "Adelard III")

INCIPITS:

Introduction: *Geometrie sicut et reliquarum facultatum usus suum antecessit artificium ...*

I. def. 1: *Punctus est illud cui pars non est.*

I.1, enunciation: *Triangulum equilaterum supra datam lineam rectam collocare.*

I.1, proof: *Esto exemplum* ab *linea data. Data dicuntur quibus equalia habitudanter invenire poterimus ...*

MANUSCRIPTS:

Boston, Medical Library, MS 24, f.110v–111r, after 1466 (introduction only)
London, BL, Burney 275, f. 308r–335r, s. 14 (IX.1–XV.2)
London, BL, Royal 15 B IV, f. 158r–167v, s. 13 (X.94–XI.37)
Oxford, Balliol College, MS 257, f. 1r–2r (X.106–107), f. 3r–98v (I.1–XV.3), s. 12ex.
Oxford, BL, Digby 174, f. 99r–132v (introduction, I–XI.1), 139r–145r (X–XI.1, repetition of 125r–132v), 146r–153r (II–V. def. 6), s. 12
Oxford, BL, D'Orville 70, f. 23v–38r, s. 13 (VII.7–X.36)
Paris, BnF, lat. 16648, f. 2r–58r, s. 13 (X–XV)
Venice, BNM, fondo antico 332 (= 1647), f. 86r–233r, s. 13[2] (I–XV, with introduction; together with Version II)
Vienna, ÖNB, Ms. 83, f. 53r–63v, s. 14 (VII.7–X.36)

EDITIONS:

Busard, 'Johannes de Tinemue'. The introduction was edited in Clagett, 'King Alfred'.

COMMENTS:

See p. 19 and Busard, 'Johannes de Tinemue'.

4.5 Commentary from the early thirteenth century

INCIPIT:

I. def. 1: *Punctus est cui non est pars.*

I.1, proof: *Ad huius demonstrationem necessaria est tertia petitio et prima et ultima ...*

MANUSCRIPTS:

Bonn, UB, S 73, f. 1r–86r, s. 13 (I–XV)
Paris, BnF, lat. 7292, f. 188r–210v, s. 15 (I–II.13)
Pistoia, Bibl. Fabron., 315, f. 2r–48v, s. 14 (I–XV)
Vatican, Regin. lat. 1268, f. 1r–69r, s. 14in. (I–XV)
Vienna, ÖNB, Ms. 5304, f. 1r–126v, s. 16 (I.1–XV)

EDITIONS:

Busard, 'Thirteenth-Century Adaptation'.

COMMENTS:

See p. 21 and Busard, 'Thirteenth-Century Adaptation'.

4.6 Another commentary from the thirteenth century (formerly called: "Adelard IIIB")

INCIPIT:

I. def. 1: *Punctus est cuius pars non est.*

I.1, enunciation: *Triangulum equilaterum supra datam lineam collocare.*

I.1, proof: *Triangulum etc. Data dicuntur quibus equalia habitudanter invenire possumus ...*

MANUSCRIPT:

Oxford, BL, Savile 19, f. 1r–36v, s. 13 (I–V)

EDITIONS: –

COMMENTS:

See p. 22.

4.7 Campanus of Novara

INCIPITS:

I. def. 1: *Punctus est cuius pars non est.*

I.1, enunciation: *Triangulum equilaterum supra datam lineam collocare.*

I.1, proof: *Esto data linea recta* a b. *Volo super ipsam triangulum equilaterum constituere ...*

MANUSCRIPTS:

Bergamo, Bibl. Civ. A. Mai, MA 279 (Δ 4.48), f. 33r–39v, s. 16 (I–III.10, enunciations only)

Berkeley, University of California, MS UCB 79 (formerly MAR 35), f. 2r–165v, s. 15 (I–XV)

Berlin, SB, Ms. lat. fol. 852, f. 85r–89v, 90v–92r, s. 14 (I–I.32, I.41–II.8)

Berlin, SB, Ms. lat. qu. 514, f. 1r–56v, s. 14 (I–VI)

Berne, Burgerbibliothek, Ms. A. 50, f. 1r–168r, a. 1412 (I–XV)

Boston, Medical Library, MS 24, f. 111v–147r, after 1466 (I–IX)

Cambridge, UL, Addit. 6866, f. 9r–94v, s. 13/14 (I–XV)

Cambridge, UL, Gg. I. 18 (= 1413), f. 1r–123r, s. 14 (I–XV)

Cambridge, UL, Gg. VI. 3 (= 1572), f. 287v–288r, s. 14 (I. def., post.)

Copenhagen, KB, Ny kgl. S. 100 8vo, f. 2r–86r, s. 15 (I–XV, enunciations only)

Cracow, BJ, 552, f. 98r–125r (p. 195–249), s. 15 (I–III)

Cracow, BJ, 567, f. 1r–127v, s. 14 (I–XV)

Cracow, BJ, 568, f. 5r–67v (p. 9–134), a. 1464/65 (I–XV)

Cracow, BJ, 569, f. 147r–198v (p. 299–402), s. 14 (I–XI.30)

Cracow, BJ, 1844, f. 41v–82r (p. 82–163), s. 15 (I–III)

Cracow, BJ, 1859, f. 139v–172v (p. 276–342), s. 15 (I–III)

Cracow, BJ, 1918, f. 3r–39v (p. 5–78), a. 1447 (I–III; on f. 1v–2v = pp. 2–4 commentary on the definitions, axioms, postulates of book 1)

Cracow, BJ, 1927, f. 20r–58r (p. 41–117), s. 15 (I–IV.5)

Dijon, Bibl. Mun., MS 441 (266), f. 1r–122v, s. 14 (I–XV)
Düsseldorf, UB, F 13, f. 21r, s. 16in. (I. ax., I.1–6 without proofs)
Emden, Bibliothek der Großen Kirche, Hs. 4° 14, f. 1r–229v, a. 1469 (I–XV)
Erfurt, UB, Amplon. fol. 37, f. 1r–36r, s. 13/14 (I–VII)
Erfurt, UB, Amplon. fol. 375, f. 18r–84v, s. 14 (I–VI)
Erfurt, UB, Amplon. qu. 22, f. 3r–42v, s. 14 (I–VI.26; the proofs in book 1 (= f.3r–15r) in a different version; see Busard, 'Campanus', p. 19)
Erfurt, UB, Amplon. qu. 344, f. 25r–67r, s. 14 (I–V.11)
Erfurt, UB, Amplon. qu. 385, f. 90r–91v, s. 14 (I–I.5)
Erfurt, UB, Amplon. qu. 387, f. 9v–23r, s. 14 (I–III)
Florence, BLaur., Plut. 29.27, f. 1r–7v, s. 15 (I–II.4, enunciations only)
Florence, BLaur., San Marco 212, f. 1r–132v, s. 15 (I–XV)
Florence, BLaur., Strozzi 62, f. 1r–60v, s. 14 (I–IX.19)
Florence, BLaur., Strozzi 71, f. 1r–187r, s. 14 (I–XV)
Florence, BNC, Conv. soppr. A V 2587, f. 1r–233r, s. 15 (I–XV)
Florence, BNC, Conv. soppr. C VII 2820, f. 1r–40v, s. 14 (I–V)
Florence, BNC, Conv. soppr. J I 18, f. 1r–135v, s. 14 (I–XV)
Florence, BNC, Conv. soppr. J IV 24, f. 1r–177v, a. 1408 (I–XV)
Florence, BNC, Magliabecch. XI 112, f. 1r–160r, a. 1259 (I–XV)
Florence, Bibl. Ricc., MS 129, f. 1r–179v, s. 14 (I–XV)
Glasgow, UL, Ms. Gen. 1115 (formerly BE8-y.18), f. 8r–172v, a. 1480 (I–XV)
Halle, UB, Za 74, f. 175r–186v, s. 15 (VII.39–X.64)
Hildesheim, Dombibliothek, MS 743a, f. 55r–69v, s. 14 (I–II.12)
Jena, UB, El. fol. 67, f. 1r–73v, s. 14 (I–XV)
Kues, Stiftsbibliothek, Ms. 205, f. 134r–188v, s. 14 (I–XV)
Leeuwarden, Prov. Bibliotheek van Friesland, nr. 57 HS, f. 1r–58v, s. 15 (I–XIV.9)
Leiden, Bibl. RU, Perizon. O.47, f. 83r–96r, s. 18 (I; copied from Leeuwarden, nr. 57 HS)
Leipzig, UB, Hs. 1461, f. 1r–79r, 80r–98r, a. 1440 (I–IX.4, IX.8–X.36)
Leipzig, UB, Hs. 1478, f. 89r–90v, s. 15 (I. def.–I.21)
Lincoln, Cathedral Library, MS 238, f. 18v, s. 14 (III.9–14; proofs only of III.9, 13, 14)
London, BL, Addit. 22783, f. 1r–102v, s. 15 (I–VII.39, VIII.6–XV)
London, BL, Arundel 84, f. 1r–113v, s. 14 (I–XV)
London, BL, Harley 5266, f. 1r–126v, s. 14 (I–XV)
London, BL, Harley 5404, f. 1r–139v, s. 14 (I–XII)
Lübeck, Bibliothek der Hansestadt, Ms. math. 1, f. 1r–44v, s. 15 (I–VII.12)
Madrid, BN, Ms. 8989, f. 1r–89v, s. 13 (I–XV)
Melk, Stiftsbibliothek, 781 (181), p. 1–56, s. 15 (I)
Milan, Bibl. Ambr., C 241 inf., f. 33r–124r, a. 1401 (I–XV)

Milan, Bibl. Ambr., M 28 sup., f. 127v–141r (I–III.12), f. 143r–177v (III–V), s. 14
Milan, Bibl. Naz. Braidense, AF. IX. 51, f. 63r–65v, s. 15 (V. def. 1–16)
Modena, Biblioteca Estense, Ms. lat. 567 (= α.W.8.12), f. 1r–176v, a. 1452 (I–XV)
Mühlhausen, Stadtarchiv, 60/2, f. 121r–190r, s. 14/15 (I–XIII.3)
Munich, BSB, Clm 247, f. 1r–223r, s. 15 (I–XV)
Munich, BSB, Clm 4377, f. 196r–200v, s. 15 (I–II.4)
Munich, BSB, Clm 11305, f. 1r–79v, s. 14 (VII–XV)
Munich, BSB, Clm 14111, f. 202v–203r, s. 15 (I–I.1 beginning)
Munich, BSB, Clm 14448, f. 70r–100r, s. 14 (I–VI.32)
Munich, BSB, Clm 28209, f. 1r–89v, s. 13/14 (I–XV)
Naples, BN, VIII C 20, f. 1r–17v, 58r–67v, 18r–57v, 68r–80v, s. 15 (I–X.1. From book 7 on, most proofs are missing)
Naples, BN, VIII C 21, f. 2r–106v, s. 15 (I–XV)
Naples, BN, VIII G 86, f. 158v–178r, 179r–220r, s. 15 (I–III.12, III. def. 1–V.34)
New York, CU, MS Plimpton 156, f. 1r–164v, ca. 1300 (I–XV)
New York, CU, MS Plimpton 157, f. 1r–14v, s. 15 (I)
New York, CU, MS Plimpton 158, f. 1rv, s. 14 (II.3 end – II.7 beginning)
New York, CU, MS Plimpton 159, f. 4v–6v, s. 14 (I.1–III.7, enunciations only)
New York, CU, MS Plimpton 160, f. 1r–16v, 17r–31v, s. 14 (I–III.29 beginning, III.8 end – V.12)
New York, CU, MS Plimpton 165, f. 29v–140v, s. 13ex. (I–XV)
Nuremberg, Stadtbibliothek, Cent. VI. 13, f. 1r–162r, s. 15 (I–XV; the beginning reworked and written by Johannes Regiomontanus. See Folkerts, 'Regiomontanus' Approach to Euclid')
Oxford, BL, Arch. Selden. B 13, f. 3r–65v, s. 13 (I–XV)
Oxford, BL, Canon. class. lat. 309, f. 1r–89v, s. 15 (I.2 end – XV)
Oxford, BL, Canon. misc. 200, f. 37r–78v, s. 15 (I–V. def. 1)
Oxford, BL, Canon. misc. 554, f. 1r–73r, s. 15 (I–VI)
Oxford, BL, Digby 191, f. 1r–66v, s. 14 (I–XV)
Oxford, BL, Savile 19, f. 36v–156r, s. 13 (VI–XV)
Padua, BU, Ms. lat. 1077, f. 1r–15v, s. 14 (I–I.38)
Paris, Bibl. Mazar. 3635, f. 1r–112r, s. 14 (I. def. 2 end – XV; missing are: I.40 end – I.49 beginning, III.35 end – IV.2 beginning, VII.39 end – X.35 beginning, XII.8 end – XIII.12 beginning)
Paris, BnF, lat. 3411, f. 92r–99v, s. 15 (I–I.31)
Paris, BnF, lat. 7213, f. 1r–173r, s. 14 (I–XV)
Paris, BnF, lat. 7214, f. 1r–207r, s. 14 (I–XV)
Paris, BnF, lat. 7216A, f. 1r–190v, s. 15 (I–XV)
Paris, BnF, lat. 7366, f. 51r–84v, s. 14 (I–VI.2)
Paris, BnF, lat. 7373, f. 2r–v, s. 14 (I. def., ax., post.)

Paris, BnF, lat. 7375, f. 1r–54r, s. 14 (I–V. def. 16)
Paris, BnF, lat. 14068, f. 5r–8v, s. 15 (I–II.11, enunciations only)
Paris, BnF, lat. 14735, f. 1r–127v, s. 15 (I–XV)
Paris, BnF, lat. 16197, f. 2r–134v, s. 13 (I–XV)
Paris, BnF, lat. 16198, f. 2r–73v, s. 14 (I–XV)
Perugia, Bibl. Comun. 387 (F.61), f. 94(95)r–99(100)r, s. 15 (I–I.9)
Pommersfelden, Graf von Schönbornsche Schloßbibliothek, Nr. 230 (2852), f. 1r–149r, a. 1398 (I–XV)
Prague, Knih. Metrop. Kapit., Ms. 1459 (M 100), f. 53v–60r, s. 14 (I–III.29, enunciations only, somewhat different from the usual text)
Prague, Národní Muzeum, Nostitz a 4 (151q), f. 1r–151v, s. 14/15 (I–XV)
Prague, UK, IV.D.5 (= 659), f. 1r–157r, s. 13/14 (I–XV; I.1 end – 12 beginning and IV.1 – 16 beginning are missing. Reworking of the Campanus text)
Prague, UK, VIII.G.27 (= 1609), f. 135r–145v, s. 15 (I.1–VI, enunciations only)
Princeton, UL, Garrett 104, f. 1r–144r, s. 15 (I–XV)
Ravenna, Bibl. Class., 136, 81ff. without foliation, s. 15 (I–IX)
Reims, Bibl. Mun., Ms. 897, f. 66r–70r, a. 1451/55 (I–I.6)
Rome, Bibl. Angelica, Ms. 127, f. 3r–5r, s. 15 (I–I.2 beginning)
Salamanca, BU 221, f. 1r–177r, s. 14 (I–XV)
Salamanca, BU 2624, f. 3r–96r, s. 13/14 (I–XV)
Stuttgart, Württembergische LB, HB XI 24, f. 2r–98v, 107r, a. 1468–72 (I–VI.32, with special introduction, some additions and commentaries: see also Vatican, Palat. lat. 1352)
Toledo, Bibl. Catedral, Ms. 98-27, f. 98r–119v, s. 15 (I–II.8)
Vatican, Archivio S. Pietro, H 35, f. 1r–142v, s. 14in. (I–XV)
Vatican, lat. 2223, f. 1r–106v, s. 15 (I–XV)
Vatican, lat. 2224, f. 1r–222r, a. 1457 (I–XV)
Vatican, Barb. lat. 103, f. 1r–60r, s. 15 (X.19–XV)
Vatican, Ottob. lat. 1551, f. 1r–130v, s. 14 (I–XV)
Vatican, Palat. lat. 1348, f. 1r–37v, s. 14 (I–V.28)
Vatican, Palat. lat. 1349, f. 1r–24r, s. 14 (I–VII.21, enunciations only, but with proof of V.1 and commentaries to V. deff.)
Vatican, Palat. lat. 1351, f. 1r–195v, s. 14 (I.3 end – XV)
Vatican, Palat. lat. 1352, f. 1r–89r, s. 15 (I–VIII, with special introduction; see also Stuttgart, Württembergische LB, HB XI 24)
Vatican, Palat. lat. 1354, f. 98v–99v, 120r–v, ca. 1464 (excerpts from books 1, 2, 3, 7, 8, 10, 9)
Vatican, Regin. lat. 1186, f. 5r–158r, s. 15 (I–XV; the proofs in book 1 in a different version)
Vatican, Regin. lat. 1261, f. 61r–197v, s. 14 (I–XV)
Vatican, Regin. lat. 1267, f. 1r–56v, s. 14 (I–VII.3)

Vatican, Urb. lat. 506, f. 2r–159v, s. 15 (I–XV)
Vatican, Urb. lat. 507, f. 1r–143r, s. 14 (I–X.3 beg., X–XV)
Venice, BNM, fondo antico 329 (= 1843), f. 12r–34r, s. 15 (I–XV, enunciations only)
Venice, BNM, Cod. lat. VIII 39 (= 3271), f. 1r–135r, s. 14 (I–XV)
Venice, BNM, Cod. lat. VIII 77 (= 3223), f. 20r–98r, s. 15 (I.24 end – VII.30 beg.; X.20 end – XV)
Vienna, ÖNB, Ms. 2367, f. 198r–247v, s. 14 (I–VI)
Vienna, ÖNB, Ms. 2465, f. 1r–142v, s. 14in. (I–XV)
Vienna, ÖNB, Ms. 4770, f. 45r–50v, s. 14 (I–I.37. The text of the first proofs differs from the usual Campanus text)
Vienna, ÖNB, Ms. 4953, f. 126r–143v, s. 15 (I)
Vienna, ÖNB, Ms. 5304, f. 127r, s. 16 (XV.7–13, enunciations only)
Vienna, ÖNB, Ms. 5417, f. 56r–74v, s. 14/15 (I–II.12)
Vienna, ÖNB, Ms. 5507, f. 1r–82r, a. 1457 (I–XV)
Wolfenbüttel, HAB, Cod. Guelf. 65 Aug. fol., f. 9r–75v, s. 15 (I–VI)

EDITIONS:

First edited by E. Ratdolt, Venice, 1482. Later editions by Henricus Stephanus, Paris, 1516; J. Hervagius, Basel, 1537, 1546, 1558 and others; some of them were published together with Zamberti's translation. See Steck, 'Bibliographia Euclideana', pp. 47f. Modern edition: Busard, 'Campanus'.

COMMENTS:

Most enunciations are as found in Version II (= Robert of Chester). Campanus has enlarged the text using, besides others, the Nayrīzī commentary and Jordanus de Nemore, *De arithmetica*. See p. 23 and Busard, 'Campanus'. – The numeration of the enunciations given here is according to the *editio princeps* of 1482 (and agrees with that in Busard, 'Campanus'.) – In manuscripts having enunciations only and ending before book 5 it is difficult or impossible to decide whether they present a Campanus or a Robert text. If such manuscripts are from the fourteenth or fifteenth centuries, they are listed here. – Reworkings of the Campanus text are to be found in Section 5.1.

4.8 Other reworkings and commentaries from the thirteenth and fourteenth centuries

4.8.1 COMMENTARY OXFORD, D'ORVILLE 70

INCIPIT:

I 1, enunciation: *Punctus est cui pars non est.*
I 1, proof: *Sit* ab *linea data, duo eius capita centra sint circulorum eam occupantium* ...

MANUSCRIPTS:
Oxford, BL, D'Orville 70, f. 1r–23v, s. 13 (I–VII.6)
Vienna, ÖNB, Ms. 83, f. 39r–53r, s. 14 (I–VII.6)

EDITIONS: –

COMMENTS:
The enunciations are in the form of Version II (= Robert of Chester). See Busard, 'Early Adaptations', p. 129, and Busard, 'Campanus', p. 27.

4.8.2 REWORKING PARIS, BNF, LAT. 7374

INCIPIT:
I.47, proof: *Sit triangulus* abc, *itaque quadratum huius lateris sit tanquam quadrata aliorum duorum laterum. Dico ergo quod iste angulus* bac *sit rectus ...*

MANUSCRIPT:
Paris, BnF, lat. 7374, f. 13v–109v, s. 13^{1} (I.47–IX.38)

EDITIONS: –

COMMENTS:
This text uses the definitions, postulates, axioms, enunciations from Version II (= Robert of Chester), but has different proofs, which seem to be connected with Adelard's translation. See Clagett, 'Medieval Latin Translations', p. 29 note 31(2), and Busard, 'Early Adaptations', p. 136.

4.8.3 REWORKING LONDON, BL, SLOANE 285

INCIPIT:
I.1, proof: *Facta dispositione figure per secundam petitionem bis sumptam, 'a puncto et cetera,' per primam conceptionem constat propositum, quia 'que uni et eidem et cetera.' ...*

MANUSCRIPT:
London, BL, Sloane 285, f. 1r–14v, s. 13 (I–VI.4)

EDITIONS: –

COMMENTS:
This text uses the definitions, postulates, axioms and enunciations from Robert, but has different proofs, which are abridged in the manner of Robert. See Clagett, 'Medieval Latin Translations', p. 29 note 31(1).

4.8.4 COMMENTARY OF ALBERTUS MAGNUS (?)

INCIPITS:
Introduction: *Sicut triplex est phylosophia ut dicit Aristoteles in sexto philosophie prime ...*
I. def. 1: *Punctum est cui pars non est.*

I.1, enunciation: *Supra datam rectam lineam equilaterum triangulum constituere.*

I.1, proof: *Sit enim data recta linea* ab. *Per tertiam autem petitionem pono pedem circini immobilem in* a . . .

MANUSCRIPT:

Vienna, Dominikanerkloster, MS 80/45, f. 105r–145r, s. 14 (I–IV)

EDITIONS:

Edition of book 1 by Tummers, 'Albertus (Magnus)' commentaar', deel II, pp. 1–102. English translation of book 1 in Lo Bello, 'Commentary of Albertus Magnus', pp. 2–147.

COMMENTS:

See p. 25. This text is connected with the anonymous commentary (see p. 21). For further details, see Hofmann, 'Euklid-Bearbeitung'; Tummers, 'Albertus (Magnus)' commentaar', deel I, and Tummers, 'Commentary of Albert'.

4.8.5 Commentary ascribed to Roger Bacon

INCIPIT:

Nulla linea nisi tantum una residuo coniungi potest ut sint ambe sub termino et specie earum que erant ante separationem scilicet quando factum erat residuo et hoc potest patere sumpta consideratione 25 premisse . . .

MANUSCRIPT:

Florence, BNC, Conv. soppr. J IX 26, f. 46r–55r, s. 15

EDITION:

Busard, 'Roger Bacon' (edition on pp. 204–217).

COMMENTS:

See p. 26 and Busard, 'Roger Bacon'.

4.8.6 Reworking Vat. Regin. lat. 1268

INCIPIT:

Executus euclides in superioribus libris quasdam linearum et angularium circulariumque figurarum naturas . . .

MANUSCRIPT:

Vatican, Regin. lat. 1268, f. 72r–91v (V–VI), f. 92r–112v (X), s. 14in.

EDITIONS:

Some propositions are edited in Busard, 'Adelard of Bath', *Addendum 3* (pp. 400–413).

COMMENTS:

On f. 72r–91v there is an unknown version of books 5 and 6; the author used Robert, Johannes de Tinemue, Gerard and Hermann. In book 10 the author used Robert, Gerard and Hermann. See Busard, 'Early Adaptations', pp. 133f., and Busard, 'Campanus', pp. 24f.

4.8.7 Nicole Oresme, "Quaestiones super geometriam Euclidis"

Incipit:
Circa librum Euclidis queritur primo circa quoddam dictum Campani dicens quod magnitudo decrescit in infinitum ...

Manuscripts:
Greifswald, UB, Ms. 742, f.16v–33r, s. 14
Seville, Bibl. Colomb., 7-7-13, f. 102v–112r, s. 14
Vatican, lat. 2225, f. 90r–98v, s. 15
Vatican, Chigi F IV 66, f. 22v–40r, s. 14

Edition:
Busard, 'Oresme'. One additional *quaestio*, which is missing in both Vatican manuscripts, is edited in Murdoch, 'Oresme', pp. 88–91.

Comments:
See p. 26; Busard, 'Oresme'; and Murdoch, 'Oresme'.

* * *

5. Humanistic translations from Greek into Latin; early translations into the vernacular

5.1 Reworkings and commentaries mostly from the fifteenth century; not identified texts

Manuscripts:

Admont, Stiftsbibliothek, MS 442, f. 1r–5r, s. 14. Contains: definitions principally taken from Euclid I (f. 1r–v); Euclid I–II.13 (Version II or Campanus), enunciations only (f. 1v–4r); proofs of Euclid I.32 and of "secondary propositions" (f. 4r–5r). See Tummers, 'Geometry and Theology'.

Cambridge, UL, Ii. I. 13 (= 1705), f. 55v–56v, s. 14 (V.1–34. Commentary on book 5, as it seems, by Walter of Evesham. The enunciations are from Campanus. The author gives numerical examples)

Cracow, BJ, 569, f. 37v–47r (p. 80–99), s. 14 (XI–XV. Reworking of the Campanus text. The enunciations and some addenda are identical with the Campanus text)

Erfurt, UB, Amplon. qu. 344, f. 68r–87v, s. 14 (*Questiones super primo Euclidis*)

Florence, BLaur., Ashburnham 1234, f. 9r–15v, s. 14 (reworking and commentary)

Jena, UB, El. fol. 73, f. 26v–30v, s. 15/16 (mathematical texts among which some excerpts from Euclid in the Campanus tradition. The excerpts seem to be from a university lecture)

Jena, UB, Prov. oct. 254, f. 11r–240v, s. 15/16 (I–XIV. Special commented version, as it seems, a university lecture. See also Kassel, Ms. math. 8° 8)

Kassel, Murhardsche Bibl., Ms. math. 8° 8, f. 2r–105r, s. 15 (I–IX. Special commented version, as it seems, a university lecture. See also Jena, Prov. oct. 254)

Leipzig, UB, Hs. 1478, f. 80r–88v, s. 15 (I–V. def. 1, with commentary, as it seems, from a university lecture)

London, BL, Addit. 24071, f. 70r–71v, s. 15 (I.2–XIII.10, enunciations only, not in the Campanus version. Only those enunciations are given which are necessary for the proofs in the treatise *De sinibus et cordis ac eorum arcubus*. See Vienna 5268)

London, BL, Sloane 1478, f. 8r–42r, s. 17 (I–VI, enunciations only with some notes. The text is based upon the Campanus version, but not a pure Campanus text)

Munich, BSB, Clm 657, f. 82r–95v, s. 15 (I–VI, diagrams only)

Munich, BSB, Clm 18800, f. 121r–123r, s. 14/15 (excerpts from Euclid, mostly book 1, and texts concerning *latitudines formarum*)

Oxford, BL, Digby 98, f. 130v–131v, s. 15 (I.1–6: reworking of the Campanus text; enunciations in the Campanus tradition. On f. 131v diagrams of I.7–47)

Oxford, BL, Digby 178, f. 88r–107r, s. 14 (I–XV. Enunciations in the Campanus version. There are only some shortened proofs which are based on the Campanus text)

Oxford, CCC 234, f. 1r–94r, 95v–96v, 98r–170r, s. 15 (I–XV. Enunciations in the Campanus version. The author of the proofs, which differ from the Campanus text, used different manuscripts, among them Campanus, Version II and the text in Paris, BN lat. 10257)

Paris, BnF, lat. 7219, f. 1r–26v, s. 15 (I–VI, diagrams only)

Paris, BnF, lat. 7292, f. 210v–245v, s. 15 (II.14–VI.32)

Perugia, Bibl. Comun. 322, s. 15 (Brani di Vitruvio e d'Euclide)

Piacenza, Bibl. Comun., Ms. Pallastrelli 211, f. 3r–32v, s. 15ex. (XIV–XV, translation by G. Valla; f. 1r–2v: dedication letter to Petrus Barocius. See p. 28.)

Prague, Knih. Metrop. Kapit., Ms. 1459 (M 100), f. 21r–48r, s. 14 (*Quaestiones in Euclidis geometriam.* On f. 49v–50r, 52v–53r there are notes, commentaries and small extracts from Euclid)

Prague, Knih. Metrop. Kapit., Ms. 1585 (O 1), f. 166r–174v, 176r–178r, s. 15 (Commentary on I–VI, as it seems, from a university lecture)

Vatican, lat. 2114, f. 257v–258r, s. 14

Vatican, Ottob. lat. 2273, f. 5r–49v, s. 16 (XIV–XV, translation by G. Valla; f. 1r–4v: dedication letter to *princeps Albertus pius Carpensis.* See p. 28.)

Vatican, Regin. lat. 1208, f. 1r–32r, s. 16 (Michael Stifel, Commentary on book 10)

Vienna, ÖNB, Ms. 5268, f. 98r–v, s. 15 (I.2–XIII.10, enunciations only, in the Campanus version. Only those enunciations are given which are necessary for the proofs in the treatise *De sinibus et cordis ac eorum arcubus*. See London, Add. 24071)

5.2 Translation of B. Zamberti

MANUSCRIPTS:

Munich, BSB, Clm 6, f. 1r–438v, a. 1539 (together with his translation of the commentary by Proclus)

New York, CU, MS Plimpton 214, f. 2r–153r, s. 16 (I–III)

Turin, BN, H.VII.46, f. 1r–134v, s. 16 (I.1–VIII; enunciations only)

Vatican, Barb. lat. 3, p. 1–355, s. 16 (I–VI; contains both Campanus and Zamberti texts)

EDITIONS:

First printed, Venice 1505. For later editions, often together with Campanus' reworking, see Steck, 'Bibliographia Euclideana', pp. 47f.

COMMENTS:

Translation from the Greek. It is unknown which Greek manuscript Zamberti used. See p. 29.

5.3 Early translations into Italian

MANUSCRIPTS:

Bologna, BU, Ms. 535 Miscell. L 4, a. 1553, 205ff. (I–XV; translated by Fabrizio Sanseverini of Ferrara)

Brescia, Biblioteca dell'Università, Fondo Carlo Viganò, P.6.3, s. 15, f.1r–34v (I–III.7)

Florence, BNC, II-383, ca. 1550, f. 1r–15v (I; translated from the Greek by Benedetto Varchi)

Florence, BNC, Conv. soppr. A V 1443, ca. 1550, f. 1r–21r (I; translated from the Greek by Benedetto Varchi)

Florence, BNC, Magliabecch. XI 12, s. 16, 62ff. (I–VI, enunciations only)

Florence, Bibl. Ricc., MS 2180, s. 15, 100ff.

London, BL, Harley 3400, ca. 1500, 121ff. (I–III)

Milan, Bibl. Ambros., O 222 sup., a. 1541, 98ff. (I–XV; translated by "Costanzo bolognese")

Milan, Bibl. Trivulz., Ms. 187, a. 1539, 31pp. (I–III. def.10; translated by "Costanzo bolognese", as it seems, from the Campanus text)

Pisa, BU, Cod. 556, a. 1485, 205ff. (autograph of Raffaello di Giovanni di Guglielmo Canacci, *maestro d'abbaco* in Florence)

Siena, Biblioteca degl'Intronati, L.IV.16, 155ff., s. 15/16 (I–XV; anonymous translation of the Campanus text)

Siena, Biblioteca degl'Intronati, L.IV.17, 135ff., s. 15/16 (I–XV; anonymous translation of the Campanus text)

EDITIONS: –

COMMENTS:

Only Italian manuscripts which were written before ca. 1550 are given here. See p. 30. For details, see Pagli, 'Volgarizzazioni'.

* * *

Bibliography

A. Allard, *Muḥammad ibn Mūsā al-Khwārizmī. Le Calcul Indien (Algorismus). Histoire des textes, édition critique, traduction et commentaire des plus anciennes versions latines remaniées du XIIe siècle*, Paris / Namur 1992.

R. O. Besthorn et al., *Codex Leidensis 399,1. Euclidis Elementa ex interpretatione al-Hadschdschadschii cum commentariis al-Narizii. Arabice et latine ediderunt*, Copenhagen. I.1, 1893; I.2, 1897; II.1, 1900; II.2, 1905; III.1, 1910; III.2, 1932.

A. Birkenmajer, "La bibliothèque de Richard de Fournival", in *Études d'histoire des sciences et de la philosophie du moyen âge*, vol. 1, Wrocław etc. 1970, pp. 117–210.

B. Bischoff, *Mittelalterliche Studien*, Bd. 3, Stuttgart 1981.

S. Brentjes, "Textzeugen und Hypothesen zum arabischen Euklid in der Überlieferung von al-Ḥaǧǧāǧ b. Yūsuf b. Maṭar (zwischen 786 und 833)", *Archive for History of Exact Sciences* 47 (1994), 53–92.

S. Brentjes, "Remarks about the proof sketches in Euclid's *Elements*, Book I as transmitted by Ms Paris, B.N., fonds latin 10257", in M. Folkerts (ed.), *Mathematische Probleme im Mittelalter: der lateinische und arabische Sprachbereich*, Wiesbaden 1996, pp. 115–137.

S. Brentjes, "The relevance of non-primary sources for the recovery of the primary transmission of Euclid's *Elements* into Arabic", in F. J. Ragep, S. P. Ragep (ed.), *Tradition, Transmission, Transformation. Proceedings of two Conferences on Pre-Modern Science Held at the University of Oklahoma*, Leiden / New York / Cologne 1996, pp. 201–225.

S. Brentjes, "Aḥmad al-Karābīsī's Commentary on Euclid's 'Elements'", in M. Folkerts, R. Lorch (ed.), *Sic itur ad astra. Studien zur Geschichte der Mathematik und Naturwissenschaften. Festschrift für den Arabisten Paul Kunitzsch zum 70. Geburtstag*, Wiesbaden 2000, pp. 31–75.

S. Brentjes, "Observations on Hermann of Carinthia's Version of the *Elements* and its Relation to the Arabic Transmission", *Science in Context* 14 (1/2) (2001), 39–84.

Ch. Burnett, "A group of Arabic-Latin translators working in northern Spain in the mid-12th century", *Journal of the Royal Asiatic Society* (1977), 62–108.

Ch. Burnett (ed.), *Hermann of Carinthia, De essentiis. A Critical Edition with Translation and Commentary*, Leiden / Cologne 1982.

Ch. Burnett (ed.), *Adelard of Bath. An English Scientist and Arabist of the Early Twelfth Century*, London 1987.

Ch. Burnett, M. Folkerts, R. Lorch, "Addendum to Adelard's Versions of Euclid's *Elements*", in Ch. Burnett (ed.), *Adelard of Bath. An English Scientist and Arabist of the Early Twelfth Century*, London 1987, pp. 65–68.

Ch. Burnett, "The Latin and Arabic influences on the vocabulary concerning demonstrative argument in the versions of Euclid's *Elements* associated with Adelard of Bath", in *Aux origines du lexique philosophique européen. L'influence de la* latinitas, Louvain-la-Neuve 1997, pp. 117–135.

H. L. L. Busard, *Nicole Oresme, Quaestiones super geometriam Euclidis*, Leiden 1961.

H. L. L. Busard, "Über einige Euklid-Scholien, die den Elementen von Euklid, übersetzt von Gerard von Cremona, angehängt worden sind", *Centaurus* 18 (1974), 97–128.

H. L. L. Busard, "Ein mittelalterlicher Euklid-Kommentar, der Roger Bacon zugeschrieben werden kann", *Archives Internationales d'Histoire des Sciences* 24 (1974), 199–218.

H. L. L. Busard (ed.), *The Translation of the Elements of Euclid from the Arabic into Latin by Hermann of Carinthia (?)*: books I–VI, *Janus* 54 (1967), 1–140, and published separately, Leiden 1968; books VII–IX, *Janus* 59 (1972), 125–187; books VII–XII, Amsterdam 1977.

H. L. L. Busard (ed.), *The First Latin Translation of Euclid's* Elements *Commonly Ascribed to Adelard of Bath*, Toronto 1983.

H. L. L. Busard (ed.), *The Latin translation of the Arabic version of Euclid's* Elements *commonly ascribed to Gerard of Cremona*, Leiden 1983.

H. L. L. Busard, "Some Early Adaptations of Euclid's *Elements* and the Use of its Latin Translations", in M. Folkerts, U. Lindgren (ed.), *Mathemata. Festschrift für Helmuth Gericke*, Stuttgart 1985, pp. 129–164.

H. L. L. Busard (ed.), *The Mediaeval Latin Translation of Euclid's Elements Made Directly from the Greek*, Stuttgart 1987.

H. L. L. Busard / M. Folkerts (ed.), *Robert of Chester's (?) Redaction of Euclid's* Elements: *the so-called Adelard II Version*, 2 vols., Basel / Boston / Berlin 1992.

H. L. L. Busard (ed.), *A Thirteenth-Century Adaptation of Robert of Chester's Version of Euclid's* Elements, Munich 1996.

H. L. L. Busard (ed.), "A Latin Translation of an Arabic Commentary on Book X of Euclid's *Elements*", *Mediaeval Studies* 59 (1997), 19–110.

H. L. L. Busard (ed.), *Johannes de Tinemue's Redaction of Euclid's Elements, the so-called Adelard III Version*, Stuttgart 2001.

H. L. L. Busard (ed.), *Campanus of Novara's Edition of Euclid's* Elements, Stuttgart 2005.

M. Clagett, "The Medieval Latin Translations From the Arabic of the *Elements* of Euclid, With Special Emphasis on the Versions of Adelard of Bath", *Isis* 44 (1953), 16–42.

M. Clagett, "King Alfred and the Elements of Euclid", Isis 45 (1954), 269–277.

M. Clagett, *Archimedes in the Middle Ages*, vol. 1, Madison 1964; vols. 2–5, Philadelphia 1976–1984.

M. Curtze, "Ueber eine Algorismus-Schrift des XII. Jahrhunderts", *Abhandlungen zur Geschichte der Mathematik* 8 (1898), 1–27.

M. Curtze (ed.), *Anaritii in decem libros priores Elementorum Euclidis commentarii ex interpretatione Gherardi Cremonensis in codice Cracoviensi 569 servata*, Leipzig 1899 (= I. L. Heiberg, H. Menge (ed.), *Euclidis Opera Omnia, Supplementum*).

G. De Young, "The Arithmetic Books of Euclid's Elements in the Arabic Tradition", unpublished Ph.D. dissertation, Harvard University 1981.

G. De Young, "The Arabic Textual Traditions of Euclid's Elements", *Historia Mathematica* 11 (1984), 147–160.

G. De Young, "New traces of the lost al-Ḥajjāj Arabic translations of Euclid's *Elements*", *Physis* 28 (1991), 647–666.

G. De Young, "The Latin Translation of Euclid's Elements Attributed to Gerard of Cremona in Relation to the Arabic Transmission", *Suhayl* 4 (2004), 311–383.

B. G. Dickey, *Adelard of Bath. An Examination Based on Heretofore Unexamined Manuscripts*, unpublished Ph.D. dissertation, University of Toronto 1982.

J. W. Engroff, *The Arabic Tradition of Euclid's Elements: Book V*, unpublished Ph.D. dissertation, Harvard University 1980.

M. Folkerts (ed.), *"Boethius" Geometrie II, ein mathematisches Lehrbuch des Mittelalters*, Wiesbaden 1970.

M. Folkerts, *Ein neuer Text des Euclides Latinus. Faksimiledruck der Handschrift Lüneburg D 4° 48, f. 13r–17v*, Hildesheim 1970.

M. Folkerts, "Anonyme lateinische Euklidbearbeitungen aus dem 12. Jahrhundert", in *Österreichische Akademie der Wissenschaften, mathematisch-naturwissenschaftliche Klasse, Denkschriften*, 116. Band, 1. Abhandlung, Vienna 1971, pp. 1–42.

M. Folkerts, "Probleme der Euklidinterpretation und ihre Bedeutung für die Entwicklung der Mathematik", *Centaurus* 23 (1980), 185–215. (Reprinted in this volume, item IV.)

M. Folkerts, "The Importance of the Pseudo-Boethian *Geometria* During the Middle Ages", in M. Masi (ed.), *Boethius and the Liberal Arts. A Collection of Essays*, Bern / Frankfurt / Las Vegas 1981, pp. 187–209. Reprinted in M. Folkerts, *Essays on Early Medieval Mathematics. The Latin Tradition*, Aldershot 2003, item VII.

M. Folkerts, "Die Altercatio in der Geometrie I des Pseudo-Boethius. Ein Beitrag zur Geometrie im mittelalterlichen Quadrivium", in G. Keil (ed.), *Fachprosa-Studien. Beiträge zur mittelalterlichen Wissenschafts- und Geistesgeschichte*, Berlin 1982, pp. 84–114. Reprinted in M. Folkerts, *Essays on Early Medieval Mathematics. The Latin Tradition*, Aldershot 2003, item VIII.

M. Folkerts, "Adelard's Versions of Euclid's *Elements*", in Ch. Burnett (ed.), *Adelard of Bath. An English Scientist and Arabist of the Early Twelfth Century*, London 1987, pp. 55–65.

M. Folkerts, "Regiomontanus' Approach to Euclid", this volume, item VII.

M. Folkerts, "Leonardo Fibonacci's Knowledge of Euclid's *Elements* and of Other Mathematical Texts", this volume, item IX.

B. Geyer, "Die mathematischen Schriften des Albertus Magnus", *Angelicum* 35 (1958), 159–175.

M. Geymonat (ed.), *Euclidis Latine facti fragmenta Veronensia*, Milano, Varese 1964.

M. Geymonat, "Nuovi frammenti della geometria 'boeziana' in un codice del IX secolo?", *Scriptorium* 21 (1967), 3–16.

G. D. Goldat, *The Early Medieval Traditions of Euclid's Elements*, unpublished Ph.D. dissertation, University of Wisconsin 1957.

Ch. H. Haskins, *Studies in the History of Mediaeval Science*, 2nd ed., Cambridge (Mass.) 1927.

I. L. Heiberg, "Prolegomena critica", in I. L. Heiberg (ed.), *Euclidis Elementa*, vol. V, Leipzig 1888, pp. XXIII–CXIII.

J. L. Heiberg, "Beiträge zur Geschichte Georg Valla's und seiner Bibliothek", *Centralblatt für Bibliothekswesen*, Beiheft 16, Leipzig 1896.

J. E. Hofmann, "Über eine Euklid-Bearbeitung, die dem Albertus Magnus zugeschrieben wird", in *Proceedings of the International Congress of Mathematicians, 14–21 Aug. 1958*, Cambridge 1960, pp. 554–566.

P. Hossfeld, "Zum Euklidkommentar des Albertus Magnus", *Archivum Fratrum Praedicatorum* 52 (1982), 115–133.

G. Junge, "Das Fragment der lateinischen Übersetzung des Pappus-Kommentars zum 10. Buche Euklids. (Nr. 7377A, Fol. 68–70 der Bibliothèque Nationale zu Paris.)", *Quellen und Studien zur Geschichte der Mathematik, Astronomie und Physik*, Abt. B, 3, Studien (1934), 1–17.

M. Klamroth, "Über den arabischen Euklid", *Zeitschrift der Deutschen Morgenländischen Gesellschaft* 35 (1881), 270–326, 788.

W. R. Knorr, "The medieval tradition of Archimedes' *Sphere and Cylinder*", in E. Grant, J. E. Murdoch (ed.), *Mathematics and its applications to science and natural philosophy in the Middle Ages. Essays in honor of Marshall Clagett*, Cambridge etc. 1987, pp. 3–42.

P. Kunitzsch, "Findings in Some Texts of Euclid's *Elements* (Mediaeval Transmission, Arabo-Latin)", in M. Folkerts, U. Lindgren (ed.), *Mathemata. Festschrift für Helmuth Gericke*, Stuttgart 1985, pp. 115–128.

P. Kunitzsch, "Gerard's translations of astronomical texts, especially the Almagest", in *Gerardo da Cremona, a cura di P. Pizzamiglio* (= *Annali della Biblioteca Statale e Libreria Civica di Cremona* XLI, 1990), Cremona 1992, pp. 71–84.

P. Kunitzsch, "'The Peacock's Tail': On the names of some theorems of Euclid's *Elements*", in M. Folkerts, J. P. Hogendijk (ed.), *Vestigia Mathematica. Studies in medieval and early modern mathematics in honour of H. L. L. Busard*, Amsterdam 1993, pp. 205–214.

R. Lemay, "De la scolastique à l'histoire par le truchement de la philologie: Itinéraire d'un médiéviste entre Europe et Islam", in *La diffusione delle scienze islamiche nel medio evo europeo. Convegno internazionale ... Roma, 2–4 ottobre 1984*, Rome 1987, pp. 399–535.

R. Lemay, "Gerard of Cremona", in *Dictionary of Scientific Biography* 15, New York 1978, pp. 173–192.

A. Lo Bello, *The Commentary of al-Nayrizi on Book I of Euclid's* Elements of Geometry, Boston / Leiden 2003.

A. Lo Bello, *The Commentary of Albertus Magnus on Book I of Euclid's* Elements of Geometry, Boston / Leiden 2003.

A. Lo Bello, *Gerard of Cremona's Translation of the Commentary of al-Nayrizi on Book I of Euclid's* Elements of Geometry, Boston / Leiden 2003.

R. Lorch, "Some Remarks on the Arabic-Latin Euclid", in Ch. Burnett (ed.), *Adelard of Bath. An English Scientist and Arabist of the Early Twelfth Century*, London 1987, pp. 45–54.

J. E. Murdoch, "Euclides Graeco-Latinus. A Hitherto Unknown Medieval Latin Translation of the *Elements* Made Directly from the Greek", *Harvard Studies in Classical Philology* 71 (1966), 249–302.

J. E. Murdoch, "The Medieval Euclid: Salient Aspects of the Translations of the *Elements* by Adelard of Bath and Campanus of Novara", in *XIIe Congrès International d'Histoire des Sciences, Colloques* (= *Revue de synthèse*, 3e série, nos 49–52), Paris 1968, pp. 67–94.

J. E. Murdoch, "Euclid: Transmission of the Elements", in *Dictionary of Scientific Biography* 4, New York 1971, pp. 437–459.

J. E. Murdoch, "Nicole Oresme, *Quaestiones super geometriam Euclidis*. Edited by H. L. L. Busard" (review), *Scripta Mathematica* 27 (1964), 67–91.

R. A. B. Mynors (ed.), *Cassiodori Senatoris Institutiones*, Oxford 1937; 2nd ed. 1961.

A. Nagl, "Ueber eine Algorismus-Schrift des XII. Jahrhunderts und über die Verbreitung der indisch-arabischen Rechenkunst und Zahlzeichen im christl. Abendlande", *Zeitschrift für Mathematik und Physik, hist.-lit. Abth.*, 34 (1889), 129–146, 161–170.

P. Pagli, "Volgarizzazioni manoscritte degli *Elementi* di Euclide", in R. Franci, P. Pagli, A. Simi (ed.), *Il sogno di Galois. Scritti di storia della matematica dedicati a Laura Toti Rigatelli per il suo 60o compleanno*, Siena 2003, pp. 51–90.

P. L. Rose, *The Italian Renaissance of Mathematics. Studies on Humanists and Mathematicians from Petrarch to Galileo*, Genève 1975.

F. Sezgin, *Geschichte des arabischen Schrifttums. Band V: Mathematik, bis ca. 430 H.*, Leiden 1974.

M. Steck, *Bibliographia Euclideana. Die Geisteslinien der Tradition in den Editionen der "Elemente"* (ΣΤΟΙΧΕΙΑ) *des Euklid (um 365–300). Handschriften – Inkunabeln – Frühdrucke (16. Jahrhundert). Textkritische Editionen des 17.–20. Jahrhunderts. Editionen der Opera minora (16.–20. Jahrhundert). ... Nach dem Tode des Verfassers herausgegeben von M. Folkerts*, Hildesheim 1981.

H. Suter, "Über einige noch nicht sicher gestellte Autorennamen in den Übersetzungen des Gerhard von Cremona", *Bibliotheca Mathematica*, 3. Folge, 4 (1903), 19–27.

H. Suter, "Über den Kommentar des Muhammed ben 'Abdelbâqî zum zehnten Buche des Euklides", *Bibliotheca Mathematica*, 3. Folge, 7 (1906–1907), 234–251.

W. Thomson, *The Commentary of Pappus on Book X of Euclid's Elements. Arabic Text and Translation*, Cambridge (Mass.) 1930.

P. M. J. E. Tummers, "The Commentary of Albert on Euclid's Elements of Geometry", in J. A. Weisheipl (ed.), *Albertus Magnus and the Sciences. Commemorative Essays 1980*, Toronto 1980, pp. 479–499.

P. M. J. E. Tummers, "Geometry and Theology in the XIIIth century", *Vivarium* 18 (1980), 112–142.

P. M. J. E. Tummers, *Albertus (Magnus)' commentaar op Euclides' Elementen der geometrie*, Nijmegen 1984.

P. M. J. E. Tummers (ed.), *The Latin Translation of Anaritius' Commentary on Euclid's* Elements *of Geometry, Books I–IV*, Nijmegen 1994. (= *Artistarium*, supplementa, IX.)

H. Weißenborn, *Die Übersetzungen des Euklid durch Campano und Zamberti. Eine mathematisch-historische Studie*, Halle/S. 1882.

IV

Probleme der Euklidinterpretation und ihre Bedeutung für die Entwicklung der Mathematik*

0. Einleitung

Euklids Hauptwerk, die *Elemente,* gilt als dasjenige wissenschaftliche Werk, das am häufigsten bearbeitet und benutzt wurde; es war über 2000 Jahre lang nicht nur das mathematische Lehrbuch schlechthin, sondern es beeinflußte auch die Entwicklung anderer wissenschaftlicher Disziplinen, z.B. der Physik, aber auch der Philosophie. Dieses erstaunliche Buch hat seit seinem Entstehen stets große Bedeutung für die Geschichte der Mathematik und der übrigen Wissenschaften gehabt. Das darin geschaffene System der euklidischen Geometrie wird auch jetzt noch in allen Schulen der Welt studiert.

Es ist naheliegend, daß ein Standardwerk wie Euklids *Elemente* oft herausgegeben wurde. Jedoch beschränkte sich die Tätigkeit der Gelehrten nicht darauf, den schon bekannten Editionen und Übersetzungen jeweils eine weitere hinzuzufügen; vielmehr bemühte man sich, die wesentlichen Gedanken in Euklids Werk zu verdeutlichen und fruchtbar zu machen. So ergab sich im Laufe der Geschichte eine erstaunliche Vielfalt von Versionen, Bearbeitungen, abgekürzten Fassungen, Kommentaren, Scholien und Sonderfassungen für spezielle Zwecke.

Dem Mathematikhistoriker stellt sich nun die Aufgabe, diese verwirrende Vielfalt innerhalb der Euklidtradition aufzuhellen, zu registrieren, welche Versionen und Kommentare es gibt und was sie ent-

*Universität Oldenburg, Postfach 2503, D-2900 Oldenburg, BRD. Überarbeitete Fassung eines Vortrags, der am 1.12.1978 am Institut für Geschichte der Naturwissenschaften der Universität Aarhus gehalten wurde. Den Mitarbeitern des Institutes danke ich für freundliche Hinweise.

halten. Schon dies ist nicht einfach, vor allem, weil viele Texte nur handschriftlich erhalten und bisher kaum oder nicht erfaßt sind. So sind heute noch keine endgültigen Aussagen möglich. Aber glücklicherweise hat gerade das Problem der Euklidtradition und -rezeption seit Beginn intensiverer mathematikhistorischer Forschung, also seit etwas mehr als 100 Jahren, stets die Forscher beschäftigt, und in den letzten Jahrzehnten sind grundlegend neue Erkenntnisse über den arabischen und lateinischen Euklid gewonnen worden. So sind heute die Überlieferungslinien im großen und ganzen bekannt.

Dieser Aufsatz setzt sich nicht das Ziel, einen möglichst vollständigen Überblick über die Euklidtraditionen zu geben, sondern will vielmehr anhand der Überlieferung untersuchen, in welcher Hinsicht die Euklidbearbeitungen die Entwicklung mathematischer Gedanken förderten – oder auch hemmten. Die Arbeit gliedert sich in drei Teile: Zunächst werde ich die Hauptwege aufzeigen, auf denen Euklids Schrift von der Antike bis in die Gegenwart gelangte. Im 2. Teil versuche ich dann, relativ pauschal die wichtigsten Probleme des Euklid zu nennen, mit denen sich die Gelehrten in den einzelnen Zeiten beschäftigt haben. Abschließend werde ich an einem Beispiel sozusagen exemplarisch die Geschichte spezieller Probleme verfolgen, die aus der Beschäftigung mit den *Elementen* erwuchsen.

1. Hauptlinien der Überlieferung der Elemente

Zunächst ein globaler Überblick über die Tradition der *Elemente.*[1] Wie man an der großen Anzahl der Bearbeitungen und Kommentare erkennt, bemühte man sich besonders in vier Epochen um das Verständnis dieses Werks: in der Antike, bei den Arabern, im christlichen Mittelalter, im 16. Jahrhundert.

Euklid verfaßte sein Werk kurz vor 300 v. Chr. Schon wenig später setzten die Bemühungen anderer Gelehrter um die *Elemente* ein: Wir wissen von mehreren Kommentaren aus der Spätantike. Der älteste, den Heron von Alexandria verfaßte (um 60 n. Chr.), ist verloren. Wichtiger war der Kommentar des Pappos von Alexandria (Ende des 3. Jahrhunderts), dessen griechischer Text zwar ebenfalls nicht mehr vorhanden ist, von dem wir aber die Bemerkungen zum 10. Buch in arabischer Übersetzung besitzen.[2] Am bekanntesten ist zweifellos der

Abhängigkeit der wichtigeren Euklid-Traditionen

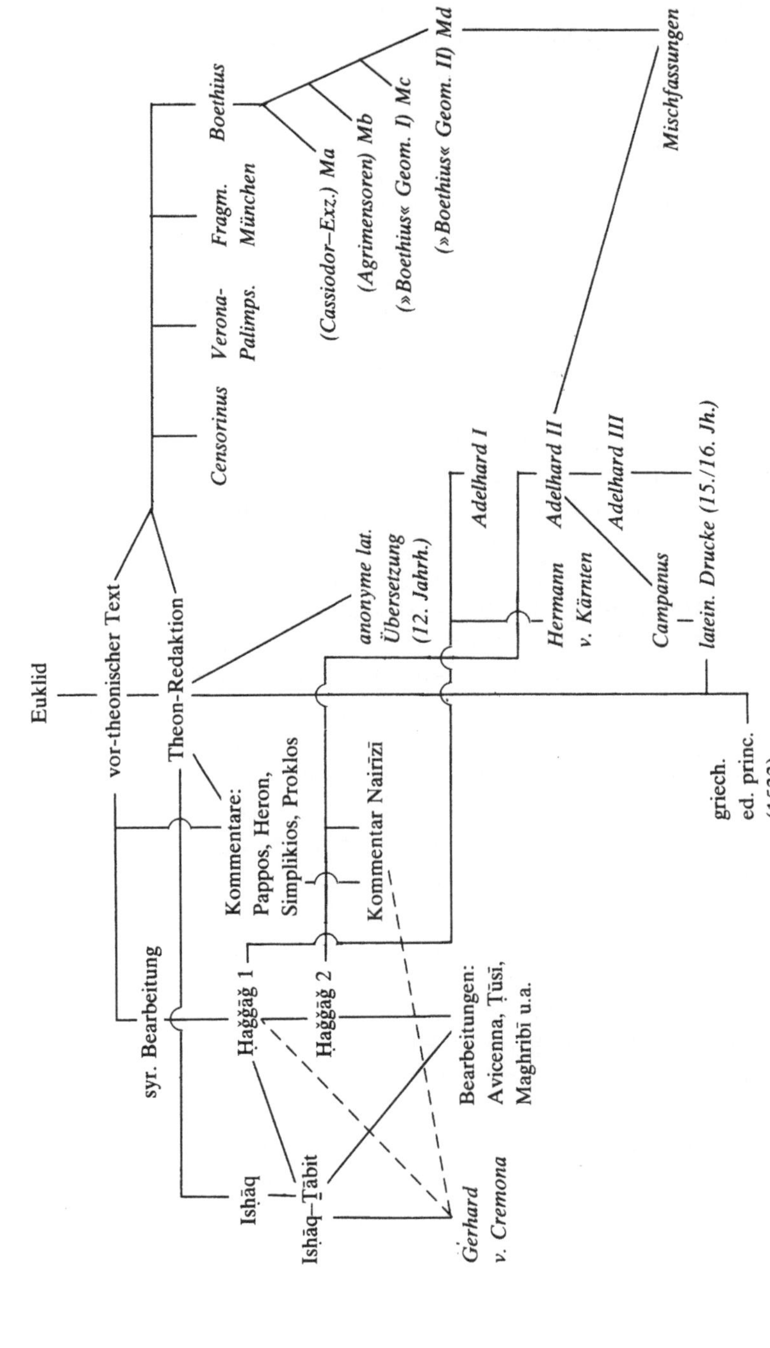

Kommentar zum Buch 1, den Proklos Diadochos (410–485) verfaßte.[3] Im 6. Jahrhundert schrieb Simplikios, der durch seine Kommentare zu Aristoteles bekannt ist, einen *Kommentar zur Einleitung des Buches von Euklid.* Teile dieser Schrift hat der Araber an-Nairīzī bewahrt. Diese Werke werden ergänzt durch die große Zahl griechischer Scholien. Viele gehen auf Proklos und Pappos zurück, andere sind viel jüngeren Datums; sie stammen zum Teil erst aus dem 14. Jahrhundert.

Das herausragende Ereignis in der griechischen Phase der Überlieferung der *Elemente* war die Edition dieses Werkes, die Theon von Alexandria im 4. Jahrhundert ausführte. Ähnlich wie bei den Werken anderer griechischer Autoren, entstand auch hier durch philologische Arbeit aus dem schon verwilderten Text eine gereinigte Fassung, die fortan bis zum 19. Jahrhundert als der maßgebliche griechische Euklidtext angesehen wurde. Erst als 1808 François Peyrard eine griechische Handschrift fand, die den vortheonischen Text aufweist, konnte man die Überlieferungsgeschichte weiter zurückverfolgen. Heute weiß man, daß auch einige arabische Euklidübersetzungen auf eine vortheonische Fassung zurückgehen.

Der griechische Urtext der *Elemente* wurde erst wieder in der Renaissance in Westeuropa bekannt. Im Mittelalter waren im Abendland lediglich lateinische Euklidübersetzungen verbreitet. Man unterscheidet zwei Phasen: die griechisch-lateinische und die arabisch-lateinische. Die bedeutendste Übersetzung aus dem Griechischen wurde um 500 von Boethius angefertigt, ist aber nur fragmentarisch in mehreren Texten auf uns gekommen.[4] Bis zum Anfang des 12. Jahrhunderts standen den Gelehrten nur diese fast unbrauchbaren Exzerpte zur Verfügung, die nicht einmal Beweise enthielten. So ist es verständlich, daß diese Bruchstücke nicht dazu beitrugen, die mathematischen Kenntnisse zu erweitern. Wir können diese Phase im folgenden vernachlässigen.

Während die griechischen Euklidhandschriften nach dem Ausgang der Antike in Westeuropa nicht mehr bekannt waren und in der byzantinischen Welt zwar aufbewahrt und bisweilen kopiert, sonst aber wenig beachtet wurden, verlagerten sich die Bemühungen um Euklids Werk im Mittelalter in die arabisch sprechenden Länder. Eine griechische Handschrift der *Elemente* gelangte aus Byzanz zum Kalifen al-Manṣūr (754–775), wurde allerdings damals nicht ins Arabische übersetzt.

Dies geschah erstmals durch al-Ḥaǧǧāǧ (um 786–833), der gleich zwei Übersetzungen anfertigte: unter Kalif Hārūn al-Rashīd (786–809) und unter al-Ma'mūn (813–833). Die zweite Fassung, von der die Bücher 1–6 und (vermutlich) 11–13 erhalten sind,[5] war kürzer als die erste, von der keine Handschrift bekannt ist. Vielleicht wurden die Kürzungen erreicht, indem man auf die benutzten Sätze zurückverwies, statt sie zu wiederholen.

Eine neue Übertragung ins Arabische veranstaltete Isḥāq b. Ḥunain († 910/911), der berühmteste Übersetzer griechischer Werke. Auch hier gab es eine zweite Rezension, die von dem bekannten islamischen Mathematiker Ṯābit b. Qurra (um 826–901) angefertigt wurde. Während die erste Version verloren ist, gibt es von der Ṯābit-Bearbeitung zahlreiche Handschriften, die aber noch nicht untersucht sind. Sicher ist, daß Ṯābit für seine Bearbeitung griechische Handschriften benutzte. Sezgin vertritt die Ansicht,[6] Isḥāqs Übersetzung habe eine theonische Handschrift benutzt, und Ṯābit habe diesen Isḥāq-Text unter Heranziehung der al-Ḥaǧǧāǧ-Version verbessert. Der Ḥaǧǧāǧ-Text seinerseits geht nach Sezgin über eine syrische Übersetzung, die wahrscheinlich im 5./6. Jahrhundert ausgeführt wurde, auf eine vortheonische Quelle zurück. Auf den sehr unterschiedlich eingeschätzten Wert der besagten syrischen Übersetzung möchte ich nicht eingehen.[7]

Die genannten zwei bzw. vier vollständigen Übersetzungen bezeichnen nur den Anfang einer erstaunlichen Aktivität, mit der sich die islamischen Gelehrten um den Euklidtext bemühten. Man unterscheidet Auszüge, Bearbeitungen und Kommentare. Der bekannteste Auszug ist derjenige des persischen Philosophen Avicenna (Ibn Sīnā, 980–1037). Er nahm in den Abschnitt Geometrie seiner umfangreichen philosophischen Enzyklopädie *Kitāb aš-Šifā'* alle 15 Bücher der *Elemente* auf, aber mit abgekürzten Beweisen. Avicenna will nicht eine vollständige Bearbeitung schaffen, sondern einen Abriß, eine Zusammenfassung, die sich in Verbindung mit mündlicher Unterweisung zur schnellen Aneignung mathematischer Kenntnisse eignete.[8]

Unter den verschiedenen Euklidbearbeitungen ist diejenige am verbreitetsten, die Naṣīraddīn aṭ-Ṭūsī (1201–1274) im Jahre 1248 abschloß. Von ihr gibt es zahlreiche Handschriften und Editionen.[9] aṭ-Ṭūsī versuchte, den alten Text des Euklid aufgrund der Übersetzung des al-Ḥaǧǧāǧ wiederherzustellen. Er benutzte auch den Isḥāq-Ṯābit-Text, gab aber in der Regel der älteren Version den Vorzug. Für die Textgestaltung wandte er philologische, historische und mathematische Kriterien an – ein Vorgehen, das ganz modern anmutet.[10] Eine zweite Bearbeitung, die nur die ersten 13 Bücher umfaßt, läuft unter Ṭūsīs Namen, ist aber vermutlich unecht, da sie erst 1298 beendet wurde. Obwohl es nur zwei Handschriften gibt, ist diese Fassung bekannter als der echte Ṭūsī-Text, da die kürzere Version schon 1594 in Rom gedruckt wurde.

Die Zahl der arabischen Kommentare zu Euklid ist Legion. Aus arabischen Quellen kennen wir die Namen von etwa 60 Kommentatoren,[11] von deren Schriften mehr als die Hälfte erhalten sind. Es wäre müßig, all diese Namen zu nennen; es reicht die Bemerkung, daß viele der bekanntesten arabischen Mathematiker in dieser Reihe stehen. Wegen des großen Umfangs der *Elemente* ist es verständlich, daß fast alle Autoren nur Teile des Werks kommentierten, zumal Euklids Schrift aufgrund ihres Aufbaus eine solche Auswahl erleichert.

Die Phase der arabischen Übersetzungen und Bearbeitungen zeigt nicht nur die größte Vielfalt in der Geschichte der *Elemente,* sondern sie war auch in mathematischer Hinsicht die schöpferischste. Demgegenüber treten Übersetzungen in andere östliche Sprachen zurück; die meisten davon gehen auf arabische Quellen zurück. Bekannt sind Übersetzungen ins Persische, Hebräische, Syrische und Armenische.

Seit dem 12. Jahrhundert nimmt das mathematische Wissen im Abendland schlagartig zu. Dies hängt nicht zuletzt mit der Tatsache zusammen, daß im Rahmen der Übersetzertätigkeit aus dem Arabischen auch Euklids Hauptwerk ins Lateinische übertragen wird und somit erstmals seit der Antike wieder ein vollständiger Text vorliegt. Genauer gesagt, ist die Überlieferung zweigleisig, da die Übersetzungen teils auf den Ḥaǧǧāǧ-Text, teils auf Isḥāq-Ṯābit zurückgehen. Einen Isḥāq-Ṯābit-Text benutzte Gerhard von Cremona (um 1114–1187), dessen Bearbeitung von allen arabisch-lateinischen Texten der griechischen Tradition am nächsten kommt.[12] Das Werk enthält aber auch Beweise nach al-Ḥaǧǧāǧ I und hat vermutlich auch Anleihen an Gerhards eigene Übersetzung des Kommentars von an-Nairīzī gemacht. Während von Gerhards Schrift immerhin acht Handschriften bekannt sind, existiert von einer anderen Übertragung, die vermutlich Hermann von Kärnten (um 1140–50) zuzuschreiben ist, lediglich eine Handschrift. Hermann benutzte den arabischen Text von al-Ḥaǧǧāǧ I.[13]

Die bei weitem wichtigste Bedeutung innerhalb der mittelalterlichen arabisch-lateinischen Euklidtradition kommt dem englischen Übersetzer, Mathematiker und Philosophen Adelhard von Bath zu (um 1116–1142). Adelhard war nicht nur der Autor von mindestens drei Versionen der *Elemente,* sondern seine Übersetzung war auch der Ausgangspunkt für zahlreiche weitere Bearbeitungen. Diese Texte wollen wir unter dem Namen "Adelhard-Tradition" zusammenfassen. Seit den grundlegenden Arbeiten von Marshall Clagett[14] unterscheidet man die Versionen Adelhard I, II und III sowie die Bearbeitung durch Campanus. Dabei waren Adelhard I und III wenig verbreitet, während die beiden übrigen Fassungen außerordentlich oft kopiert wurden. Adelhard I ähnelt der Übersetzung des Hermann und benutzte wie dieser eine Ḥaǧǧāǧ-I-Vorlage. Es handelt sich um eine relativ getreue Übersetzung. Dagegen ist Adelhard III eine Bearbeitung des Adelhard-II-Textes. Viel wichtiger sind die Fassungen Adelhard II, von der 50 Handschriften existieren, und der Campanus-Text (über 100 Handschriften). Adelhard II entstand um 1120, Campanus' Fassung kurz vor 1260. Adelhard II ist keine Übersetzung, sondern eine abgekürzte Bearbeitung, die auf al-Ḥāǧǧāǧ II beruht. Typisch sind die Beweise, die nicht ausgeführt werden, sondern bei denen nur erwähnt wird, nach welchen Propositionen der Beweis geführt werden könnte. – Der Campanus-Text der Lehrsätze stimmt wörtlich mit Adelhard II überein; die Beweise jedoch sind weitgehend unabhängig. Hier hat Campanus teilweise beachtliche Erweiterungen vorgenommen. Er verwandte große Sorgfalt darauf, unklare Abschnitte zu erläutern. Um das Werk auf den mathematischen Stand seiner Zeit zu bringen, ergänzte er es mit Material seiner Zeitgenossen; insbesondere benutzte er die Arithmetik des Jordanus Nemorarius.

Wenn wir versuchen, die Geschichte der *Elemente* in der Renaissance und der Neuzeit zu verfolgen, so müssen wir zunächst vier zentrale Ereignisse erwähnen: die Veröffentlichung des Campanus-Textes (erstmals 1482 in Venedig), eine neue Übersetzung vom Griechischen ins Lateinische durch Bartolomeo Zamberti (1505), die *editio princeps* des griechischen Textes durch Simon Grynaeus (Basel 1533), eine weitere lateinische Übersetzung durch Federigo Commandino (1572).

Mit der Übersetzung, die Zamberti aus dem Griechischen anfertigte, stand erstmals wieder ein vollständiger Euklidtext zur Verfügung, der relativ rein war. Trotzdem war auch Zambertis Text nicht frei von eigenwilligen Interpretationen und Zusätzen, so daß in der Folgezeit Parallelausgaben Zamberti–Campanus veröffentlicht wurden (seit 1516). – Die Ausgabe des griechischen Textes im Jahre 1533 – übrigens gemeinsam mit dem Kommentar des Proklos – blieb bis zum 18. Jahrhundert die einzige vollständige griechische Edition. Es ist erstaunlich, daß trotzdem die lateinischen Ausgaben des Campanus und Zamberti bedeutender blieben als die Edition des griechischen Textes. – Ich kann nicht alle weiteren lateinischen Bearbeitungen auch nur des 16. Jahrhunderts erwähnen.[15] Viele umfaßten nur die ersten sechs Bücher, die ja in gewisser Weise eine Einheit bilden.[16] In dieser Zeit entstanden auch die ersten Übersetzungen in die Nationalsprachen: 1543 durch N. Tartaglia ins Italienische, 1564 durch P. Forcadel ins Französische, 1558 bzw. 1562 durch Johann Scheubel bzw. Wilhelm Holtzmann ins Deutsche, 1570 durch Henry Billingsley ins Englische. All diese beruhten auf den lateinischen Versionen des Campanus und Zamberti.

Ein Markstein war die Veröffentlichung einer neuen Übersetzung durch Federigo Commandino (Pesaro 1572). Commandino (1509–1575), der auch andere griechische mathematische Werke ins Lateinische übersetzte, zog neben der Ausgabe von 1533 mindestens eine weitere griechische Handschrift für seine Euklidübersetzung heran. So entstand ein philologisch genauer und zugleich mathematisch zuverlässiger Euklidtext. Praktisch alle späteren Ausgaben und Übersetzungen bis 1808 beruhen auf Commandinos Text.[17]

Die Euklidkommentare dieser Zeit dienten wie die mittelalterlichen Bearbeitungen oft dazu, die Strukturen im Euklid logisch klarzulegen;[18] andere versuchten, das Werk für Praktiker nutzbar zu machen. Hierzu gehört etwa die erste deutsche Übersetzung durch Holtzmann (Xylander), der sich an Maler, Goldschmiede und Baumeister wendet (1562). Die wohl bekannteste gelehrte Bearbeitung stammt vom Jesuiten Christopher Clavius (1537–1612; Rom 1574ff.). Diese ist eher als eine Übersetzung ein Kommentar, der mit großem Fleiß die Arbeiten von Campanus, Zamberti und Commandino benutzt. Der Hauptwert liegt in der großen Menge des Hilfsmaterials, das sie enthält. So fügte Clavius zu den 486 Propositionen des griechischen Euklid 671 eigene hinzu.

Man könnte die Euklidarbeiten im 16. Jahrhundert unter dem Aspekt subsummieren, ein umfassendes Hand- oder Schulbuch herzustellen. Im 17. Jahrhundert dagegen bemüht man sich, den Text zu vereinfachen. Man verkürzt Beweise, ändert bisweilen die Reihenfolge, läßt Sätze oder sogar Bücher aus. Hier sind vor allem die Arbeiten von André Tacquet (seit 1654) und Isaac Barrow (seit 1655) zu nennen. Noch kritischer geht man im 18./19. Jahrhundert mit Euklid um: Die Mathematiker versuchen, Unvollkommenheiten zu beseitigen. Auf der anderen Seite wenden sich Philologen und Historiker dem griechischen Text zu: Im 19. Jahrhundert schaffen Peyrard und Heiberg maßgebliche Textausgaben.

2. *Problematische Teile in den Elementen als Objekt für Kommentatoren*

Im folgenden soll aufgezeigt werden, welche Teile in Euklids Werk in den einzelnen Zeiten die Aufmerksamkeit der Bearbeiter vor allem auf sich lenkten. Eine Antwort auf diese Frage setzt voraus, daß man nicht nur die Titel der Euklidbearbeitungen und -kommentare kennt, sondern auch ihren Inhalt. Dies ist leider nicht immer der Fall. So kann ich nur einen groben Überblick geben; weitere Arbeiten, die zu erwarten sind, können das Bild in vielen Punkten ändern.

Was wissen wir aus griechischen Euklidausgaben, Bearbeitungen, Kommentaren, Scholien und Erwähnungen in anderen Werken über die Bemühungen der Griechen um Probleme, die mit dem Euklidtext verbunden sind? Thomas L. Heath hat das Wesentliche zusammengefaßt:[19] Herons Kommentar, dessen Inhalt wir zum Teil aus Proklos, zum Teil aus an-Nairīzīs Kommentar kennen, enthielt ein paar allgemeine Bemerkungen, etwa, daß nur drei Axiome zugelassen sind; ferner bei verschiedenen Beweisen in Buch 1, 3, 6, 7 zusätzliche Fallunterscheidungen; weiterhin alternative Beweise, etwa zu II 1–10, wo statt der geometrischen die heute übliche algebraische Beweismethode gewählt wird, oder in I 19.25, wo Heron die indirekten Beweise

Euklids durch direkte ersetzte. Ferner bewies Heron die Umkehrungen einiger Sätze (II 12.13; VIII 27) und erweiterte den Euklidtext an einigen Stellen (I 1.37.47; III 17.22; IV 16; VII 3). All diese Ergänzungen oder Änderungen enthalten substantiell wenig Neues. Ebenfalls nicht bedeutend waren offenbar die Änderungen, die Porphyrios (232–304) ausführte, soweit die wenigen Bemerkungen bei Proklos ein Urteil zulassen. Viel wichtiger war der Kommentar, den Pappos im 3. Jahrhundert verfaßte. Ein Teil davon, nämlich seine Bemerkungen zum 10. Buch, ist erhalten. Hier liegt erstmals eine Erklärung zu diesem schwierigsten Buch des Euklid vor. Auch das, was wir über andere Teile seines Kommentars, vor allem wiederum durch Erwähnungen bei Proklos, wissen, zeigt, daß seine Erörterungen wesentliche Fragen anschnitten. Ein Beispiel möge genügen: Pappos stellte fest, daß die Umkehrung des 4. Postulats (Alle rechten Winkel sind gleich) nicht gelte. Die Aussage, daß alle Winkel, die einem rechten Winkel gleich seien, ebenfalls rechte Winkel seien, stimme dann nämlich nicht, wenn es sich um Kreiswinkel handle. Hier schneidet Pappos ein Problem an, das in der Folgezeit die Mathematiker immer wieder beschäftigte: das des hornförmigen Winkels. – Auch das Werk des letzten bekannten griechischen Euklidkommentators, Simplikios (6. Jahrhundert), betraf wesentliche Punkte, nämlich die Definitionen, Postulate und Axiome des 1. Buches und hier vor allem das berühmte 5. Postulat, das Parallelenpostulat. A. I. Sabra hat diesen Abschnitt kürzlich aus arabischen Quellen ediert und analysiert.[20]

Der einzige vollständig erhaltene griechische Kommentar stammt von Proklos und betrifft nur Euklids 1. Buch. Proklos' Interesse an philosophischen Fragen zeigt sich in seinem Kommentar, wo er nicht nur im Vorwort eine Philosophie der Mathematik gibt, sondern auch später oft philosophische Betrachtungen einfließen läßt. Ihn interessieren nicht zusätzliche Beweise, Fallunterscheidungen und ähnliches, sondern die grundsätzlichen Fragen, etwa nach dem Wesen der Definitionen, Axiome und Postulate und nach ihrer philosophischen Begründung. So ist es kein Wunder, daß er sich besonders ausführlich mit dem Parallelenpostulat beschäftigt und hier auch – ausnahmsweise – einen Beweisversuch liefert. Im übrigen beschränkt er sich im allgemeinen darauf, die Ansichten seiner Vorgänger (Eudemos, Geminos, Apollonios) darzustellen und auf Probleme hinzuweisen.

Kommen wir jetzt zu den Arabern. Man kann die Mathematik der Araber in drei Perioden einteilen: die Periode der Rezeption und Assimilation der Mathematik der Vorgänger (bis ca. 850), die schöpferische Periode (ca. 850–1200) und die Spätzeit.[21] In der Periode der Assimilation wird auch der Euklidtext übersetzt, der fortan allen Gelehrten selbstverständlich bekannt ist. Für die schöpferische Periode ist charakteristisch, daß die Wissenschaftler in der Lage sind, die Ergebnisse ihrer Vorgänger zu verbessern, deren Lehren zu erweitern und zu korrigieren. In diese Zeit fallen die wichtigsten Arbeiten zu Euklid. Ich möchte nur den Inhalt einiger hier nennen.

Schon in der Rezeptions- und Assimilationsperiode versucht man, einen Beweis für das Parallelenpostulat zu bringen. Der älteste Ansatz dieser Art bei den Arabern ist bei al-Ǧauharī (Anfang 9. Jahrhundert) zu finden. Nur wenig später gab al-Māhānī (825–888) eine kritische Analyse der Proportionenlehre des Eudoxos und Euklid in zwei Kommentaren zum 5. und 10. Buch; außerdem verfaßte er eine Schrift über 26 Sätze des 1. Buches von Euklid, die ohne den indirekten Beweis bewiesen werden können.[22] Auch im Abendland sehr bekannt wurde an-Nairīzī (um 900), der nicht nur einen Kommentar zum Gesamtwerk verfaßte, sondern zusätzlich noch eine Schrift über das 5. Postulat. Die zentrale Figur in der Euklidkommentierung ist im 11. Jahrhundert Ibn al-Haiṯam († 1041). Er verfaßte verschiedene Schriften zu den Postulaten des Euklid und erwarb sich vor allem in der Geschichte des Parallelenpostulats Ruhm. – Nur noch wenige Bemerkungen zu späteren Kommentaren: 'Umar al-Ḫaiyām († 1123), der durch seine Methode, kubische Gleichungen geometrisch zu lösen, bekannt ist, beschäftigte sich auch mit den Lehren von den Proportionen, den Parallelen und dem Zahlbegriff. Er verfaßte einen dreiteiligen Kommentar zu den schwierigsten Stellen und Postulaten der *Elemente,* wobei der erste Teil die Parallelenlehre, die letzten beiden die Verhältnislehre behandeln.[23] Die Bearbeitung der *Elemente* durch aṭ-Ṭūsī († 1274) haben wir schon erwähnt. Sie hat zusammen mit speziellen Schriften desselben Verfassers einen direkten Einfluß auf die Entwicklung der Parallelenlehre in Westeuropa ausgeübt, insbesondere die Arbeiten von J. Wallis (1663) und G. Saccheri (1733) beeinflußt.

Um etwas Ordnung in diese verwirrende Fülle zu bringen: Wenn man statt der Autoren die Themen betrachtet, die sie behandeln, so kann man ohne große Schwierigkeiten trotz der unzureichenden Vorarbeiten die Hauptinteressen der Kommentatoren ermitteln. Man kommentierte im allgemeinen nicht das ganze Werk, sondern Teile, und hier interessierten vor allem die Prämissen, d.h. Definitionen, Postulate, Axiome. Unter ihnen wiederum steht das Parallelenpostulat mit Abstand an erster Stelle. Eine einfache Durchmusterung des Inhalts derjenigen Euklidkommentare, die Sezgin erwähnt, ergibt je etwa 15 Arbeiten zum Parallelenpostulat und zum 10. Buch und knapp 10 Schriften zum 5. Buch des Euklid. Spezielle Kommentare zu anderen Büchern oder Themen sind dagegen nur vereinzelt vertreten (etwa zum 3., 7., 12., 13. Buch). Man wird kaum überrascht sein über die außerordentlich häufige Behandlung dieser drei Themen, die übrigens nicht auf die arabische Welt beschränkt ist: Die Unübersichtlichkeit in der Behandlung irrationaler Größen in Buch 10 machte Erläuterungen erforderlich. Auf der anderen Seite spielt die Theorie der Proportionen, die in Buch 5 dargestellt wird, eine zentrale Rolle in der Mathematik. Da es einige Schwierigkeiten gab, um die eudoxische Definition gleicher Verhältnisse in Buch 5 zu erfassen, wurde dieses Buch bevorzugtes Objekt für Kommentatoren. Das Bemühen um das Parallelenpostulat schließlich hängt mit der Vermutung zusammen, daß man dieses besonders komplizierte Postulat vielleicht beweisen könne. Schon aus der Antike sind verschiedene Beweisversuche bekannt.

Mit dem Parallelenpostulat, der Proportionenlehre und der Lehre von den Irrationalitäten hatten die Araber die drei Themenkreise angegeben, die in Zukunft – im Mittelalter wie in der Neuzeit – die europäischen Mathematiker hauptsächlich beschäftigen sollten. Aber natürlich änderten sich die Interessen und damit auch die Schwerpunkte der Arbeiten im Laufe der Zeit. Vergleichen wir einmal den arabischen mit dem arabisch-lateinischen Euklid: In beiden Traditionen findet man eine große Menge verschiedener Versionen, wobei aber wesentlich mehr arabische als lateinische Kommentare verfaßt wurden. Vermutlich gibt es nur einen original in Latein geschriebenen Kommentar, der – mit Fragezeichen – dem Dominikaner Albertus Magnus zugeschrieben wird; er benutzt früher übersetztes Material, etwa den Kommentar von an-Nairīzī.[24] Zwar findet man im Abendland bisweilen *Questiones super Euclidem,* etwa im 14. Jahrhundert von Nicole Oresme, aber diese behandeln nicht so sehr Sätze des

Euklid, sondern eher sehr allgemeine Probleme der Geometrie, Mathematik oder Philosophie. Dies führt uns zu einem anderen Unterschied zwischen der arabischen und lateinischen Tradition: Die lateinischen Texte wurden offenbar großenteils mit der Absicht verfaßt, philosophische Fragen zu behandeln, während die arabischen mehr auf die reine Mathematik ausgerichtet blieben. So findet man in den lateinischen Texten nirgends etwas, das mit den arabischen Diskussionen über das Parallelenpostulat vergleichbar wäre. Dagegen ist die Proportionenlehre auch im Abendland während des Mittelalters stark beachtet worden, sicherlich in Verbindung mit den Bewegungsgesetzen, die vor allem in der Oxforder und Pariser Schule seit dem 14. Jahrhundert aufgestellt wurden. Auch die Frage des krummlinig begrenzten Winkels trat in den Vordergrund.

Charakteristisch für die Adelhard–Campanus-Texte, die im lateinischen Mittelalter vorherrschten, sind ihre zahlreichen Zusätze. Sie zeigen das Bestreben, Euklids Werk zu einem didaktisch aufgebauten Lehrbuch umzuarbeiten, ein Trend, der schon in der arabischen Epoche festzustellen war. Man stellt Erörterungen über den Aufbau eines Beweises an, verweist auf ähnliche Propositionen und fügt Axiome oder Definitionen ein, um alle nur denkbaren Lücken in der Beweisführung zu schließen. Diese didaktischen Bemühungen richteten sich einerseits an den Studenten, der innerhalb der Artes auch Teile des Euklid verarbeiten mußte; andrerseits konnte man dadurch auch außerhalb der Mathematik liegende, vor allem philosophische, Probleme in den Griff bekommen. In dieser Hinsicht die wichtigsten Fragen sind die nach der Inkommensurabilität, nach dem hornförmigen Winkel zwischen Kreisumfang und Tangente und nach der Teilbarkeit von Größen. All diese Ideen beziehen sich letztendlich auf Probleme des Unendlichen und der Stetigkeit, zentrale Fragen innerhalb der mittelalterlichen Philosophie. So lieferte die Adelhard-Tradition ein Lehrbuch Euklid, das hervorragend zu den scholastischen Interessen paßte.

Während die lateinischen Euklidtexte im späten Mittelalter von Darstellungsweise und Zielsetzung her einen relativ homogenen Eindruck machen und sich verhältnismäßig gut charakterisieren lassen, sind die Euklidausgaben und -bearbeitungen in Renaissance und Neuzeit, wie erwähnt, wesentlich uneinheitlicher. Eine zusätzliche Schwierigkeit entsteht dadurch, daß die große Masse der gedruckten

Texte höchstens in Bibliographien titelmäßig erfaßt, aber inhaltlich noch nicht ausgewertet ist; die handschriftlichen Bearbeitungen sind praktisch völlig unbekannt. So sind allgemeine Angaben über wesentliche Probleme noch nicht möglich; vermutlich werden sich bei den sehr unterschiedlichen Zielen der einzelnen Ausgaben auch keine einheitlichen Gesichtspunkte finden lassen.

3. Die Proportionenlehre in Buch 5

Wir haben gesehen, daß es vor allem drei Problemgruppen gibt, die man zu allen Zeiten für erklärenswert hielt: das Parallelenpostulat zu Beginn von Buch 1, die eudoxische Definition der Verhältnisgleichheit in Buch 5 und die Klassifikation der Irrationalitäten in Buch 10. Im folgenden möchte ich einige Ansätze vorführen, die im Lauf der Geschichte gemacht wurden, um diese Probleme in den Griff zu bekommen. Der unterschiedliche Umfang historischer Vorarbeiten zu diesen Themen zwingt mich dabei, eine Auswahl zu treffen: Lediglich die Weiterentwicklung der Parallelenlehre seit Euklid ist bisher recht eingehend untersucht worden, so daß ich sie hier vernachlässigen kann;[25] dagegen werden die Probleme, die mit Euklids 5. Buch zusammenhängen, in mathematikhistorischen Darstellungen meist nur gestreift, und über die Arbeiten zum 10. Buch liest man praktisch nichts. Dies ist sehr bedauerlich, da gerade das 10. Buch mit seiner Lehre von den inkommensurablen Größen und der Theorie der Irrationalitäten insbesondere bei den Arabern, aber auch im 16. Jahrhundert in Europa nachhaltiges Interesse gefunden hat. So kann ich auf Probleme, die mit dem 10. Buch zusammenhängen, an dieser Stelle nicht eingehen. Stattdessen will ich versuchen, Entwicklungslinien darzustellen, die von der eudoxischen Definition der Verhältnisgleichheit in Buch 5 ausgehen.

3.1 Antike

Schon vor den Griechen hat man mit Verhältnissen gerechnet, aber es gab noch keine allgemeine Erklärung des Begriffs oder eine Theorie der Proportionen. Diese nun haben die Griechen geschaffen, und zwar in verschiedenen Ansätzen: Zunächst bauten die Pythagoreer, ausgehend von der Musiktheorie, eine Theorie der Verhältnisse mit Zahlen, also mit diskreten Größen, auf; erst Jahrhunderte später, durch Eudoxos (4. Jahrhundert), entstand eine entsprechende Lehre mit stetigen Größen. Die ältere Theorie ist bei Euklid im 7. Buch, die jüngere im 5. Buch wiedergegeben.

Euklid gibt im 7. Buch keine Definition des Verhältnisses; hier müßte man stillschweigend die alte, von Nikomachos überlieferte[26] Erklärung ergänzen:

> Verhältnis (λόγος, *proportio*) ist eine Beziehung (σχέσις, *relatio, habitudo*) zweier Terme zueinander.

Dabei ist σχέσις der Name für eine allgemeine Relation, λόγος für das Verhältnis im heute noch üblichen Sinn.

Die Zahlentheorie in Buch 7 geht von der Erklärung aus, was man unter »Teil« und »Menge von Teilen« einer Zahl versteht (Def. 3–5). Sinngemäß wird gesagt:

> Sind a und b zwei (natürliche) Zahlen, so ist jede eine Menge von Teilen der anderen: $a = m \cdot \frac{b}{n}$ und $b = n \cdot \frac{a}{m}$.[27]

Als 20. Definition folgt dann die Definition der Verhältnisgleichheit:

> Zahlen stehen in Proportion, wenn die erste von der zweiten Gleichvielfaches oder derselbe Teil oder dieselbe Menge von Teilen ist wie die dritte von der vierten.
> (Also $a{:}b = c{:}d$, wenn es teilerfremde m, n gibt mit $a = m \cdot \frac{b}{n}$ und $c = m \cdot \frac{d}{n}$.)

Mit dieser Definition können Rechengesetze für Proportionen bewiesen werden, allerdings nur für natürliche Zahlen.

In Buch 5 stellt Euklid im Anschluß an Eudoxos eine Proportionenlehre für beliebige Größen auf. Zunächst definiert er, was ein Verhältnis ist:

> V Def. 3: Verhältnis (λόγος) ist das gewisse Verhalten (σχέσις) zweier gleichartiger Größen der Abmessung nach.

Dies entspricht der Definition bei Zahlen mit dem Unterschied, daß jetzt die Begriffe *gleichartig* und *der Abmessung nach* hinzugekommen sind; letzteres bedeutet den *Wert* = *Zahlengröße*.

Um auszuschließen, daß man heterogene Größen miteinander vergleicht, etwa Strecken und Flächen, aber auch krummlinig und gradlinig begrenzte Winkel, formuliert Euklid zunächst das berühmte Axiom des Messens:

> V Def. 4: Daß sie ein Verhältnis haben, sagt man von Größen, die vervielfältigt einander übertreffen können.

Es gibt zwei äquivalente Formulierungen dieses Axioms in der griechischen Mathematik: zunächst

> Euklid X 1: Nimmt man bei Vorliegen zweier ungleicher (gleichartiger) Größen von der größeren ein Stück größer als die Hälfte weg und vom Rest ein Stück größer als die Hälfte und wiederholt dies immer, dann muß einmal eine Größe übrig bleiben, die kleiner als die kleinere Ausgangsgröße ist. (Es wird angemerkt, daß der Beweis auch zu führen ist, wenn man stets nur die Hälften fortnimmt.)

Die dritte Formulierung stammt von Archimedes in der Schrift *Über Kugel und Zylinder,* wo er unter Berufung auf Eudoxos folgendes fordert:

> Die größere von zwei gegebenen Größen, sei es Linie, Fläche oder Körper, überragt die kleinere um eine Differenz, die, genügend oft vervielfacht, jede der beiden gegebenen Größen übertrifft.[28]

Dies ist in geometrischer Gestalt genau das Axiom von Archimedes–Eudoxos, das heute der Begründung der Theorie der irrationalen Zahlen zugrunde gelegt wird:

> Sind a und b zwei Elemente aus der Menge der reellen Zahlen, die beide >0 sind, so soll es stets mindestens eine natürliche Zahl n geben, so daß $nb>a$ ausfällt.

Auf diesem Axiom beruht bekanntlich ganz wesentlich die Vollständigkeit des Systems der reellen Zahlen.

Das von Eudoxos–Archimedes formulierte Axiom des Messens ist für die weitere Entwicklung der griechischen Mathematik grundlegend; in der divisiven Form (Euklid X 1) wird es etwa bei infinitesimalen Betrachtungen herangezogen und ist der wichtigste Schritt bei der Exhaustionsmethode.

Die äquivalente Definition V 4 bereitet dann die zentrale Definition der Verhältnisgleichheit für beliebige Größen vor:

> V Def. 5: Es gilt $A:B = C:D$, wenn für je zwei Zahlen m, n
> stets zugleich $nA > mB$ und $nC > mD$
> oder $nA = mB$ und $nC = mD$
> oder $nA < mB$ und $nC < mD$ gilt.

Diese geniale Definition schließt die herkömmliche Definition in Buch 7 ein und ist auch auf inkommensurable Größen anwendbar, da jetzt statt des Gleichheitszeichens die Größer-Relation zugelassen ist. Mit Hilfe dieser zugegebenermaßen etwas komplizierten Definition stellt Euklid dann eine Proportionenlehre für allgemeine Größen auf; auf Einzelheiten kann ich hier nicht eingehen.

Die Notwendigkeit einer neuen Proportionenlehre ergab sich für die Griechen mit der Entdeckung irrationaler Größen durch die Pythagoreer. Man weiß nicht, ob diese Entdeckung am Verhältnis Diagonale zu Seite beim Fünfeck oder Quadrat gemacht wurde. Man ist sich aber einig, daß die Existenz inkommensurabler Größen mit Hilfe der Wechselwegnahme (ἀνθυφαίρεσις) gemacht wurde. Dieses Verfahren ist uns heute als *Euklidischer Algorithmus* aus den Elementen der Zahlentheorie wohlbekannt. Für Zahlen ausgesprochen findet es sich in VII 1.2, für Größen in X 2.3 der *Elemente*, wo auch die Konstruktion des größten gemeinsamen Teilers angegeben wird:

> Seien A, B zwei Größen. Dann nimmt man von der größeren die kleinere so oft wie möglich weg, von der kleineren den Rest usw.:
> $A = p_1 B + R_1$
> $B = p_2 R_1 + R_2$
> ....

Wenn ein gemeinsames Maß existiert, so bricht das Verfahren ab, und

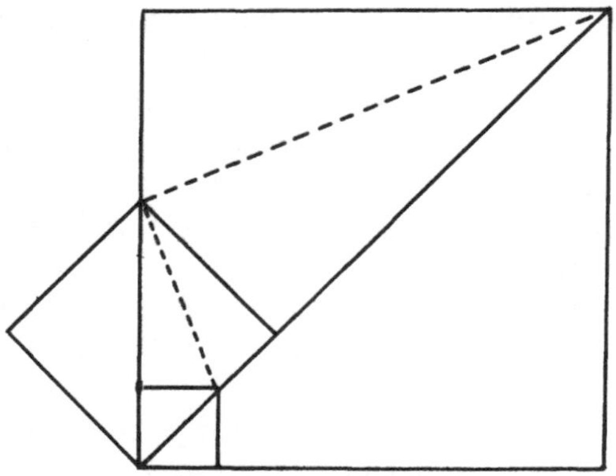

Fig. 1.

der letzte Rest ist der größte gemeinsame Teiler. Bei nicht abbrechendem Verfahren sind die Größen inkommensurabel, und die p_i bilden die Koeffizienten der Kettenbruchentwicklung. Wendet man dies Verfahren geometrisch auf Diagonale und Seite des Quadrats an, so sieht man leicht, daß sich folgende Kettenbruchdarstellung ergibt:[29]

$$\sqrt{2} = 1 + \cfrac{1}{2 + \cfrac{1}{2 + \cfrac{1}{2 + \cfrac{1}{\cdots}}}}$$

Hieraus läßt sich eine Definition der Gleichheit zweier Proportionen herleiten: Zwei Größenpaare (A, B) und (C, D) stehen im gleichen Verhältnis, wenn das Verfahren der Wechselwegnahme dieselben p_i liefert, d.h., wenn B ebensooft in A enthalten ist wie D in C. Diese Definition wurde in der Antike nicht explizit formuliert, aber es gibt Anklänge daran.[30] Jedenfalls ist diese Definition, die ebenfalls auf inkommensurable Größen anwendbar ist, anschaulicher als die des Eudoxos, die eigentlich zu allen Zeiten als schwer verständlich galt.

Man sieht, daß die pythagoreische Theorie der Proportionen für rationale Zahlen Ähnliches leistet wie die eudoxische für allgemeine Größen. Das Verhältnis ganzer Zahlen entspricht einem rationalen

Bruch, das Verhältnis zweier kontinuierlicher Größen entspricht in gewissem Sinne einer reellen Zahl. Die Griechen haben jedoch die Unterscheidung von Zahlen und Verhältnissen kontinuierlicher Größen beibehalten. Dabei wurde die allgemeine Theorie offenbar nicht für praktische Zwecke benutzt, sondern vor allem für die Exhaustionsmethode. Da die Praxis des Operierens mit Irrationalitäten in Griechenland nicht weit entwickelt war, ist es verständlich, daß die in Buch 5 vorhandene Theorie nicht ganz vollständig ist: Die Zusammensetzung der Verhältnisse, die äquivalent mit der Multiplikation irrationaler Zahlen ist, wird nur für Spezialfälle (z.B. doppeltes, dreifaches Verhältnis) betrachtet. Ein zweiter Mangel: Es gibt keinen allgemeinen Satz über die Existenz einer vierten Proportionalen; vielmehr zieht Euklid in Buch 11 und 12 für allgemeine Größen ein solches Theorem stillschweigend heran. Auch die Frage, ob die eudoxische Definition der Verhältnisgleichheit Beziehungen zur Theorie der Wechselwegnahme aufweist, hat die Griechen offenbar nicht interessiert.

3.2 Araber

Das Verdienst, auf diese Lücken hingewiesen zu haben, gebührt den islamischen Mathematikern. Ihnen ist die Erweiterung des Zahlbegriffs bis zur irrationalen Zahl allmählich gelungen. Hierfür sind vor allem zwei Momente verantwortlich: eine umfangreiche Rechenpraxis, insbesondere in Verbindung mit Trigonometrie und Astronomie, sowie die Kritik und Weiterentwicklung der griechischen Theorie.[31]

Irrationale Größen waren für die Araber nichts Besonderes; sie rechneten bei der Bestimmung von Gleichungswurzeln oder trigonometrischen Größen selbstverständlich mit Näherungswerten, so daß man die Irrationalitäten wie Zahlen betrachtete.

Mit dieser konkreteren Beziehung zu den irrationalen Größen hängt auch die kritische Analyse zusammen, der man die euklidische Proportionenlehre bei den Arabern unterzieht. E. B. Plooij hat in seiner Dissertation[32] erstmals die Arbeiten mehrerer islamischer Mathematiker[33] zur Proportionenlehre untersucht und ist dabei zu überraschenden Ergebnissen gekommen, die Juschkewitsch und Rosenfeld dann präzisiert haben:[34]

Die Araber bestreiten nicht die Richtigkeit der Definition der Verhältnisgleichheit in Buch 5, meinen aber, daß diese Definition das Wesen der Proportion, nämlich seine messende Funktion, nicht aufdecke. Denn bei kommensurablen Größen erkennt man im euklidischen Algorithmus den Prozeß der Messung einer Größe mit Hilfe einer anderen; dagegen werden in der allgemeinen Definition des 5. Buches willkürliche Vielfache der 1. und 2. Größe mit entsprechenden der 3. und 4. Größe verglichen, ohne daß man ein Messen erkennt. Die Mathematiker der arabischen Länder haben sich bemüht, gerade diese Maßfunktion herauszustellen, die die Theorie der Zahlenverhältnisse mit der allgemeinen Theorie der Proportionen verbindet. Allmählich wurde daher die eudoxische Definition der Proportion durch eine andere ersetzt, die im wesentlichen mit der voreudoxischen Theorie der Wechselwegnahme übereinstimmte, also die Gleichheit der Kettenbruchentwicklung als Kriterium für die Gleichheit zweier Verhältnisse annahm. Zunächst ging man nur von dieser Definition aus; später, vor allem seit Ibn al-Haiṯam, untersuchte man den Zusammenhang zwischen dieser Definition und der des Eudoxos und bewies ihre Äquivalenz. Man erkannte auch Lücken in der griechischen Theorie: z.B. das Fehlen eines Satzes über die 4. Proportionale oder einer Lehre von den zusammengesetzten Verhältnissen, die man vor allem für die Trigonometrie benötigte.

Die neue Theorie ist besonders gut bei al-Ḫaiyām dargelegt. Er definiert nicht nur die Gleichheit zweier Verhältnisse, sondern auch das größere Verhältnis mit Hilfe der Kettenbruchdarstellung und gibt dabei auch ein Kriterium an, um irrationale und rationale Zahlen zu vergleichen. al-Ḫaiyām beweist die Äquivalenz seiner Definitionen der Gleichheit und Ungleichheit von Verhältnissen mit den Definitionen Euklids. Dabei beruhen seine Beweise auf dem Satz über die Existenz einer vierten Proportionalen zu drei gegebenen Größen. al-Ḫaiyām betont die Wichtigkeit dieses Satzes und beweist ihn mit Hilfe eines Stetigkeitsprinzips, das er nach eigenen Worten Aristoteles entnommen hat. Es besteht in der Behauptung, daß man Größen bis ins Unendliche teilen kann. Sein Beweis ist allerdings nicht ganz exakt.

Die Ideen al-Ḫaiyāms wurden 150 Jahre später von aṭ-Ṭūsī wiederaufgenommen und weiterentwickelt. Er spricht deutlich aus, daß jedes Verhältnis als Zahl bezeichnet werden kann, die durch die Einheit gemessen wird. So werden auch alle irrationalen Verhältnisse zu Zahlen, die näherungsweise durch rationale Brüche ausgedrückt werden können.

3.3 Westeuropäisches Mittelalter

Ich habe erwähnt, daß in der arabischen Welt nicht wie bei den griechischen Mathematikern das Verfahren der anthyphairesis durch die eudoxische Definition der Verhältnisgleichheit ersetzt war, sondern daß beide Ideen bekannt waren und man sich bemühte, Beziehungen zwischen beiden Kriterien aufzuzeigen. Im lateinischen Mittelalter waren diese Ansätze weitgehend unbekannt, weil sie sich überwiegend in speziellen Arbeiten vorfanden, die nicht ins Lateinische übersetzt wurden; in den arabischen Euklidausgaben des al-Ḥaǧǧāǧ und in der Isḥāq-Ṯābit-Tradition wird wie im griechischen Original lediglich die Definition des Eudoxos (ohne Erklärung) erwähnt, während die Methode der anthyphairesis nur in VII 1.2 und X 2.3 in Verbindung mit der Frage nach dem gemeinsamen Maß zweier Zahlen bzw. Größen benutzt wird. Zwei spezielle arabische Arbeiten allerdings waren auch in lateinischer Übersetzung im Mittelalter bekannt: die Schrift von Aḥmad b. Yūsuf (Ametus filius Josephi) über die Proportionenlehre[35] und der Euklidkommentar des an-Nairīzī,[36] beides von Gerhard von Cremona im 12. Jahrhundert übersetzt. Wir wissen, daß Campanus das Werk des Aḥmad für die Kommentierung von Buch 5 des Euklid benutzte und es auch Leonardo von Pisa sowie Thomas Bradwardine bekannt war.[37]

Aḥmad wendet sich aus methodischen Gründen gegen die eudoxische Definition: Sie sei zwar gültig, aber kein erstes Prinzip im aristotelischen Sinne, da es nicht einfacher als alle anderen sei. Stattdessen gibt er folgende Definition:[38]

> Verhältnis ist die Beziehung zwischen zwei Dingen, die etwas gemeinsam haben hinsichtlich der Art der Quantität oder Qualität, durch die sie gleich sind oder sich unterscheiden.

Mit ihrer Hilfe sucht er die eudoxische Definition zu beweisen. Aḥmad erwähnt nicht die anthyphairetische Methode; er bemüht sich nur, die Gültigkeit der eudoxischen Definition durch ein Prinzip zu bestätigen, das intuitiv als richtig angesehen werden konnte und das einfacher als jene Formulierung war.

Das zweite arabische Werk zur Proportionenlehre, das, abgesehen von den arabischen Euklidübersetzungen, im Abendland in lateinischer Übersetzung bekannt war, ist der Kommentar von an-Nairīzī zu Euklid I–X.[36] an-Nairīzī erwähnt Definition 3[39] und den unechten Zusatz nach dieser Definition[40] und nimmt diesen Text zum Anlaß, um das Vorgehen der anthyphairesis ausführlich und klar zu erläutern. Jedoch bringt an-Nairīzī diesen Begriff nicht in Verbindung mit den grundlegenden eudoxischen Definitionen. Vielmehr sind seine Bemerkungen zu V Def. 4 und 5 nur kurz und geben lediglich wieder, was er auf den vorhergehenden Seiten gesagt hatte; er geht in keiner Weise auf die Beziehungen zwischen den beiden Begriffen ein. Zu Definition 5 bemerkt er, daß Beweisversuche mißlungen sind;[41] möglicherweise bezieht sich dies auf Aḥmad.

Fassen wir zusammen: Was konnten die westeuropäischen Gelehrten im Mittelalter von den arabischen Bemühungen um die eudoxische Definition wissen? Die Arbeiten von Aḥmad b. Yūsuf und an-Nairīzī vermittelten wenigstens in Andeutungen eine Vorstellung von den arabischen Versuchen um die Proportionenlehre. Die Idee der anthyphairesis war bekannt, und man hätte sie als Ausgangspunkt für weitere Überlegungen nehmen können. Aber dies geschah nicht.

Im folgenden wollen wir uns auf die wesentlichen lateinischen Euklidübersetzungen des Mittelalters beschränken: auf Adelhard II und Campanus sowie hin und wieder auf die weniger verbreitete Fassung des Gerhard von Cremona. Bei Adelhard und Gerhard fehlt ein Hinweis auf die anthyphairesis. Dies ist verständlich, da zumindest in Buch 5 die eudoxische Definition jene verdrängt hatte. Aber auch in anderen Versionen und Einzelhandschriften, die zusätzliche, manchmal methodische, Kommentare aufweisen, wird jenes Verfahren nicht erwähnt. Dies ist besonders bei Campanus überraschend, da er Aḥmads Werk über die Proportionen und wohl auch an-Nairīzīs Euklidkommentar kannte. Einiges spricht dafür, daß Campanus so sehr mit der antiken Lehre der Zahlenverhältnisse verhaftet war, daß er die Alternative der allgemeinen Theorie nicht akzeptierte, vielleicht sogar ihre volle Reichweite nicht einmal verstand.[42]

Bekanntlich war während des ganzen Mittelalters Boethius' Arithmetik die Hauptautorität, was die Verhältnisse ganzer Zahlen betraf. Auf Boethius beruht die Arithmetik des Jordanus Nemorarius, in der ebenfalls arabische Elemente verarbeitet sind. Auf Boethius dürfte auch die etwas überraschende Terminologie in der mittelalterlichen Proportionenlehre zurückgehen: Die Übersetzungen von λόγος (= ursprünglich Verbindung, später Verhältnis zwischen zwei Größen) und ἀναλογία (= Verhältnisgleichheit zwischen mehr als zwei Größen) sind nicht, wie zu erwarten, *ratio* und *proportio,* sondern vielmehr *proportio* (für λόγος) und *proportionalitas* (für ἀναλογία) gemäß Boethius. Vor allem A. G. Molland hat gezeigt, daß in der Proportionenlehre des Mittelalters die eudoxische Definition der Gleichheit von Proportionen oft durch arithmetische Kriterien ersetzt wurde.[43] Es bildete sich in Anlehnung an Ideen des Boethius eine Lehre der Denominationen heraus, wobei dieses Wort bisweilen die verschiedenen Arten rationaler Proportionen bezeichnet (etwa in Boethius' Arithmetik), bisweilen aber auch mit *Bruch* wiedergegeben werden kann. Eine Arithmetisierung der Proportionenlehre setzt ein: Anstelle der abstrakten Definition des Eudoxos findet man in Euklidbearbeitungen und anderen mathematischen Schriften oft Zahlenbeispiele. Dabei beschränkt man sich meistens auf rationale Verhältnisse; erst Bradwardine läßt in der *Geometria speculativa* auch irrationale Proportionen zu.

Es gibt mehrere Gründe, warum man davon abkam, die allgemeine eudoxische Theorie der Proportionen zu benutzen, und mehr die arithmetischen Anwendungen betonte (etwa die Anwendbarkeit der Theorie auf Probleme der Physik, Musik und Astronomie). Zu den wichtigsten Gründen gehörten jedenfalls Verständnisschwierigkeiten, die aus Überlieferungs- und Übersetzungsproblemen resultierten. Sie betreffen die entscheidenden Definitionen V Def. 4 und 5.

Alle lateinischen Euklidtexte enthalten zwischen Def. 3 und 4 die nichtauthentische Definition *Proportionalitas est similitudo proportionum,* einen Text, der auch in einigen guten griechischen Handschriften am Rande anzutreffen ist und wohl auch in arabischen Codices steht. Diese zusätzliche Definition, die sich an unpassender Stelle befindet – die Gleichheit von Proportionen wird ja erst in Def. 5 erklärt –, dürfte zur Verwirrung der mittelalterlichen Mathematiker beigetragen haben, insbesondere zur Umdeutung der überlieferten Definition 4.

Das Hauptproblem entstand durch den Text der Def. 4. Diese Definition wird zwar in Gerhards Euklidübersetzung und in seiner Übertragung des an-Nairīzī-Kommentars richtig wiedergegeben, ist aber bei Hermann und vor allem in den Adelhard-Fassungen durch eine andere, mißverständliche Definition ersetzt. Die originale Def. 4 besagt bekanntlich, daß Größen zueinander im Verhältnis stehen, wenn sie vervielfältigt einander übertreffen können; dies ist das eudoxisch-archimedische Axiom des Messens. Dagegen enthalten alle drei Adelhard-Versionen, die Hermann-Übersetzung und der Campanus-Text stattdessen eine Feststellung, die keineswegs den Inhalt der Def. 4 trifft:

> *Quantitates que dicuntur continuam proportionalitatem habere, sunt quarum eque multiplicia aut equa sunt aut eque sibi sine interruptione addunt aut minuunt.*
> (Größen, von denen man sagt, sie stehen in stetiger Proportionalität, sind solche, deren Gleichvielfache entweder gleich sind oder sich in gleicher Weise ohne Unterbrechung übertreffen bzw. geringer sind.)

Diese Feststellung geht nicht auf das Wesen der ursprünglichen Def. 4 ein, nämlich auf das Prinzip der Vergleichbarkeit bei Proportionen, sondern führt vielmehr den hier unpassenden Begriff der »stetigen Proportion« $A:B = B:C$ ein. Es ist bis heute unklar, woher diese unechte Def. 4 stammt. Es liegt nahe, sie in einer arabischen Vorlage zu suchen, doch hat man sie bisher weder in al-Ḥaǧǧāǧ II noch in Handschriften des Isḥāq-Ṯābit-Textes noch in den beiden aṭ-Ṭūsī-Versionen gefunden.

Die unechte Def. 4 wird in den betrachteten Übersetzungen in Verbindung gesetzt mit der folgenden Def. 5, der erwähnten eudoxischen Definition der Gleichheit von Verhältnissen. Diese Definition wurde i.a. korrekt übersetzt:

> *Quantitates que dicuntur esse secundum proporcionem unam . . . sunt quarum prime et tercie multiplicaciones equales multiplicacionibus secunde et quarte equalibus fuerint simul* (schlechter: *similes*) *vel addicione vel diminucione vel equalitate eodem ordine sumpte.*

> (Größen, von denen man sagt, sie stehen im selben Verhältnis, sind solche, bei denen die Gleichvielfachen des ersten und dritten den Gleichvielfachen des zweiten und vierten hinsichtlich der Addition, Subtraktion oder Gleichheit gleich sind, wenn man sie in derselben Reihenfolge nimmt.)

Man erkennt, daß die Terminologie der unechten Def. 4 derjenigen der Def. 5 ähnelt, so daß man sie im Mittelalter in Beziehung setzte. Auch in der nichtauthentischen Def. 4 werden ja Gleichvielfache zur Erklärung verwendet; so könnte der Eindruck entstehen, es handle sich um die Anwendung des eudoxischen Kriteriums auf den Spezialfall der stetigen Proportionalität. Aber ein Wort läßt diese Interpretation nicht zu: Die Gleichvielfachen sollen einander *in gleicher Weise* (*eque*) übertreffen oder untereinander geringer sein. Dies widerspricht Eudoxos' Idee: Es wird bei ihm nur gefordert, *daß* ein Überschuß oder Defizit in gleicher Weise vorhanden ist; daß der Überschuß oder das Defizit hinsichtlich des Betrages gleich groß sein soll, ist sinnlos. So war nicht mehr die Möglichkeit gegeben, den Begriff Proportion auf inkommensurable Größen anzuwenden.

Beim Übergang von der unechten Def. 4 zur echten Def. 5 wandten die lateinischen Kommentatoren das (falsche) Kriterium des *gleichen* Überschusses oder Defizits auf die eudoxische Definition der Verhältnisgleichheit an und verfälschten dadurch ihren wahren Charakter. Dies tritt besonders deutlich bei Campanus hervor, der sich seitenlang über die Definitionen des Buchs 5 ausläßt.[44] Für ihn ist alles klar: Die Proportionalität wird in zwei Arten eingeteilt: die stetige (behandelt in der unechten Def. 4) und die nicht stetige (in Def. 5). Campanus interpretiert – absolut uneudoxisch – die unechte Def. 4 so:

> In stetiger Proportion stehen solche Größen, bei denen alle Gleichvielfachen in stetiger Proportion stehen.
> (D.h.: $A{:}B = B{:}C$ genau dann, wenn $(nA){:}(nB) = (nB){:}(nC)$.)

Natürlich ist diese Aussage bei den Gleichvielfachen nA, nB, nC trivial. Tatsächlich gibt Campanus zu, daß seine Deutung der Def. 4 ein Zirkelschluß sein könnte. – Um den Überschuß bzw. das Defizit in die Definition einzubeziehen, wendet Campanus den Kunstgriff an, daß

Überschuß bzw. Defizit *quantum ad proportionem* und nicht *quantum ad quantitatem* genommen werden muß: Bei 12:6 = 6:3 übertrifft 6 die 3 und 12 die 6 in gleicher Weise *quantum ad proportionem*, da 12:6 = 6:3, nicht aber *quantum ad quantitatem*, da 12–6 ≠ 6–3 ist.

Nachdem diese Interpretation der unechten Def. 4 gefunden war, lag es nahe, eine ähnliche Deutung auch für Def. 5 zu versuchen. Zunächst schreibt Campanus beim Vergleich der Verhältnisse $A:B$ und $C:D$ folgendes:

> Wenn E (= $2A$) größer als G (= $3B$) ist und in gleicher Weise F (= $2C$) größer als H (= $3D$), oder wenn E kleiner als G ist und F in gleicher Weise kleiner als H oder wenn $E = G$ ist und $F = H$, dann ist $A:B = C:D$.

Dies entspricht aber nur scheinbar der eudoxischen Definition, denn etwas später betont Campanus, daß Gleichheit beim Überschuß oder Defizit hier dasselbe bedeutet wie bei der stetigen Proportion, nämlich nicht *quantum ad quantitatem*, sondern *quantum ad proportionem*. Das heißt: $A:B = C:D$ genau dann, wenn $(nA):(mB) = (nC):(mD)$. Somit hat Campanus das Wesen der eudoxischen Definition nicht erfaßt, weil er die unechte Def. 4 im Auge hatte.

Am Beispiel des Campanus haben wir gesehen, daß es durch die korrupte Textüberlieferung den westeuropäischen Mathematikern im 13. Jahrhundert unmöglich war, die griechischen Ideen der Verhältnisgleichheit auch für inkommensurable Größen zu begreifen. So ist es nicht verwunderlich, daß Roger Bacon Euklid tadelt, weil er in Buch 5 »außerordentlich dunkle Worte« gebraucht, die »in keinem Teil der Mathematik, Philosophie oder Theologie« benutzt werden.[45] Als Alternative fand man schließlich die folgende im Spätmittelalter übliche Definition:

> Proportionen sind gleich, wenn ihre Denominationen gleich sind.

Dabei ist die Denomination diejenige ganze oder gebrochene Zahl, die die Proportion in den kleinsten Größen ausdrückt. Diese Definition findet man etwa bei Roger Bacon, bei Bradwardine in der *Geometria speculativa* und im *Tractatus de proportionibus.*[46] Campanus übernahm mit anderen Definitionen aus Jordanus' Arithmetik auch diese Erklärung der Gleichheit von Proportionen in Buch 7 der *Elemente;* dort begegnet sie uns als Nr. 20 unter den von Campanus zugefügten Definitionen und ersetzt die euklidische Definition von der Proportionalität von Zahlen. In den scholastischen Abhandlungen des 14. und 15. Jahrhunderts wird diese Definition Allgemeingut.

Diese – noch unzulängliche – Definition der Verhältnisgleichheit mit Hilfe der Denominationen genügte für die Zwecke des Spätmittelalters. Bradwardine benutzte die Denomination als ein Kriterium, um rationale und irrationale Proportionen zu unterscheiden: Eine rationale Proportion ist direkt (*immediate*) durch eine gegebene Zahl benannt (so 6:2 durch 3 usw.), während inkommensurable Größen nur indirekt (*mediate*) durch eine Zahl benannt werden können, weil sie direkt benannt werden durch ein gegebenes Verhältnis und dieses wiederum direkt durch eine Zahl.[47] Diese etwas spitzfindige Erklärung wurde von anderen Mathematikern der Scholastik im allgemeinen akzeptiert und modifiziert, etwa in der Art, daß $(4:2)^{3:2}$ direkt benannt wird durch die Proportion 3:2 und indirekt durch die Zahl $1\frac{1}{2}$, die wiederum 3:2 direkt benennt.[48] Es ist klar, daß diese Deutungen problematisch waren und die Anwendung der Denominationstheorie auf das Gleichheitskriterium für Proportionen unbefriedigend bleiben mußte.

Fassen wir zusammen: Dadurch, daß es nicht gelang, die eudoxischen Grundideen in Buch 5 zu begreifen, verlor das Abendland eines der beiden allgemeingültigen Kriterien für die Verhältnisgleichheit, die in der Antike bekannt waren. Obwohl den Gelehrten die zweite denkbare Definition (durch anthyphairesis) in der lateinischen Übersetzung des an-Nairīzī zugänglich war, nutzten sie nicht die Möglichkeit, die sich ihnen bot. Stattdessen behandelten sie Proportionen mit der Theorie der Denominationen, einer Theorie, die nicht genügend durchgearbeitet und nicht recht auf inkommensurable Größen anwendbar war. Diese Theorie führte in gewisser Weise zu einer Arithmetisierung der Proportionenlehre. Sie behinderte nicht den Ausbau dieses Gebiets, insbesondere die Ausdehnung auf das Rechnen mit gebrochenen Exponenten. Man kam zu Rechenregeln über die Zusammensetzung von Verhältnissen: $A:C = (A:B)\cdot(B:C)$, die man etwa in der mathematischen Astronomie für den Transversalensatz auf der Kugel benötigte. In der Algebra wandte man die Proportionenlehre etwa auf Lösungen von Gleichungen mittels des einfachen oder doppelten falschen Ansatzes an. Am wichtigsten wurde die Theorie der Verhältnisse aber in ihren physikalischen Anwendungen, etwa in Bradwardines Bewegungsgesetz

$$F_2:R_2 = (F_1:R_1)^n \text{ mit } n = V_2:V_1.$$

Dies zeigt, daß im Mittelalter trotz des Unverständnisses wesentlicher Teile der eudoxischen Proportionenlehre neue mathematische Ideen wirksam werden konnten. Beides – das Mißverständnis der genialen griechischen Gedanken und die Originalität der Gelehrten des Spätmittelalters – sind bezeichnend für die Lehre von den Proportionen und wohl auch für die Mathematik im europäischen Mittelalter schlechthin. Dabei muß man noch zugute halten, daß die falsche Interpretation hauptsächlich durch den Zustand eines Textes im mittelalterlichen Corpus der arabisch-lateinischen Übersetzungen bewirkt worden war.

3.4 Ausblick

Erst im 16. Jahrhundert begann man das Wesen der griechischen Proportionenlehre wieder zu verstehen, nachdem griechische Handschriften des Euklid zur Verfügung standen. Auch neue Ideen arabischer Gelehrter wurden seit dieser Zeit vereinzelt in Westeuropa bekannt. So erfuhren europäische Gelehrte nach 1594, als aṭ-Ṭūsīs *Darlegung des Euklid* in Rom erschien, daß islamische Mathematiker den Zahlbegriff in Richtung auf die reellen Zahlen erweitert hatten. Mit dem Aufschwung der Rechentechnik im 17. Jahrhundert begann man, die abstrakte eudoxische Theorie der Proportionen zu kritisieren und zu arithmetisieren. So erinnert die Theorie der Verhältnisse bei Gregorius a Sancto Vincentio im 17. Jahrhundert an die arabischen Vorstellungen, und Andreas Tacquet kritisierte Euklid von einem Standpunkt aus, der al-Ḫaiyām ähnelt. Isaac Newton schließlich deutete die Zahl nicht mehr in griechischer Manier als Menge von Einheiten, sondern als abstraktes Verhältnis irgendeiner Größe zu einer anderen gleichartigen Größe, die als Einheit gewählt wurde:

Unter »Zahl« verstehen wir nicht sowohl eine Menge von Ein-

> heiten, sondern vielmehr das abstrakte Verhältnis irgendeiner Größe zu einer anderen Größe derselben Gattung, die als Einheit angenommen wird. Sie ist dreifacher Art: ganz, gebrochen und irrational; ganz, wenn die Einheit sie mißt; gebrochen, wenn ein Teil der Einheit, dessen Vielfaches die Einheit ist, sie mißt; irrational, wenn die Einheit mit ihr inkommensurabel ist.[49]

Damit ist ein vorläufiger Abschluß erreicht.[50]

Unter denjenigen Mathematikern, die im 19. Jahrhundert neue Definitionen der reellen Zahl fanden, ist natürlich vor allem Dedekind zu erwähnen. Er hat wohl als erster die Konstruktion des Eudoxos in aller Tiefe erfaßt. Zwischen seiner Theorie, nach der die reellen Zahlen als Schnitt im Gebiet der rationalen Zahlen eingeführt werden, und derjenigen des Eudoxos besteht eine so tiefe Analogie, daß R. Lipschitz Dedekind in einem Brief fragte, was er eigentlich im Vergleich zu den Alten Neues gemacht habe. Tatsächlich wird auch in der Definition des Eudoxos eine Klasseneinteilung der Paare ganzer (bei Dedekind der rationalen) Zahlen hervorgerufen: Sei a, b das gegebene Paar archimedischer Größen. Dann gehört ein Paar ganzer Zahlen m, n zur Klasse A, wenn $ma > nb$, und zur Klasse B, wenn $ma \leq nb$ ist. Dadurch zerfallen alle Paare ganzer Zahlen in zwei Klassen, die sich weder schneiden noch leer sind. Dieser Schnitt definiert wie bei Dedekind eine (nicht notwendig rationale) Zahl, und aus der eudoxischen Definition folgt, daß bei Verhältnisgleichheit $a{:}b = c{:}d$ das Paar c, d denselben Schnitt hervorruft. Es gibt zwischen beiden Theorien aber auch schwerwiegende Unterschiede: So operiert Dedekind auf dem Gebiet der rationalen Zahlen, Eudoxos aber mit Paaren ganzer Zahlen, für die weder eine arithmetische Operation noch eine Ordnungsrelation definiert ist.

Trotz der bruchstückhaften Auswahl hat diese Arbeit vielleicht erkennen lassen, welch vielfältige Anregungen die Lektüre des Euklid in den verschiedenen Zeiten gegeben hat, daß es auch in dem sonst so verschrienen Mittelalter interessante mathematische Gedanken gegeben hat und daß manche Ideen, die uns erstmals in Euklids Werk begegnen, auch heute noch nicht völlig überholt sind.

ANMERKUNGEN

1. Einen guten Überblick gibt J. E. Murdoch im *Dictionary of Scientific Biography,* Band 4, New York 1971, S. 437–459. – Die hier beigefügte Tabelle versucht, die Zusammenhänge zwischen den wichtigsten Bearbeitungen zu veranschaulichen.
2. Ediert von Thomson, W., und Junge, G., *The commentary of Pappus on book X of Euclid's Elements,* Cambridge (Mass.) 1930.
3. Ediert von G. Friedlein. Leipzig 1883.
4. Edition der erhaltenen Auszüge bei Folkerts, M., »*Boethius*« *Geometrie II, ein mathematisches Lehrbuch des Mittelalters,* Wiesbaden 1970.
5. Buch 1–6 wurden ediert von J. L. Heiberg, R. O. Besthorn und anderen, *Codex Leidensis 399, 1: Euclidis Elementa ex interpretatione al-Hadschdschadschii cum commentariis al-Narizii,* Kopenhagen 1893–1932.
6. Sezgin, F., *Geschichte des arabischen Schrifttums,* Band 5, Mathematik, Leiden 1974 (hier S. 95f.).
7. Siehe Sezgin,[6)] S. 87–90.
8. Lokotsch, K., *Avicenna als Mathematiker, besonders die planimetrischen Bücher seiner Euklidübersetzung,* Erfurt 1912, S. 25.
9. Sezgin,[6)] S. 111–113.
10. Sezgin,[6)] S. 95.
11. Sezgin,[6)] S. 105–120.
12. H. L. L. Busard bereitet eine kritische Ausgabe dieses Textes vor.
13. Siehe Busard, H. L. L., *The translation of the Elements of Euclid from the Arabic into Latin by Hermann of Carinthia (?),* Books I–VI: Leiden 1968, Books VII–XII, Amsterdam 1977.
14. Vor allem »The medieval Latin translations from the Arabic of the Elements of Euclid, with special emphasis on the versions of Adelard of Bath«, *Isis* 1953, *44,* 16–42.
15. Einen recht umfassenden Überblick gibt die Euklidbibliographie von Max Steck, die in Hildesheim erscheinen wird.
16. Z. B. Oronce Fine (seit 1536) oder Jacques Peletier (seit 1557).
17. Etwa David Gregory (1703); Robert Simson (1756).
18. Etwa bei Pierre de la Ramée (seit 1545).
19. *The thirteen books of Euclid's Elements,* Band 1, New York (Dover) 1956, S. 19–45.
20. Sabra, A. I., »Simplicius's proof of Euclid's parallels postulate«, *Journal of the Warburg and Courtauld Institutes* 1969, *32,* 1–24.
21. Sezgin,[6)] S. 25f.
22. Sezgin,[6)] S. 105.
23. Sezgin,[6)] S. 51.
24. Siehe Hofmann, J. E., »Uber eine Euklid-Bearbeitung, die dem Albertus Magnus zugeschrieben wird«. *Proceedings of the International Congress of Mathematicians,* Cambridge 1958, S. 554–566.
25. Siehe insbesondere Bonola, R., *Die nichteuklidische Geometrie,* 3. Auflage, Leipzig/Berlin 1921; Stäckel, P., *Die Theorie der Parallellinien von Euklid bis auf Gauss,* Leipzig 1895. Beide Bücher berücksichtigen kaum die damals noch weitgehend unbekannten Arbeiten der Araber. Hierzu siehe: Juschkewitsch, A. P., *Geschichte der Mathematik im Mittelalter,* Leipzig 1964, S. 277–288; Juschkewitsch, A. P., und Rosenfeld, B. A., *Die Mathematik der*

Länder des Ostens im Mittelalter, Berlin 1960, S. 149–151; Rosenfeld, B. A., »The theory of parallel lines in the medieval East«, *Actes du XIe Congrès International d' Histoire des Sciences, 1965,* Band *3,* Wroclaw usw. 1968, S. 175–178; Sabra, A. I., »Thābit ibn Qurra on Euclid's parallels postulate«, *Journal of the Warburg and Courtauld Institutes* 1968, *31,* 12–32 und die in Anmerkung 20) genannte Arbeit desselben Verfassers.

26. Arithmetik II 21 (= S. 120, 5f. Hoche).
27. Euklid braucht Fallunterscheidungen, weil $m = 1$ oder $n = 1$ nicht zugelassen ist.
28. Postulat 5; siehe *Archimedis opera omnia,* ed. J. L. Heiberg, Band 1, Leipzig 1910, S. 8.
29. Die Figur wurde von H. G. Zeuthen rekonstruiert; siehe Becker, O., *Das mathematische Denken der Antike,* 2. Auflage, Göttingen 1957, S. 73.
30. Siehe Szabó, Á., *Anfänge der griechischen Mathematik,* München/Wien 1969, S. 134f.
31. Juschkewitsch,[25] S. 248; Juschkewitsch/Rosenfeld,[25] S. 132.
32. *Euclid's conception of ratio and his definition of proportional magnitudes as criticized by Arabian commentators,* Rotterdam 1950.
33. al-Māhānī, an-Nairīzī, Ibn al-Haiṯam, al-Ḫaiyām, al-Ǧaiyānī, aṭ-Ṭūsī.
34. Juschkewitsch,[25] S. 248–256; Juschkewitsch/Rosenfeld,[25] S. 126–136.
35. Ediert, ins Englische übersetzt und erläutert von Schrader, D., *The Epistola de proportione et proportionalitate of Ametus Filius Josephi.* Diss. Univ. of Wisconsin 1961.
36. Ediert von M. Curtze als Supplement zu Heibergs Euklidausgabe: *Anaritii in decem libros priores Elementorum Euclidis commentarii,* Leipzig 1899.
37. Schrader,[35] S. 33–39.
38. Schrader,[35] S. 78.
39. *Et proportio est aliqua relatio quantitatis, que est inter duas res unius generis,* S. 156, 9f. Curtze.
40. *Proportionalitas est similitudo proportionum,* S. 157, 28f. Curtze.
41. S. 163, 18–24 Curtze.
42. Über das 5. Buch in den Euklidübersetzungen des 12. Jahrhunderts aus dem Arabischen unterrichtet ausführlich Cunningham, T. J., *Book V of Euclid's Elements in the twelfth century. The Arabic-Latin traditions.* Diss. Univ. of Wisconsin 1972.
43. »The denomination of proportions in the Middle Ages«, *Actes du XIe Congrès International d'Histoire des Sciences,* 1965, Band *3,* Wroclaw usw. 1968, S. 167–170; »The geometrical background to the »Merton School««, *British Journal of the History of Science* 1968, *4,* 108–125; »An examination of Bradwardine's Geometry«, *Archive for History of Exact Sciences* 1978, *19,* 113–175. Einen sehr guten Überblick gibt Murdoch, J. E., *The medieval language of proportions.* Scientific Change, ed. A. C. Crombie, London 1963, S. 237–271.
44. In der Campanus-Ausgabe Paris 1516 umfaßt dieser Abschnitt 11 Druckseiten.
45. *Communia mathematica,* ed. R. Steele, Oxford 1940, S. 86.
46. Siehe Murdoch,[43] S. 257.
47. *Tractatus de proportionibus,* ed. H. Lamar Crosby, Madison 1955, S. 66, 8–19.
48. Siehe Murdoch,[43] S. 259–261.
49. *Arithmetica universalis,* Ausgabe Leiden 1732, S. 4.
50. Über den Weg von der antiken Definition der Zahl bis hin zur modernen Fassung des Zahlbegriffs unterrichtet Gericke, H., *Geschichte des Zahlbegriffs,* Mannheim 1970 (hier vor allem S. 68–72).

Addendum to IV

[pp. 186–193] Since this article appeared, research has shown that the transmission of Euclid in the Middle Ages is more complicated. In particular this is true of the Arabic translations and reworkings of Ḥajjāj, Isḥāq and Thābit and of the Latin texts here called "Adelard II" and "Adelard III". The diagram on p. 187 must be modified accordingly. Our present knowledge of the transmission of Euclid in the Middle Ages and of the various texts involved is outlined in item III: "Euclid in Medieval Europe", pp. 2–31.

V

DIE MATHEMATISCHEN STUDIEN REGIOMONTANS IN SEINER WIENER ZEIT

Wer in ZINNERS Regiomontan-Biographie[1] blättert in der Absicht, Auskunft über dessen frühe mathematische Studien zu erhalten, wird enttäuscht feststellen, daß ZINNER hierüber fast nichts aussagt. Dies liegt nicht daran, daß ZINNER sich nur für die astronomischen Leistungen dieses großen Gelehrten interessierte; vielmehr ist die Quellenlage recht ungünstig. In der Bibliographie der Handschriften und frühen Drucke, die mit Regiomontan in Verbindung gebracht werden[2], findet man nämlich eine ganze Anzahl astronomischer Texte, aber, abgesehen von der Handschrift Wien 5203, keine einzige mathematische Schrift, die Regiomontan in seiner Wiener Zeit, also bis 1461, selbst verfaßte oder kopierte. Bekanntlich sind die größeren mathematischen Werke, insbesondere die Dreieckslehre, die Streitschrift gegen Nikolaus von Kues über die Kreisquadratur, die trigonometrischen Tafeln, die Euklidbearbeitung, der Briefwechsel mit Bianchini, Jakob von Speyer und Christian Roder, erst in den Jahren ab 1462 entstanden[3]. Aus dieser Tatsache könnte man folgern, daß Regiomontan sich in den Wiener Jahren überwiegend mit astronomischen Fragestellungen beschäftigte. Dies mag zutreffen, jedoch zeigt die Vertrautheit mit verschiedenen mathematischen Teilgebieten, die wir in den in Italien entstandenen Schriften konstatieren müssen, daß Regiomontan sich in der vorangehenden Zeit, insbesondere also in Wien, ein umfassendes mathematisches Wissen angeeignet haben muß. Ich möchte im folgenden diese an sich triviale Feststellung konkretisieren, indem ich diejenigen mathematischen Schriften im engeren Sinne benenne, die Regiomontan mit Sicherheit in seinen Wiener Jahren kennenlernte oder selbst schrieb. Dadurch wird man erkennen, daß man durch mühselige Kleinarbeit noch

---

[1] ERNST ZINNER, *Leben und Wirken des Joh. Müller von Königsberg genannt Regiomontanus*. (2. A. Osnabrück 1968).

[2] Abgedruckt bei ZINNER a. O. 297—371.

[3] Siehe MENSO FOLKERTS, Regiomontanus als Mathematiker. *Centaurus 21* (1977) 14—45.

viele Mosaiksteinchen finden kann, die das Bild vom Mathematiker Regiomontan zwar nicht völlig verändern, aber doch sehr viel deutlicher machen können. Dabei werde ich einerseits die schon genannte Wiener Handschrift etwas näher analysieren und andrerseits auf andere mathematische Texte hinweisen, die Regiomontan während des Aufenthaltes in Wien kennenlernte.

1. Regiomontans Wiener Rechenbuch (Wien, Ms. 5203)

Der Codex Wien 5203 ist die einzige bei Zinner erwähnte mathematische Handschrift Regiomontans, die in Wien geschrieben wurde. Dieser Codex ist als das *Wiener Rechenbuch* Regiomontans seit Ende des letzten Jahrhunderts bekannt, aber erstaunlicherweise nur zum Teil ausgewertet. Für die Mathematik hat dies wie so oft Maximilian Curtze in Ansätzen versucht[4]. Er hat dabei auch hier nur die Rosinen herausgepickt, allerdings manche interessante Einzelheit übersehen. Das Wiener Rechenbuch enthält eigene und fremde Arbeiten, die Regiomontan in den Jahren 1454 bis etwa 1462 schrieb[5], wobei die meisten Abschriften und Entwürfe vor 1458 liegen[6]. Zinner geht bei der Datierung offenbar von der Form der Schriftzüge aus[7], ein Verfahren, das vermutlich nicht immer zu unanfechtbaren Ergebnissen führt. Wenn man Zinner folgt, so stammt nur eine mathematische Schrift in diesem Rechenbuch nicht aus der Wiener Zeit, nämlich Regiomontans Erörterung über die Berechnung von Sinustafeln[8], die dieser um 1462 geschrieben haben soll[9]. Diese Abhandlung, die Schöner 1541 offenbar nach dieser Handschrift publizierte[10], steht in Verbindung mit Peuerbachs *Tractatus sinuum et*

---

[4] Maximilian Curtze, Die Quadratwurzelformel des Heron bei den Arabern und bei Regiomontan und damit Zusammenhängendes. *Zeitschrift für Mathematik und Physik, hist.-lit. Abt. 42* (1897) 145—156; weitere Auszüge in: Eine Studienreise. *Centralblatt für Bibliothekswesen 16* (1899) 257—306.

[5] Zinner a. O. 307.

[6] Zinner a. O. 64.

[7] Zinner a. O. 67f.

[8] Auf f. 28r—32v.

[9] Zinner a. O. 307.

[10] In: *Tractatus Georgii Peurbachii super Propositiones Ptolemaei de Sinubus et Chordis. Item Compositio Tabularum Sinuum per Ioannem de Regiomonte. Adiectae sunt et Tabulae Sinuum duplices per eundem Regiomontanum* (Nürnberg 1541) f. B2r—Cr. Der von Schöner abgedruckte Text

chordarum, der, von Regiomontan geschrieben, ebenfalls in dieser Handschrift vorhanden ist[11] und der wiederum auf Johannes von Gmundens Traktat *De sinibus, chordis et arcubus* zurückgeht[12]. Da Regiomontans Sinustraktat nicht mehr in die Wiener Zeit fällt, können wir ihn hier vernachlässigen.

Welche anderen nichtastronomischen Schriften sind nun in dieser Handschrift enthalten? Zu erwähnen ist ein Musiktraktat auf f. 128 v—131 r. Es handelt sich um die *Musica speculativa* des Johannes de Muris. Dabei ist der Text fast wörtlich mit der von M. GERBERT edierten Fassung identisch[13]. Diese Schrift war unter den Muris zugeschriebenen Musiktraktaten am verbreitetsten[14]. Regiomontan kopierte hier eine gekürzte Fassung der *Musica speculativa*, die in engem Zusammenhang mit einer Melker Handschrift steht[15].

Zwei andere Texte am Schluß der Handschrift berühren mechanische Fragen; sie behandeln Waagebalken und Gewichte. Dies sind der auf ein hellenistisches Original zurückgehende *Liber karastonis* des Tābit b. Qurra[16] und eine mechanische Schrift des Jordanus Nemorarius[17]. Nach den Forschungen von F. BUCHNER[18], deren Ergebnisse M. CLAGETT

---

ist wörtlich identisch mit der korrigierten Fassung des Regiomontan in Wien Nat. Bibl. Cod. 5203, f. 28r—32v. Regiomontan hat vor allem in Proposition 6 große Teile verändert.

[11] Auf f. 124r—128r. Dieser Traktat ist ebenfalls von Schöner in seiner Ausgabe (A. 10) publiziert worden (auf f. A2r—Bv). Auch hier stimmt Schöners Abdruck mit dem Text in der Handschrift Wien 5203 überein. Nur Regiomontans kritische Schlußbemerkung (f. 128r, 9 Zeilen) wurde nicht wiedergegeben.

[12] Johannes von Gmundens Schrift wurde erstmals publiziert von H. L. L. BUSARD, Der Traktat De sinibus, chordis et arcubus von Johannes von Gmunden. *Österreichische Akademie der Wissenschaften, math.-nat. Kl., Denkschriften 116/3* (1971).

[13] MARTIN GERBERT, *Scriptores ecclesiastici de musica sacra potissimum, t. III* (St. Blasien 1784 = Hildesheim 1963) 249—255. Der Text beginnt mit *Prima petitio* (GERBERT, 249) und endet mit *augmentari* (GERBERT, 255). Für nähere Hinweise danke ich Herrn DR. F. ZAMINER, Berlin.

[14] Von ihr sind 48 Handschriften bekannt. Siehe ULRICH MICHELS, *Die Musiktraktate des Johannes de Muris* (Wiesbaden 1970) 17—24. Die Wiener Handschrift hat bei MICHELS die Signatur Wi_3; er vermutet — zu Unrecht — deutsche Herkunft: S. 124.

[15] MICHELS a. O. 23.

[16] Auf f. 172r—173v.

[17] Auf f. 174r—180v.

[18] Die Schrift über den Qarastun von Thabit b. Qurra, in: *Sitzungsberichte der physikalisch-medizinischen Sozietät in Erlangen 52* (1920) 142—147.

übernahm[19], weicht der erste Text von der gängigen Version des *Liber karastonis* ab. Er ähnelt der Handschrift Thorn, R. 4⁰ 2, ist aber klarer und besser. Bei der zweiten Schrift handelt es sich um einen anonymen Kommentar zu dem Jordanus zugeschriebenen Traktat *De ratione ponderis*[20].

Die mathematischen Texte im engeren Sinne gehören teils der Arithmetik, teils der Geometrie an. Zunächst zur Arithmetik: Regiomontan hat mit großer Geduld eine Quadrat-[21] und eine Kubikzahlentafel[22] berechnet, aber die Rechnungen nicht zu Ende geführt: Die Quadratzahlen, die bis 6140^2 reichen sollten, enden schon bei 1130^2, die Kubikzahlen statt bei 651^3 schon bei 387^3. Dabei hat Regiomontan zunächst offenbar jede zehnte Quadratzahl bestimmt, wie man aus den vorhandenen Werten 1130^2 bis 1680^2 folgern kann, und dann die dazwischenliegenden Zahlen mit Hilfe der Differenzen ermittelt, die zwischen zwei aufeinanderfolgenden Quadrat- oder Kubikzahlen stets angegeben sind. — Diese Tafeln benötigte Regiomontan sicherlich auch für die Bestimmung von Quadrat- und Kubikwurzeln. Ein eigener Traktat über Quadratwurzeln ist auf f. 167v—168r vorhanden; da er von CURTZE ediert und analysiert wurde[23], brauche ich nur kurz darauf einzugehen. Regiomontan bestimmt nach dem alten griechischen und letztlich babylonischen Verfahren

$$\sqrt{n} = \sqrt{a^2 + b} \approx a + \frac{b}{2a} = n_1$$

und bildet dann $(\frac{n}{n_1} + n_1) : 2 = n_2$ als zweite Näherung. Diese bei Heron nachzuweisende Methode dürfte Regiomontan über arabische Vermittler in lateinischen Texten kennengelernt haben[24]. — Seine Verfahren, Quadrat- und Kubikwurzeln aus Brüchen rational zu bestimmen,

---

19 E. A. MOODY / M. CLAGETT, *The medieval science of weights* (Madison 1960) 77—86, vor allem 84f.

20 Siehe RON B. THOMSON, Jordanus de Nemore: Opera. *Mediaeval Studies 38* (Toronto 1976) 97—144. Nach THOMSON, 106, enthält der Wiener Codex eine verkürzte Fassung der *Aliud commentum*-Version.

21 Auf f. 33r—41v.

22 Auf f. 48r—50r.

23 CURTZE a. O. (A. 4, 1897) 148—150. 146f.

24 Zur Geschichte dieser Methode vergleiche man die zusammenfassende Darstellung bei JOHANNES TROPFKE, *Geschichte der Elementar-Mathematik in systematischer Darstellung 2* (3. A. Berlin/Leipzig 1933) 170—176.

indem man die Wurzeln durch geschicktes Erweitern der Nenner beseitigt:

$$\sqrt{\frac{a}{b}} = \frac{1}{bc}\sqrt{abc^2},\quad \sqrt[3]{\frac{a}{b}} = \frac{1}{bc}\sqrt[3]{ab^2c^3},\quad \sqrt{\frac{a}{b}} = \frac{ac}{\sqrt{abc^2}},\quad \sqrt[3]{\frac{a}{b}} = \frac{ac}{\sqrt[3]{a^2bc^3}}$$

findet man in vielen mittelalterlichen Rechenbüchern. Auch sie gehen auf arabische Methoden zurück und wurden mit den Algorismusschriften seit dem 12. Jahrhundert im Abendland bekannt[25].

Interessant ist ferner ein Abschnitt über vollkommene Zahlen, der von Regiomontan selbst stammt[26]. Der Begriff der vollkommenen Zahl, die gleich der Summe ihrer Teiler ist (einschließlich der Eins), wurde von den Griechen formuliert und wird bei Euklid als selbstverständlich vorausgesetzt. Bei Euklid findet man ebenfalls den Satz, daß die Zahl $(2^n—1) \cdot 2^{n-1}$ vollkommen ist, sofern nur $2^n—1$ eine Primzahl ist[27]. Auf diese Weise ergeben sich leicht die ersten vier vollkommenen Zahlen 6, 28, 496 und 8128, die sich bei Nikomachos (um 100 n. Chr.) nachweisen lassen[28]. Die vollkommenen Zahlen, die in der Spätantike durch die Neupythagoreer zu neuem Ruhm gelangten, blieben im ganzen Mittelalter bekannt; bei Isidor[29], Jordanus Nemorarius[30] und Leonardo von Pisa[31] finden sie sich ebenso wie in arabischen Quellen[32]. Die fünfte vollkommene Zahl allerdings, $(2^{13}—1) \cdot 2^{12} = 33\,550\,336$, scheint erst im 15. Jahrhundert bekannt geworden zu sein. Regiomontan erwähnt sie im genannten Traktat und dürfte sie wohl als erster gefunden haben, wenn sie sich auch in einer deutschen Abhandlung aus etwa derselben

---

25 Siehe A. P. Juschkewitsch / B. A. Rosenfeld, Die Mathematik der Länder des Ostens im Mittelalter. *Sowjetische Beiträge zur Geschichte der Naturwissenschaft* (Berlin 1960) 102, 105.

26 Auf f. 167r. Dieser Abschnitt ist bei Curtze a. O. (A. 4, 1899) 288f. ediert.

27 *Elemente* IX 36.

28 In Buch I, Kapitel 16. Siehe *Nicomachus of Gerasa, Introduction to arithmetic, translated into English by* Martin Luther d'Ooge (New York 1926) 209—212. Zur Geschichte der vollkommenen Zahlen siehe Leonard Eugene Dickson, *History of the theory of numbers 1* (Washington 1919 = New York 1971) 3—33.

29 *Origines*, Buch 3, Kapitel 5, § 11.

30 *De elementis arismetice artis*, Buch VII 55f. in der Ausgabe von Jacques Lefèvre d'Etaples (Paris 1496).

31 *Liber Abbaci*, 283, in der Ausgabe von B. Boncompagni, *Scritti 1* (Rom 1857).

32 Zum Beispiel Ṯābit b. Qurra. Siehe Fuat Sezgin, *Geschichte des arabischen Schrifttums*, Band 5: Mathematik. Leiden 1974, 270, Nr. 13.

Zeit (1461) ebenfalls nachweisen läßt[33]. Es ist erwähnenswert, daß fast 100 Jahre später M. Stifel die fünfte vollkommene Zahl noch nicht kennt, sondern stattdessen 256 · 511 angibt, obwohl 511 keine Primzahl ist[34]. Regiomontan nennt sogar noch die Zahlen, die zur sechsten vollkommenen Zahl $(2^{17}-1) \cdot 2^{16} = 8\,589\,869\,056$ führen, rechnet allerdings die Multiplikation nicht mehr durch; das Ergebnis findet sich erstmals 1555 bei Scheybl in seiner Euklidübersetzung[35]. — Man fragt sich, woher Regiomontan diese Kenntnisse besaß. Ich glaube nicht, daß er hier einen fremden Text kopierte, zumal er bei der Ableitung der vollkommenen Zahlen sich zweimal versah, weil er zusammengesetzte Zahlen als Primzahlen ansah, später jedoch seinen Fehler korrigierte. Vielmehr wird Regiomontan die Anweisung zur Konstruktion derartiger Zahlen bei Euklid gekannt und benutzt haben. Hieraus folgt, daß er sich schon in Wien intensiv auch mit den späteren Büchern Euklids befaßte. Direkte Hinweise auf seine Bemühungen um Euklid besitzen wir erst aus den Italiener Jahren, als Regiomontan mehrere Handschriften des großen griechischen Mathematikers bearbeitete, um daraus einen zuverlässigen Text zu machen[36].

Vor der Abhandlung über vollkommene Zahlen hat Regiomontan den *Algorismus demonstratus* des Gernardus abgeschrieben, ein im Spätmittelalter recht bekanntes Rechenbuch aus dem 13. Jahrhundert[37]. Diese Schrift ist später von Schöner ediert worden[38]. Da es sich um ein

---

[33] In München, Clm 14908, f. 27v—29r. Zu diesem Text, der in St. Emmeram von Fridericus Gerhart geschrieben wurde, siehe M. Curtze, Mathematisch-historische Miscellen 5. Zur Geschichte der vollkommenen Zahlen. *Bibliotheca Mathematica, Neue Folge 9* (Stockholm/Berlin/Paris 1895) 39—42.

[34] *Arithmetica integra* (Nürnberg 1544). Vgl. J. E. Hofmann, Michael Stifel (1487 ?—1567). Leben, Wirken und Bedeutung für die Mathematik seiner Zeit. *Sudhoffs Archiv, Beiheft 9* (Wiesbaden 1968) 8f.

[35] Joh. Scheybl, *Das sibend acht und neunt buch des hochberümbten Mathematici Euclidis Megarensis* (Augsburg 1555); zu Euklid IX 36.

[36] Siehe hiezu Menso Folkerts, Regiomontans Euklidhandschriften. *Sudhoffs Archiv 58* (1974) 149—164.

[37] Siehe hiezu Gustaf Eneström, Der „Algorismus de integris" des Meisters Gernardus. *Bibliotheca Mathematica, 3. Folge 13* (1912/13) 289—332, und derselbe, Der „Algorismus de minutiis" des Meisters Gernardus. *Bibliotheca Mathematica, 3. Folge 14* (1913/14) 99—149. Die Identifikation Gernardus = Jordanus Nemorarius ist ungewiß. Diese Schrift befindet sich in Wien Cod. 5203 auf f. 148r—166v. Weitere Handschriften nennt Thomson a. O. 112.

[38] *Algorithmus demonstratus* (Nürnberg — J. Petreius 1534).

bekanntes Werk handelt, das nicht von Regiomontan stammt, ist es nicht nötig, ausführlicher darauf einzugehen. Wichtig ist eine weitere Abhandlung in dieser Handschrift[39]. Sie betrifft die Proportionenlehre, jenes mathematische Teilgebiet, das im Gefolge der mittelalterlichen Euklidinterpretation vor allem im 14. Jahrhundert durch Gelehrte in Oxford und Paris zu einer selbständigen Disziplin ausgebaut worden war. Unser Text geht auf eine frühere Zeit zurück, als die Proportionen noch nicht benutzt wurden, um Bewegungsabläufe zu studieren. Vielmehr hängt er mit dem Studium der zusammengesetzten Verhältnisse zusammen, das von arabischen Mathematikern betrieben wurde. Ausgehend von der nacheuklidischen Definition des zusammengesetzten Verhältnisses[40], hatte etwa aṭ-Ṭūsī 1260 in seiner *Abhandlung über das vollständige Vierseit* eine Theorie dieser Verhältnisse aufgestellt[41]. Die zusammengesetzten Verhältnisse wurden von Ṯābit b. Qurra[42] und Aḥmad b. Yūsuf[43] benutzt, um die 18 möglichen Schreibarten für die Formel des Menelaischen Transversalensatzes aufzustellen. Von jenen beiden arabischen Wissenschaftlern wiederum hängen die Schriften des Campanus[44] und Jordanus Nemorarius[45] über die Proportionenlehre ab[46]. Es ist seit einigen Jahren bekannt, daß es nach dem von Schöner 1534 herausgegebenen *Algorismus demonstratus*[47] einen Anhang gibt, der Jordanus' Schrift über die Proportionen ähnelt, aber im Unterschied dazu zu Beginn das arithmetische, geometrische und harmonische Mittel definiert[48]. Der Autor dieses Traktats war bisher unbekannt, weil man nicht bemerkte, daß Schöners Druckvorlage der Text f. 136r—137r in Wien 5203 ist. Jetzt ist klar, daß dieser Anhang von Regiomontan nach einer Vorlage des Jordanus bearbeitet wurde.

Kommen wir jetzt zu den geometrischen Abhandlungen im Wiener Rechenbuch. Regiomontan schrieb auf f. 142r—146r eine Arbeit über

---

39 Auf f. 136r—137r.

40 Euklid, *Elemente* VI Def. 5.

41 Siehe A. P. Juschkewitsch, *Geschichte der Mathematik im Mittelalter* (Leipzig 1964) 304—308.

42 Siehe Sezgin a. O. 265. 268, Nr. 1.

43 Siehe Sezgin a. O. 288—290.

44 *De proportione et proportionalitate.*

45 *De proportionibus.*

46 Ediert von H. L. L. Busard, Die Traktate De Proportionibus von Jordanus Nemorarius und Campanus. *Centaurus 15* (Kopenhagen 1971) 193—227.

47 Siehe die Anmerkungen 37 und 38.

48 Busard a. O. 196. Der Text ist dort auf 222—226 abgedruckt.

isoperimetrische Figuren ab. Dabei handelt es sich um den manchmal Jordanus zugeschriebenen Traktat, der auf die griechische Abhandlung des Zenodoros zurückgeht[49]. Diese Schrift über Figuren gleichen Umfangs ist offenbar ohne den Umweg über das Arabische ins Lateinische übersetzt worden; sie gehört zu jener kleinen Gruppe von Schriften, die um 1160 in Süditalien/Sizilien direkt aus dem Griechischen übertragen wurden[50]. Regiomontans Text repräsentiert nicht die übliche Fassung, sondern er gehört einer späteren Bearbeitung an, die nicht sieben, sondern acht Lehrsätze enthält. Er ähnelt der Handschrift Florenz, Bibl. Naz., Conv. soppr. J.V. 18, f. 11ra—12va, unterscheidet sich aber im Text der Beweise[51]. Somit hat Regiomontan auch hier nicht die gängige Fassung des Traktats wortgetreu kopiert, sondern sie vermutlich selbständig bearbeitet.

Ein weiterer geometrischer Text von Regiomontans Hand betrifft Exzerpte aus den *Konika* des Apollonios von Perge[52]. Die Definitionen und einige Propositionen des 1. Buches (ohne Beweise) waren im 12. Jahrhundert höchstwahrscheinlich durch Gerhard von Cremona nach einer arabischen Vorlage ins Lateinische übersetzt worden. Regiomontan hat diesen Text, den Heiberg ediert hat[53], wörtlich abgeschrieben.

Neben diesen verhältnismäßig leicht zu identifizierenden fremden Schriften gibt es noch mehrere geometrische Traktate, die vermutlich eigene Bearbeitungen des Regiomontan sind. Hierzu gehört ein kurzer Text über die Berechnung der Vieleckseite, der von Curtze ediert wurde[54]. Er lehrt ein allgemeines Verfahren, um die Seite s_n eines beliebigen

---

[49] Edition des griechischen Textes in *Pappi Alexandrini collectionis quae supersunt*, Ed. F. Hultsch, *3* (Berlin 1878) 1138—1165.

[50] H. L. L. Busard, dem ich für freundliche Hinweise danke, gibt den lateinischen Text mit wissenschaftshistorischer Einleitung heraus.

[51] Hinweis von Dr. Busard. Weitere Handschriften dieser Tradition, die nach dem Traktat *De curvis superficiebus* entstanden sein muß, sind in Florenz: Bibl. Naz., Conv. soppr. J. V. 30, f. 12v (unvollständig) und J. V. 18, f. 96rv. Zu dieser Schrift vergleiche man auch die Anmerkung 13 zu Teil III, Kapitel 2, Abschnitt II in M. Clagett, *Archimedes in the Middle Ages 3* (im Druck).

[52] Auf f. 141rv.

[53] *Apollonii Pergaei quae Graece exstant*, Ed. J. L. Heiberg, *2* (Leipzig 1893, LXXV—LXXX). Der Text in Wien Cod. 5203 endet auf LXXIX, Z. 13 Heiberg. — Andere Handschriften dieser Übersetzung sind Paris, BN lat. 9335, f. 85v—86r; Dresden, Db. 86, f. 277v; Vat. Reg. lat. 1012, f. 74; London, BM Cotton Vespasian A. II, f. 139rv; Vat. Reg. lat. 1253, f. 62r—63r.

[54] Curtze a. O. (A. 4, 1897) 150f.

in einen Kreis mit dem Durchmesser d einbeschriebenen regulären Polygons zu ermitteln:

$$s_n = \sqrt{\frac{4{,}5\,d^2}{(n-1)\frac{n}{2} + 3}}\text{, also } s_n = \frac{6r}{\sqrt{n(n-1)+6}}.$$

Mit dieser Formel ergeben sich die richtigen Werte $s_3 = r\sqrt{3}$, $s_4 = r\sqrt{2}$, $s_6 = r$ und der Näherungswert $s_7 = \frac{r}{2}\sqrt{3} = \frac{1}{2}s_3$. Es ist mir nicht gelungen, die Vorlage für diesen Text zu ermitteln[54a]. Regiomontan sagt, dies sei die Methode der indischen Philosophen, und erwähnt neben Ptolemaios auch die Sarrazenen[55], so daß wiederum ein arabisches Bindeglied zu vermuten ist. Der Text ähnelt der Darstellung, die Jordanus in *De triangulis* gibt[56], doch hat Regiomontan hier wohl nicht die Schrift des Jordanus benutzt. Eine vergleichbare Vorschrift finden wir im arabischen Bereich in Abu l-Wafā's *Buch für die Schreiber*, wo ähnlich wie bei Regiomontan der Durchmesser d eines Kreises durch die Seite s_n eines einbeschriebenen regulären n-Ecks ausgedrückt wird:

$$d^2 = s_n^2\left[\frac{(n-1)n}{2} + 3\right]\cdot\frac{2}{9}\text{[57]}.$$

Diese Formel wird bei Abu l-Wafā' ebenfalls als indisch bezeichnet; ihre Herkunft ist unbekannt.

Weitere Probleme der praktischen Geometrie werden im Wiener Rechenbuch auf f. 131v—133r und 134v abgehandelt. Zunächst gibt Regiomontan eine Formel an, um die Fläche des regulären 5-, 6- und 7-Ecks zu finden[58]. Es ist die von den römischen Agrimensoren stammende Vorschrift, nach der die Flächen mit den entsprechenden Poly-

---

[54a] Ein fast gleichlautender Text steht in der aus Toledo stammenden Handschrift Madrid, Biblioteca Nacional 10053, f. 70v, aus dem 13. Jahrhundert. Demnach ist auch dieser Text nicht Eigentum des Regiomontanus.

[55] *Philosophi Indorum*: CURTZE a. O. (A. 4, 1897) 150; *more Sarrazenorum*: CURTZE a. O. (1897) 151.

[56] M. CURTZE: Jordani Nemorarii geometria vel de triangulis libri IV. *Mitteilungen des Coppernicus-Vereins für Wissenschaft und Kunst zu Thorn* 6 (Thorn 1887) 43f. = Buch IV, 23. Auch Jordanus redet von einer *questio Indorum*.

[57] JUSCHKEWITSCH a. O. 272. Siehe SEZGIN a. O. 322f., 324, Nr. 2.

[58] Auf f. 131v. Dieser Text ist ediert von CURTZE a. O. (A. 4, 1897), 152.

gonalzahlen gleichgesetzt werden[59]. Die Erwähnung des Boethius in diesem Zusammenhang zeigt, daß Regiomontan diese Formel aus der Boethius zugeschriebenen sogenannten 2. Geometrie kannte[60]. Natürlich bemerkte er den groben Fehler, den man hier begeht, und gab für das Sechseck den richtigen Weg an. Dieser Abschnitt ist sicher Eigentum des Regiomontan.

Die auf f. 131v—133r und 134v folgenden Texte zeigen, daß Regiomontan in der Wiener Zeit auch mit Archimedes' Kreismessung bekannt war[61]. Regiomontan hat zunächst zwei Abschnitte aus der Bearbeitung kopiert, die ein Gordanus vor 1390 verfaßte und die in einer Handschrift aus dem 15. Jahrhundert als Teil einer *Compilacio quorundam canonum in practicis astronomie et geometrie* vorhanden ist[62]. Der einleitende Abschnitt beschäftigt sich mit der Tatsache, daß die Kreisfläche das Produkt aus halbem Durchmesser und halbem Umfang ist[63]. Es folgt fast wörtlich der Beweis des Gordanus zur Proposition 1 von Archimedes' Kreismessung[64]. An diesen langen Abschnitt schließen sich vier kurze Bemerkungen an, die sich mit dem Verhältnis von Kreisen und Quadraten bzw. von Würfel und Kugel beschäftigen[65].

---

59 Die Fläche $F_n(a)$ des regelmäßigen n-Ecks mit der Seite a wird mit der a-ten n-Eckszahl $p_n{}^a$ gleichgesetzt:

$$F_n(a) = p_n{}^a = \frac{(n-2)a^2-(n-4)a}{2} \quad (n \geqslant 3,\ a \geqslant 1).$$

Diese Formel ist seit den römischen Agrimensoren fester Bestandteil der praktischen Geometrie im Mittelalter.

60 Siehe Menso Folkerts, „*Boethius*" *Geometrie II, ein mathematisches Lehrbuch des Mittelalters* (Wiesbaden 1970) 101f. Die Dreiecksformel, die Regiomontan erwähnt, steht in der Geometrie II in Buch II, Kapitel 5 (149, Z. 593—597 Folkerts).

61 Die folgenden Bemerkungen beruhen auf den Forschungen Marshall Clagetts, die im 3. Band seines *Archimedes in the Middle Ages* publiziert werden sollen. Ich danke Prof. Clagett für die Überlassung des betreffenden Manuskriptabschnitts.

62 Hierzu siehe M. Clagett, *Archimedes in the Middle Ages*, Band 1. Madison 1964, 142—146. Die betreffenden Teile stehen im 8. Abschnitt dieser *Compilacio*, der den Titel *In mensura figurarum* trägt.

63 Dieser Text ist Teil 8, Kapitel 18 von Gordanus' *Compilacio* entnommen; vgl. Clagett a. O., Band 1, 143, Anm. 4.

64 Er bildet Teil 8, Kapitel 23 von Gordanus' Schrift und ist bei Clagett a. O., Band 1, 148—162, ediert. Dieser Abschnitt endet in Wien Cod. 5203 oben auf f. 133r.

65 Auf f. 133r.

Auch diese Texte gehen auf Gordanus' *Compilacio* zurück[66], jedoch hat Regiomontan manches verändert oder verkürzt.

Auf f. 134v befindet sich ein weiterer geometrischer Text, in dem bewiesen wird, daß das Volumen einer Kugel gleich dem Produkt aus dem Radius und einem Drittel der Oberfläche ist[67]. Regiomontans Beweis ähnelt demjenigen, den die Banū Mūsā in der Proposition XV ihrer Schrift *Verba filiorum* bieten[68]. Dies zeigt, daß Regiomontan diese Abhandlung schon in seiner Wiener Zeit kannte[69]. Er hat den Beweis der Banū Mūsā aber in einem wesentlichen Punkt modifiziert: Während diese wie Archimedes in *De sphaera et cylindro* zwischen zwei Kugeln einen Rotationskörper einfügen, der von einem regulären Polygon erzeugt wird, ersetzt Regiomontan diesen durch ein Polyeder, das aus Pyramiden besteht. Dasselbe Verfahren hatte schon Leonardo von Pisa in seiner *Practica geometriae* gewählt. Regiomontan meistert die Schwierigkeit, die sich dadurch ergibt, daß dieses Polyeder unregelmäßig sein kann[70].

Fassen wir zusammen: Regiomontan beweist in seinem Rechenbuch eine gute Kenntnis der römischen Mathematik (Boethius, Agrimensoren), aber auch derjenigen Werke, die durch die arabisch-lateinischen Übersetzungen damals bekannt waren oder von ihnen abhängen (z. B. Jordanus' Schrift über die Proportionen, der direkt aus dem Griechischen übertragene Traktat über isoperimetrische Figuren, der *Algorismus demonstratus* des Gernardus, Texte über Quadratwurzeln und über die Seiten der einbeschriebenen Polygone). Auch die Kegelschnitte des Apollonios und auf Archimedes zurückgehende Texte waren über ara-

---

[66] Auf die Kapitel 24—26, 33 des Teils 8. Näheres bei CLAGETT in Band 3 seines *Archimedes in the Middle Ages*.

[67] *Corpulenciam sive embadum spere comprehendere. Multiplica semidiametrum in terciam partem superficiei spere, et exibit ipsa corpulencia.*

[68] Ediert bei CLAGETT a. O., Band 1, 332—334.

[69] Daß Regiomontan eine Kopie der *Verba filiorum* besaß, wird aus seinen Bücherverzeichnissen ersichtlich: Im Katalog von 1512 wird eine Handschrift *libellus trium fratrum cum multis aliis* erwähnt; der Katalog von 1522 beschreibt wohl dieselbe Handschrift etwas genauer: *Liber in quo libellus trium fratrum Moisi*; *Comensurator*; *epistola Ahmeti de proporcione*; *Menelaus de spericis*; *de ponderibus*; *Tebith de Carascove* (= *carastone*); *planisperium Jordani*; *demonstracio circulorum azimuth dicte*; *isoperimetria Jordini* (= *Jordani*); *Teodosius de speris*; *et de speculis comburentibus* (ZINNER a. O. 330). Es ist jetzt wahrscheinlich, wenn auch nicht sicher zu belegen, daß Regiomontan diesen Codex schon in seinen Wiener Jahren erwarb oder schrieb.

[70] Näheres bei CLAGETT in Band 3 seines *Archimedes in the Middle Ages*, Part III, Chap. 2, Sect. II.

bische Vermittler schon in der Wiener Zeit dem Regiomontan bekannt. Welche Vorlagen Regiomontan benutzte, wird sich schwerlich klären lassen; man wird am ehesten an mathematische Sammelhandschriften wie Paris 9335, Basel F II 33 oder Dresden Db 86 denken, die zahlreiche aus dem Arabischen übersetzte Texte enthalten[71]. Aber auch in dieser frühen Zeit zeigt sich Regiomontan schon als gelehriger Schüler, der vieles verbessert und der selbst Neues hinzufügt. Dies erkennt man etwa an den Abhandlungen über die vollkommenen Zahlen, ferner an den Texten über Wurzeln, über die regulären Polygone und die Kugelbestimmung.

2. Regiomontans Kenntnis von Visiertraktaten und deren Verbreitung

Ein weiteres damals sehr wichtiges Gebiet der angewandten Mathematik hat Regiomontan mit Sicherheit in seinen Wiener Jahren kennengelernt: die Visierkunst. Sie lehrt Verfahren, um den Rauminhalt von Fässern mit Hilfe von Visierruten mechanisch zu bestimmen. Dabei wird das Faß mit einem mittleren Zylinder gleichgesetzt, und durch besonders geeichte Meßskalen auf der Rute wird die Inhaltsbestimmung auf eine Multiplikation von abzulesenden Maßzahlen zurückgeführt. Die Technik des Visierens wurde wohl im 14. Jahrhundert im süddeutsch-österreichischen Raum entwickelt und war dort im 15./16. Jahrhundert sehr weit verbreitet, wie die zahlreichen handschriftlichen und gedruckten Visiertraktate beweisen: Allein aus den Jahren 1445 bis etwa 1520 besitzen wir über 30 oft recht ausführliche Darstellungen. Sie entstanden meistens in Süddeutschland und zeigen durch ihre unbeholfene Ausdrucksweise und den Verzicht auf mathematische Begründungen, daß sie überwiegend von Praktikern für Praktiker geschrieben wurden. Daneben gibt es einige Visiertraktate aus der Mitte des 15. Jahrhunderts, die in Erfurt entstanden und mit der dortigen Universität in Verbindung gebracht werden müssen. Ein weiterer Schwerpunkt war Österreich, speziell Wien. Offenbar wurde auch hier schon um 1450 die Visierkunst an der aufblühenden Universität betrieben, wenn mir auch nicht bekannt ist, daß diese Disziplin das Thema einer speziellen Vorlesung war[72].

---

[71] Eine Anzahl derartiger Texte standen auch in der in Anmerkung 69 erwähnten Sammelhandschrift aus Regiomontans Bibliothek. Dieser Codex ist nicht erhalten.

[72] Einen Überblick über die Entwicklung der Visierkunst mit Bibliographie der Handschriften und Drucke gibt Menso Folkerts, Die Entwicklung und Bedeutung der Visierkunst als Beispiel der praktischen Mathematik der frühen Neuzeit. *Humanismus und Technik 18* (1974) 1—41.

Aus dem Gesagten ergibt sich mit hoher Wahrscheinlichkeit, daß Regiomontan in Wien die Methoden der Visierkunst kennenlernte[73]. Es lassen sich dafür aber noch andere Indizien finden: Der Klosterneuburger Probst Georg Müstinger berechnete um 1440 eine Tafel für die Visierrute[74]; bei den engen Beziehungen zwischen Müstinger und Peuerbach dürfte dies Regiomontan nicht verborgen geblieben sein. Wichtiger noch: Peuerbach selbst schrieb eine Visierkunst, die als verloren gilt, unter dem Titel *Compositio novae virgae visoriae cum lineis et tabula nova*[75]; einer seiner Schüler mit Namen Wilhelm verfaßte eine heute noch erhaltene ähnliche Schrift[76]. So ist klar, daß auch Regiomontan als enger Mitarbeiter Peuerbachs von diesem Zweig der angewandten Mathematik wußte. Aus seiner Verlagsanzeige (1474) ersieht man, daß Regiomontan auch eine eigene Schrift folgenden Titels drucken wollte: *Radii uisorii multorum generum cum usibus suis*[77]. Der Titel läßt vermuten, daß hiermit Instrumente gemeint sind, die über die Visierkunst hinaus verwendbar sind.

Spuren eigener und fremder Arbeiten zur Visierkunst wird man in Regiomontans Bücherverzeichnissen suchen. Bekanntlich existieren solche aus den Jahren 1512, 1522 und 1563, wobei in allen Fällen auch die Bücher Bernhard Walthers († 1504) verzeichnet sind[78]. Die Beschrei-

---

[73] Auch in Regiomontans Rechenbuch, Wien 5203, befinden sich auf f. 133r neun Aufgaben der praktischen Geometrie, die sich überwiegend mit den Beziehungen zwischen Fässern verschiedener Maße beschäftigen. Regiomontan teilt nur die Probleme mit, ohne ihre Lösungen zu nennen.

[74] ZINNER a. O. 17. ZINNER gibt nicht an, woher er diese Information hat. Vermutlich bezieht er sich auf die Handschrift Klosterneuburg 683, die mit Müstinger in Verbindung gebracht werden muß. Dazu siehe S. 193.

[75] ZINNER a. O. 44.

[76] ERNST ZINNER, *Verzeichnis der astronomischen Handschriften des deutschen Kulturgebietes* (München 1925), Anmerkung zu Nr. 11708—11718. Dieser Text, der in Zürich, Zentralbibl. C 107—20, f. 108r—109v, aufbewahrt wird, stammt aus dem Jahre 1479.

[77] ZINNER (A. 1) 353. Die Verlagsanzeige ist im Faksimile wiedergegeben bei ZINNER, Tafel 26.

[78] ZINNER hat alle drei ausgewertet und das von 1563 erstmals abgedruckt; ZINNER (A. 1) 252—258. Die Liste von 1512 erstellte Christoph Scheurl für Spalatin: ZINNER (A. 1) 248f. Wenn ich ZINNERS Angaben (A. 1, 260f.) richtig interpretiere, beruht dieses (offenbar nicht mehr vorhandene) Verzeichnis auf einer Bestandsaufnahme Pirckheimers, die in der Pirckheimer-Sammlung erhalten ist. Diese Liste, die als Ganzes vermutlich unediert ist, enthält 193 Nummern Bücher und 6 Nummern Geräte. ZINNER hat die Angaben in dieser Liste für seine Bibliographie (A. 1, 297—371) herangezogen, aber nicht zusammenhängend herausgegeben. — 1522 er-

bungen sind zum Teil sehr knapp und nicht immer zuverlässig. Im Verzeichnis von 1512 sind zwei Schriften zur Visierkunst genannt[79]; einige weitere Titel könnten sich hierauf beziehen, sind aber zu ungenau[80]. In der Liste von 1522 wird ein großes Buch erwähnt, das auch eine *ars visorie Latine et Germanice* enthielt[81]:

Liber arithmetice, in quo de integris, in fractis, de proporcionibus; textus algobre; de additionibus et diminutionibus; regule algobre; de radicibus surdis; de arte bynomiali; de quadratis quadratorum et de cubicis cubicorum; numerus datorum; liber de numeris et pospositionibus eorum; arithmetica Johanns Muris; musica speculativa; ars visorie Latine et Germanice (kunst des visirens in teutsch); tractatus quadrantis etc. et usus eius; (vil exempel im rechen; regula virginum); plures propleumata in arithmetica.

Im jüngsten Verzeichnis schließlich deutet nur eine Erwähnung auf die Visierkunst hin: *Pro arte visurandi Pragmata quaedam. geschriben vngepünden. in fol.*[82].

Fassen wir zusammen: In allen drei Bücherverzeichnissen werden Visiertraktate erwähnt; dabei könnte die *ars visorie Latine et Germanice* im Verzeichnis von 1522 mit den beiden Traktaten *Compositio virge* und *visir puchleyn* (1512) identisch sein, möglicherweise auch mit der Erwähnung 1563, obwohl dies wegen der unterschiedlichen Formulierung nicht sicher ist.

Bisher ging man davon aus, daß kein Visiertraktat aus dem Besitz Regiomontans erhalten ist; ZINNER wenigstens gelang es nicht, eine Handschrift nachzuweisen. Untersucht man alle Codices[83], so findet man jedoch einige Hinweise auf Regiomontan. Ich möchte im folgenden drei

---

folgte eine neue Bestandsaufnahme der Bücher Regiomontans und Walthers: ZINNER (A. 1) 250, 260f. Dieses ausführlichere Bücherverzeichnis, das insgesamt 215 Nummern enthält, wurde von H. PETZ herausgegeben: Urkundliche Nachrichten über den literarischen Nachlaß Regiomontans und B. Walters 1478—1522. *Mitteilungen des Vereins für Geschichte der Stadt Nürnberg 7* (Nürnberg 1888) 237—262. — Zu der Liste von 1563, die 146 Bücher und 2 Instrumente recht ausführlich beschreibt, siehe ZINNER (A. 1) 250f.

[79] *Compositio virge* und *visir puchleyn radix 4*: ZINNER (A. 1) 353, Nr. 167.

[80] *Descriptiones vasorum*: ZINNER (A. 1) 360, Nr. 255; *multa geometrica*: ZINNER (A. 1) 359, Nr. 235; *Practica varignatia*: ZINNER (A. 1) 359, Nr. 231.

[81] PETZ a. O. 250, wiederabgedruckt bei ZINNER (A. 1) 318, Nr. 49. Die eingeklammerten Stellen sind im Original durchstrichen.

[82] ZINNER (A. 1) 257, Nr. 115.

[83] Ich kenne etwa 50, von denen ich 30 gesehen habe.

Handschriften nennen. Dabei wird die letzte Erwähnung mich veranlassen, auf die Verbreitung einer bisher unbekannten Sammlung von Visiertraktaten im 15. Jahrhundert einzugehen.

Der Codex München, Clm 19819, der sich zuletzt in Tegernsee befand, enthält auf f. 90r—93v zwei Traktate über Herstellung und Gebrauch der Visierrute. Zwar befindet sich in diesem Sammelband ein 1442 in Erfurt geschriebener Text, aber f. 90rv und 91v—93v scheinen bereits zu den mit Tegernsee zu verbindenden Teilen zu gehören. Die Hand von f. 91v—93v schrieb wahrscheinlich auf f. 101r den Namen des Tegernseer Priors Johannes Keck[84]. Oben auf f. 90r findet man die Buchstaben .IHS. und die Jahreszahl 1457. Da Regiomontan oft das Zeichen Jesus oder seine Abkürzung IHC an den Anfang seiner Abschriften oder Handschriften stellte[85], könnte man hier einen Zusammenhang mit Regiomontans Bibliothek vermuten. Jedoch ist die Form IHS besonders in der abweichenden Schreibung nicht mit Sicherheit für Regiomontan oder Abhängigkeit von ihm in Anspruch zu nehmen. Weitere Indizien für eine Beziehung zu Regiomontan gibt es nicht. Inhaltlich bietet der Text gegenüber anderen Visiertraktaten nichts Neues; vielmehr wird mechanisch Herstellung und Gebrauch der Visierrute dargestellt. Am unteren Rand der vorhergehenden Seite f. 89v lesen wir: *Scriptum notabile de Geometria practicum*; oben auf f. 90r sind die Worte *Virga regia* später durchgestrichen. Der Text beginnt mit: *Nota si velis facere virgam regiam fac fieri primo diametrum scilicet der do sey oben precise als weit als vnden* (dies sind übrigens die einzigen deutschen Worte im Traktat). Ein zweiter Text beginnt auf f. 91v; er trägt die Überschrift *Virgam sive baculum ad capacitates vasorum meciendas geometrice fabricare.* Ich halte es für nicht sehr wahrscheinlich, daß dieser Text Regiomontan oder seiner Bibliothek zuzuordnen ist; es ist auch nicht zu entscheiden, ob einer der beiden Traktate in einem der Bücherverzeichnisse Regiomontans gemeint ist.

Eine Handschrift der Wolfenbütteler Bibliothek wird im gedruckten Katalog mit Regiomontan in Verbindung gebracht: Der Codex 1127. Helmst. war nach O. v. Heinemann *früher doch wohl im Besitze des Schreibers (Johann Regiomontan)*[86]. Diese Handschrift enthält neben

---

[84] Diese Mitteilung verdanke ich Prof. B. Bischoff, München.

[85] Zinner (A. 1) 293.

[86] *Kataloge der Herzog-August-Bibliothek Wolfenbüttel.* Die alte Reihe: 3. Band: Die Helmstedter Handschriften. Beschrieben von Otto von Heinemann. Teil III (Wolfenbüttel 1888 = Frankfurt/Main 1965) 69f.

einigen astronomischen Traktaten auch eine Schrift zur Visierkunst[87]. Der Traktat behandelt überwiegend die Herstellung einer Quadratrute, wobei die Einteilung der Tiefenskala und der Wechsel im Vordergrund steht; der Gebrauch der Rute wird nur gestreift. Dieser relativ elementare Text hat inhaltlich keine Beziehungen zu Regiomontan, und auch äußerlich deutet nichts darauf hin, daß der Codex Regiomontan gehörte: Es fehlt sein Eignerzeichen, und die Schrift — es lassen sich übrigens mindestens zwei Hände unterscheiden — weicht stark von derjenigen Regiomontans ab. So beruht HEINEMANNS Annahme auf einem Irrtum: Es gibt keinen Grund, den Codex mit Regiomontan in Verbindung zu bringen.

Besonders interessant ist eine dritte Handschrift, ein Sammelcodex, der u. a. zwei Handschriften aus dem Besitz Regiomontans enthält. Dieser Codex, der der Regiomontan-Forschung bisher entgangen ist, gehörte GEORGE A. PLIMPTON und befindet sich heute in der Columbia University, New York (Signatur: Plimpton 188)[88]. Die Handschrift besteht aus mindestens vier Teilen: Die Blätter 1—96 stammen aus dem Jahr 1456, f. 97—198 aus dem Jahr 1501, f. 199—210 aus der Zeit nach 1450 und der Schluß (f. 211—236) etwa aus dem Jahre 1550. Uns interessieren nur der 1. und 3. Teil, die, wie ich ausführen werde, Regiomontan gehörten, aber zu seiner Zeit zwei getrennte Handschriften bildeten. Zunächst möchte ich auf Teil 3 eingehen; zu Teil 1 siehe das folgende Kapitel.

Auf f. 199r—210v befindet sich ein am Schluß unvollständiger Visiertraktat. Dieser Text ist nicht von Regiomontan geschrieben, jedoch deutet ein wichtiges Indiz auf ihn: sein Eignerzeichen, ein mit Tinte gezeichnetes Kreuz auf einem Hügel, umgeben von zwei Sternen, das sich in der linken oberen Ecke auf f. 199r befindet[89]. Nur dieses — allerdings untrügliche — Zeichen beweist, daß dieser Abschnitt Regiomontans Bibliothek angehörte und er offenbar damals den Beginn einer selbständigen Handschrift bildete; Randbemerkungen oder Korrekturen von Regiomontans Hand fehlen. Der Text selbst steht auf einem

[87] Auf f. 60r—64v. Incipit: *In nomine domini amen. Pro confectione virge visorie notandum quod in prima arte due dimensiones scilicet* . . . Der auf den Visiertext folgende Traktat, der von einer anderen Hand geschrieben wurde, trägt die Jahreszahl 1475.

[88] Der Codex ist sehr kurz beschrieben in S. DE RICCI / W. J. WILSON, *Census of medieval and renaissance manuscripts in the United States and Canada 2* (New York 1937) 1787f., und (unter verschiedenen Nummern) bei D. E. SMITH, *Rara arithmetica* (Boston/London 1908) 454—456. 468. 480. 486f.

[89] Abgebildet bei ZINNER (A. 1) 295.

hohen Niveau und unterscheidet sich dadurch von allen anderen Visiertraktaten vor Kepler, die ich kenne. Die Vermutung, daß dieser Traktat von Regiomontan selbst stammen könnte, trifft allerdings nicht zu: Es gibt nämlich noch andere Handschriften dieses Textes, die zum Teil vor Regiomontans Wirken entstanden. Da dieser Visiertraktat bisher unbekannt war, er andrerseits aber im 15. Jahrhundert eine gewisse Verbreitung erlangt hat, möchte ich auf ihn eingehen. Wir können an diesem Beispiel erkennen, welch enge wissenschaftliche Beziehungen in der 1. Hälfte des 15. Jahrhunderts zwischen verschiedenen kulturellen Zentren Süddeutschlands und Österreichs bestanden, und wir sehen, daß auch Regiomontan in dieser Tradition steht.

Die Abhandlung zur Faßmessung, die in der Handschrift Plimpton 188 enthalten ist, besteht aus mehreren in sich abgeschlossenen Texten, die sich mit Herstellung und Gebrauch verschiedener Typen von Visierruten beschäftigen, so daß man von einem Corpus von Visiertraktaten sprechen muß. Von diesem Corpus oder einzelnen Schriften aus ihm sind mir neun Handschriften bekannt. Das Corpus enthält im wesentlichen vier verschiedene Traktate, die im folgenden mit A, B, C, D bezeichnet werden sollen:

A: *De virga visoria plana et virga scripta*, in 29 bzw. 31 Kapiteln[90]
B: *De virga visoria quadrata*[91]
C: *De virga visoria trigona seu quadrante*[92]
D: *De virga visoria cubica*, in 17 Kapiteln mit Zusätzen[93].

In manchen Handschriften gibt es noch Kommentare zu A[94] und selbständige Erweiterungen[95].

---

[90] Die Kapitel 30 und 31 folgen (außer in der Handschrift Yale) nicht unmittelbar nach Kapitel 29; oft fehlen sie (Klosterneuburg, Freiburg, Klagenfurt). Der Text beginnt mit: *Virgam visoriam planam ad vasorum columnarium capacitatem inveniendam construere. Virgam visoriam dico, quia ...*

[91] Incipit: *Virgam visoriam ad vasorum columnarium capacitatem per ipsam inveniendam faciliter construere. Sit longitudo mensurae qua vas metiemur ...*

[92] Incipit: *Quadrantem seu trigonum ad capacitatem partis vasis vacuae inveniendam construere. Accipe primo virgam planam ...*

[93] Incipit: *Virgam visoriam cuius usus et factio faciles figurare. Sit mensura data qua vasa vel huius metiri volueris ...*

[94] Zwei voneinander abweichende, lange Kommentare stehen in Yale 25, f. 43r—52r, und Paris, BN lat. 10259, f. 13v—18v. Ein Kapitel aus dem letzteren findet sich in Plimpton 188, f. 172r; Vat. Pal. lat. 1354, f. 12v und Clm 14504, f. 374r. Ein Kommentar zu Proposition 6 steht in Yale 25, f. 54r—55v, Klosterneuburg 684, f. 178rv, und Vat. Pal. lat. 1354, f. 251v.

[95] Es würde zu weit führen, diese im einzelnen aufzuführen.

Die älteste Handschrift stammt aus dem Benediktinerstift St. Peter in Salzburg. Sie trug die alte Signatur X, VI 20 und wurde später als Inkunabel 800 eingereiht. Um 1924 wurde dieser Codex wie viele andere Handschriften und frühen Drucke verkauft; heute befindet er sich in der Yale Medical Library, New Haven (Nr. 25 im Supplement zum Census)[96]. Der Codex stammt aus dem Besitz von Joh. Serlinger, der wohl mit dem Bischof von Seckau identisch ist († 1511). Die Handschrift, die auf f. 1r—55v die Traktate A—D[97] und weitere kleinere Texte zur Visierkunst enthält, muß mit Nürnberg in Verbindung gebracht werden: Auf f. 10r steht am Ende von A, Kapitel 27: *et sic est finis factum Nuremberge Anno 1425°*, und in den Tabellen zu Traktat A wird zweimal Nürnberg erwähnt[98]. Auf f. 12v schließlich beginnt ein Zusatz des Schreibers nach Kapitel 31 mit den Worten: *Eadem longitudo et latitudo seu profunditas urne norembergensis que dicitur urna cubica stat in virga mea id est magistri Johannis schindel circulari signata per lineolam super cuprum transversum* . . . Johann Schindel besaß demnach also eine Visierrute, die auf den im Text angegebenen Maßen beruhte. Schindel, der um 1370 in Königgrätz geboren wurde, hielt von 1407 bis 1409 Vorlesungen in Wien, war 1423—1438 Stadtarzt in Nürnberg und lebte danach bis zu seinem Tode (1450) in Prag[99]. Während seiner Tätigkeit als Stadtarzt in Nürnberg berechnete er die Polhöhe und verschiedene Tafeln für Nürnberg; er verfaßte mehrere astronomische Arbeiten und stellte Instrumente her. Die Jahreszahl 1425 paßt gut zu Schindels Lebenslauf; ob er allerdings diesen Traktat selbst geschrieben hat, vermag ich nicht zu entscheiden.

In der aus Salzburg stammenden Handschrift befindet sich am Ende des Kommentars zum Traktat A auf f. 52r folgender Vermerk: *Item de hijs habetur una tabula postea quam fecit honorabilis et reverendus pater dominus prepositus Newburgensis.* Die Tafel selbst, die wohl auf der leeren Seite 52v stehen sollte, fehlt. Mit *prepositus Newburgensis* ist Prior Georg Müstinger aus dem Augustinerstift Klosterneuburg gemeint. Müstingers Bemühungen um die Astronomie sind bekannt; er stand in Beziehungen zu Johannes von Gmunden und ließ die wichtigsten Arbei-

---

[96] C. U. Faye / W. H. Bond, *Supplement to the Census of medieval and Renaissance manuscripts in the United States and Canada* (New York 1962) 58f.

[97] A: f. 1r—12v. 17v—24r; B: f. 13v—17r; C: f. 28r—31v; D: f. 32r—40v.41v—42r.

[98] Auf f. 18r und 21r.

[99] Siehe Ernst Zinner: *Deutsche und niederländische astronomische*

ten der Wiener Schule abschreiben[100]. Die Erwähnung in der Salzburger Handschrift bestätigt, daß sich Müstinger ähnlich wie Schindel auch mit der Visierkunst befaßte.

Zu den Handschriften, die Müstinger abschreiben ließ, gehören auch Klosterneuburg 683 und 684[101]. Der in diesem Zusammenhang besonders wichtige Codex 683 wurde um 1431 geschrieben[102]. Er enthält neben astronomischen Schriften[103] auch einen Vermessungstraktat[104] und eine interessante Sammlung mathematischer Aufgaben[105]. Auf f. 57r—78v finden wir Teile des Corpus zur Visierrechnung, und zwar die Texte B[106] und A[107]. Dieser Abschnitt entstand im Jahre 1431. Ich habe keine Anzeichen dafür gefunden, daß dieser Codex Tafeln zur Visierkunst enthält, die Müstinger berechnete[108], doch ist aus dem oben

---

Instrumente des 11.—18. Jahrhunderts. (2. A. München 1967) 501f. und 56.

100 Siehe ZINNER (A. 1) 17. 23. 70.

101 So ZINNER (A. 1) 17. Nach ZINNER (A. 99) 57, wurde dieser Codex von Nikolaus von Heybech aus Erfurt geschrieben. Heybech verfaßte Schriften über Sonnenuhren und andere astronomische Traktate: ZINNER (A. 99) 382f. Ich habe Zweifel, ob er auch den Visierteil der Handschrift Klosterneuburg 683 schrieb; wenigstens liefert die Handschrift selbst keine positiven Indizien.

102 Diese Jahreszahl steht in der subscriptio auf f. 73r: *Finitum anno domini 1431 incompleto iovis die 16ma augusti etc.* Auf f. 106v wird die Jahreszahl 1425, auf f. 212v 1433 erwähnt. Auch die Finsternistafeln auf f. 142r—184v müssen vor 1433 geschrieben sein, wie die Tempusangabe *(1433) in qua erit eclipsis solis* auf f. 182r zeigt.

103 Unter anderem die Finsternistafeln des Johannes von Gmunden für die Jahre 1422—1440 (auf f. 142r—184v).

104 Auf f. 79r—89v. Diesen Text hat MAXIMILIAN CURTZE nach der Handschrift München, Clm 56, f. 207r—218v, publiziert: De inquisicione capacitatis figurarum. Anonyme Abhandlung aus dem 15. Jahrhundert. *Abhandlungen zur Geschichte der Mathematik 8* (Leipzig 1898) 29—68. In Clm 56 befindet sich übrigens auf f. 197r—206v ein *Tractatus de quantitate trium solidorum,* der Johannes Schindel zugeschrieben wird.

105 Auf f. 118r—141v. Der Text enthält Aufgaben der Unterhaltungsmathematik, darunter Rätsel in Versen und in Prosa, und Kaufmannsrechnungen, aber auch eine Darstellung der Bruchrechnung.

106 Auf f. 57r—60v, ohne den Schluß.

107 Auf f. 61r—78v, ohne Kapitel 30+31.

108 So ZINNER (A. 1) 17. Die Tafeln in Klosterneuburg 683 decken sich mit den (anonymen) in Yale 25 aus Salzburg. Der Kommentar mit dem Hinweis auf Müstinger, der in Yale vorhanden ist, fehlt im Klosterneuburger Codex.

Gesagten klar, daß es Beziehungen zwischen dieser Handschrift und Müstinger gibt[109]. Daneben wird auch hier auf Nürnberg hingewiesen[110].

Mathematische und astronomische Kenntnisse der Wiener Schule, die zur Zeit Müstingers in Klosterneuburg zur Verfügung standen, dürften über einen Magister Reinhard aus dem Kloster Reichenbach, der in Klosterneuburg Bücher abschrieb, auch nach Regensburg gelangt sein, wo zur selben Zeit in St. Emmeram Fridericus Gerhart wirkte, der an den Naturwissenschaften sehr interessiert war[111]. Diese Beziehung erkennt man auch an zwei Handschriften aus St. Emmeram, die u. a. Texte des Visiercorpus enthalten: Clm 14504 und Vat. Pal. lat. 1354. Der Münchner Codex Clm 14504 wurde zu einem großen Teil von Fridericus geschrieben, jedoch stammt der selbständige Teil f. 358—393 nicht von seiner Hand. Diese Sammelhandschrift, die Traktate aus den Jahren 1439—1457 enthält, wurde von Vogel untersucht[112]. Auf f. 358r—393v sind fast dieselben Texte zur Visierkunst wie in Yale 25 vorhanden[113]. Auch der jetzt im Vatikan befindliche Codex Pal. lat. 1354 stammt aus St. Emmeram. Er wurde um 1464 von einem ungenannten Schreiber kopiert, dessen Hand auch in Vat. Pal. lat. 1376 nachzuweisen ist[114]. Diese Handschrift enthält auf f. 2v—30v dieselben Texte wie Clm 14504 in fast derselben Reihenfolge[115]. Hinzugefügt sind zu Beginn (f. 2r) und am Schluß (f. 31r—32r) zwei deutsche Visiertraktate.

---

[109] Bei den Tafeln über Maßstäbe auf der Visierrute (f. 78r) finden wir die Bemerkung: *pro stauffa terre Austrie.*

[110] Etwa in einem astronomischen Text auf f. 94r: *ad orizontem Nürenbergensem.*

[111] Zu Fridericus Gerhart und seinen Beziehungen zu Klosterneuburg siehe Zinner (A. 1) 68—70, und Kurt Vogel: *Der Donauraum, die Wiege mathematischer Studien in Deutschland* (München 1973) 14.

[112] Kurt Vogel: *Die Practica des Algorismus Ratisbonensis* (München 1954) 23f.

[113] Traktat A: f. 358r—368v. 373rv. 391v—393v
Traktat B: f. 383v—391r. 383r
Traktat C: f. 368v—372r
Traktat D: f. 376v—381v. 374r—376r. 382v.

[114] Siehe B. Bischoff: Studien zur Geschichte des Klosters St. Emmeram im Spätmittelalter. *Mittelalterliche Studien 2* (Stuttgart 1967) 128f. Es handelt sich um eine astronomisch-mathematische Sammelhandschrift von 251 Blättern. An mathematischen Texten sind außer den Visiertraktaten noch Auszüge aus Euklids Elementen vorhanden (auf f. 98v—99v. 120rv).

[115] Traktat A: f. 2v—9r. 12rv. 26r—30v
Traktat B: f. 23r—25r. 22v
Traktat C: f. 9r—11v
Traktat D: f. 18v—21v. 14r—18r. 13r—14r. 22rv.

Zumindest einen Teil der genannten Texte hat Regiomontan gekannt: Die Handschrift Plimpton 188, deren dritter Teil, wie erwähnt, Regiomontan gehörte, enthält auf f. 199v—201r den Visiertraktat B. Es folgen auf f. 201v—204v die 17 Kapitel des Traktats D. Auf f. 205r—206r steht Text C (ohne Tabellen), und den Schluß (f. 206v—210v) bilden die 17 ersten Kapitel von Traktat A; der Rest[116] fehlt durch Blattverlust. Ob die Handschrift ursprünglich weitere Texte enthielt, läßt sich nicht mehr feststellen. Ebenfalls unbekannt ist die genaue Entstehungszeit. Regiomontan hat zu dem von fremder Hand geschriebenen Text keine eigenen Bemerkungen hinzugefügt, so daß nicht klar ist, wie intensiv er ihn studierte. Die Überschrift *Compositio virge visorie quadrate* (auf f. 200r) könnte mit der Erwähnung *Compositio virge* im Bücherverzeichnis von 1512 identisch sein[117]; die zweite Titelzeile *De duplicj arte visorie,* die auf f. 199r auf dem sonst leeren Vorsatzblatt neben Regiomontans Eignerzeichen hinzugefügt ist, deckt sich nicht mit Erwähnungen in den Bücherverzeichnissen[118], sie stammt jedoch auch von späterer Hand.

Die genannte Handschrift Plimpton 188 enthält fast dieselben Visiertexte noch ein zweites Mal auf f. 163r—195r[119]. Dieser Teil der Handschrift wurde erst 1501 in Nürnberg[120] kopiert, wie die Erwähnungen auf f. 163r und 164r belegen. Er wurde also erst nach Regiomontans Tod zusammen mit anderen Texten[121] dem Visierteil auf f. 199—210 vorgebunden und stammt von einer anderen Handschrift.

Eine weitere späte Abschrift aller vier Visiertraktate ist Paris, BN lat. 10259[122]. Dieser Codex wurde in sehr sauberer Schrift im Jahre

---

[116] Die Kapitel 30 und 31, die auch in anderen Handschriften oft gesondert überliefert werden, sind auf f. 204v—205r vorhanden.

[117] Siehe Anm. 79.

[118] Siehe S. 187f.

[119] Es sind vorhanden:
Traktat A: f. 189r—195r. 170v—171v (nur Kap. 1—7. 30f.)
Traktat B: f. 188rv (nur der Schluß)
Traktat C: f. 166r—169v
Traktat D: f. 179v—186v. 172v—175r. 171v—172r. 187r—188r.

[120] Auf f. 188v finden wir den Zusatz: *audivi ab Walther.* Vielleicht ist hiermit der Nürnberger Kaufmann und Astronom Bernhard Walther († 1504) gemeint, der Regiomontans wissenschaftlichen Nachlaß übernahm.

[121] U. a. die *Practica geometriae* des Dominicus de Clavasio auf f. 103r—141r.

[122] Traktat A: f. 1r—13r. 27r—28r
Traktat B: f. 20r—21r
Traktat C: f. 21v—23v
Traktat D: f. 24r—27r. 28r—32v. 18v—19v.

1504 in Legnago geschrieben[123]. Er enthält neben den vier Traktaten A—D noch einen sehr ausführlichen Kommentar zur Abhandlung A mit vielen mathematischen Sätzen[124].

Es gibt noch zwei weitere Handschriften, die im wesentlichen nur den Traktat A enthalten. Dies sind die medizinisch-astronomisch-mathematische Sammelhandschrift 28 der Universitätsbibliothek Freiburg/Breisgau, die Texte aus verschiedenen Zeiten des 15. Jahrhunderts enthält[125], und der Codex XXX. b. 7 der Bischöflichen Bibliothek Klagenfurt[126]. Über die Herkunft beider Handschriften ist mir nichts bekannt.

Diese Bemerkungen über die Verbreitung der vier Visiertraktate mögen genügen. Es wird sich lohnen, die Abhandlungen auch inhaltlich zu analysieren, da sie die ältesten erhaltenen Texte zur Faßmessung bilden[127] und relativ anspruchsvoll sind: Der Autor (bzw. die Autoren) kennt die gängige wissenschaftliche Literatur seiner Zeit und erwähnt in Traktat A etwa Euklid XI 9, I 47, XII 2, die *Practica geometriae* des Dominicus de Clavasio, einen *Algorismus* und einen *tractatus de minutiis*, Johannes de Lineriis (*in quaerendo arcum per sinum*), Ptolemaios' *Almagest* und die *Geometrie* des Boethius. Er arbeitet im Traktat C mit dem *sinus rectus* und *versus* als etwas Selbstverständlichem. Die zum

---

[123] f. 32v: *MDIIII . . . porti Lignagi.*

[124] Auf f. 13v—18v. Die Sätze betreffen meist Kreise und Quadrate und sind größtenteils elementar. Daneben werden aber auch Regeln zur Ermittlung der Seite eines einbeschriebenen n-Ecks angegeben (speziell $n = 7, 8, 9, 10$).

[125] Beschreibung bei Winfried Hagenmaier, *Die lateinischen mittelalterlichen Handschriften der Universitätsbibliothek Freiburg im Breisgau* (*Hs. 1—230*) (Wiesbaden 1974) 28—32. Die Texte wurden 1428 bis ca. 1479 geschrieben. Der Text f. 167—178 ist weder durch das Wasserzeichen noch durch eine subscriptio datierbar. Der Autor des vorangehenden Traktates (f. 141r—166r) ist Nicolaus Hulleyben, Rektor der Erfurter Universität. Der Visiertraktat A steht auf f. 167r—177r.

[126] Beschreibung bei F. S. Benjamin jr. / G. J. Toomer: *Campanus of Novara and medieval planetary theory. Theorica planetarum* (Madison/Milwaukee/London 1971) 92f. Der Traktat A steht auf f. 11r—20r. Voraus gehen auf f. 2r—9v deutsche Texte zur praktischen Geometrie, die sich auch mit der Faßmessung beschäftigen. Dieser Teil der Handschrift stammt sicherlich aus dem 15. Jahrhundert. Die übrigen Texte, die z. T. jünger sind, betreffen die Astronomie. Vorhanden ist u. a. Regiomontans *Tabula primi mobilis*: Zinner (A. 1) 326.

[127] Die ältesten bisher bekannten selbständigen Texte stammen aus den Jahren nach 1443: s. Folkerts (A. 72) 36. Demgegenüber wurde Yale 25 im Jahre 1425 geschrieben oder zumindest von einer Handschrift aus dieser Zeit kopiert (siehe S. 192); auch Klosterneuburg 683 ist älter als 1443.

Teil recht komplizierten Rechnungen, die für die Teilpunkte auf der Rute erforderlich sind, werden richtig durchgeführt und die Ergebnisse in großen Tabellen zusammengefaßt. Den Autor dieses Corpus vermag ich nicht anzugeben; ich vermute, daß mehrere Mathematiker daran beteiligt waren und der Stoff an verschiedenen Stellen durch Zusätze und Kommentare erweitert wurde. Zumindest der Traktat A dürfte in Nürnberg entstanden sein; jedenfalls hat Johannes Schindel ihn um 1425 gekannt. Eine zentrale Rolle bei der Verbreitung dieser Texte spielt Klosterneuburg: Georg Müstinger kannte die Schriften, bearbeitete eine Tabelle und war indirekt dafür verantwortlich, daß das Corpus wenig später auch in St. Emmeram bekannt wurde. Von Klosterneuburg gibt es enge Verbindungen zur Wiener Universität. Daß Regiomontan die Klosterneuburger Handschrift kannte, ist möglich, aber nicht zu beweisen. Sicher ist jedoch, daß er eine andere Kopie desselben Corpus besaß (Plimpton 188) und dadurch mit dem Inhalt der Schriften vertraut war. Somit liefern diese Untersuchungen zwar keine neuen Erkenntnisse über Regiomontans eigene Arbeiten, aber sie unterstreichen die engen Verbindungen zwischen Wien, Klosterneuburg, Nürnberg und Regensburg in der ersten Hälfte des 15. Jahrhunderts und sind dadurch geeignet, das geistige Umfeld, in dem Regiomontan lebte und wirkte, zu verdeutlichen.

3. Frühe algebraische Studien Regiomontans

Es ist allgemein bekannt, daß Regiomontan die Algebra seiner Zeit sehr gut kannte und auf verschiedenen Gebieten Neues hinzufügte. Er beherrschte die Formen der quadratischen Gleichung und stellte zahlreiche Aufgaben, die auch auf Gleichungen höheren als 2. Grades führen[128]. Was wir darüber wissen, beruht allerdings fast ausschließlich auf den Aufgaben, die Regiomontan in seinem Briefwechsel behandelt, und zum geringeren Teil auf der Dreieckslehre; eine eigene Schrift zur Algebra ist nicht erhalten. Aus den genannten Texten wird auch deutlich, daß Regiomontan algebraische Symbole benutzte, und zwar speziell hochgestellte Zeichen für x (ꝛc) und x^2 (ꝗ) sowie einen langgezogenen Strich anstelle des Gleichheitszeichens[129]. Regiomontan ist neben Fri-

[128] Eine Zusammenfassung findet man bei Folkerts (A. 3) 26—30.

[129] Zur frühen algebraischen Symbolik siehe Wolfgang Kaunzner, Über die Entwicklung der algebraischen Symbolik vor Kepler im deutschen Sprachgebiet. *Kepler-Festschrift des Naturwissenschaftlichen Vereins Regensburg* (Regensburg 1971) 175—185, vor allem 178; und derselbe, Deutsche

dericus Gerhart der früheste Autor, der eine solche Symbolik für die Potenzen der Unbekannten anwendet: Während Fridericus' Ansätze mindestens bis 1461 zurückreichen, stammen die ältesten bisher bekannten Belege bei Regiomontan aus dem Jahre 1463[130]. Im Wiener Rechenbuch sind keine Aufgaben vorhanden, die eine algebraische Symbolik erforderlich machen würden, so daß man bisher nicht wußte, wie intensiv sich Regiomontan schon in Wien mit algebraischen Fragen beschäftigte. Dementsprechend findet man in mathematikhistorischen Darstellungen nur ganz allgemeine Bemerkungen zu Regiomontans algebraischen Leistungen. Allgemein wird bedauert, daß alle diesbezüglichen Schriften verloren sind.

Ganz so negativ ist die Sachlage in Wirklichkeit nicht, denn immerhin hat sich eine Handschrift erhalten, die Regiomontan gehörte und die neben fremden Schriften zur Algebra auch kleinere eigene Entwürfe sowie viele Randbemerkungen von Regiomontans Hand aufweist. Diese Handschrift hat wie so viele andere bis heute im Dornröschenschlaf gelegen; allen Mathematikhistorikern ist sie bisher entgangen. Lediglich Louis Karpinski hat sie im Zusammenhang mit seiner Edition von al-Ḫwārizmīs Algebra im Anfang dieses Jahrhunderts flüchtig studiert, ohne allerdings ihre Bedeutung voll zu erkennen[131]. Es handelt sich um den ersten Teil des schon genannten Codex Plimpton 188 in der Columbia University, New York. Dieser Teil entstand im Jahre 1456, wie die Jahreszahl oben auf f. 85r beweist.

Der genannte Abschnitt, der die Blätter 1—96 umfaßt, enthält nur zwei Traktate, die ursprünglich offenbar einen eigenen Codex bildeten: die *Flores arithmeticae* (f. 1r—70v) und al-Ḫwārizmīs Algebra mit Anhängen (f. 73r—96r). Diese Handschrift gehörte einst Regiomontans Bibliothek an; sie ist identisch mit der Erwähnung *Liber, in*

Mathematiker des 15. und 16. Jahrhunderts und ihre Symbolik. *Veröffentlichungen des Forschungsinstituts des Deutschen Museums für die Geschichte der Naturwissenschaften und der Technik*, Reihe A, Nr. 90 (München 1971), vor allem 15f.

130 In seinen Briefen an Bianchini. Dieselbe Symbolik läßt sich auch in Regiomontans Autograph der Dreieckslehre nachweisen, das etwa derselben Zeit angehört; siehe den Aufsatz von Wolfgang Kaunzner in diesem Band.

131 Robert of Chester's Latin translation of the Algebra of al-Khowarizmi. *Contributions to the history of science 1* (1915 = Ann Arbor 1930) 36.

quo Mahumethi de algebra etc; *item flores arithmetice* im Katalog aus dem Jahre 1522[132].

Der Codex beginnt mit einem arithmetisch-algebraischen Kompendium des Titels *Flores arithmeticae*, also *arithmetische Blütenlese*. Dieses Werk, das übrigens inhaltlich ebenfalls noch nicht untersucht wurde, beruht im wesentlichen auf zwei Quellen: dem Prosaabschnitt von Johannes de Muris' mathematischem Hauptwerk *Quadripartitum numerorum* (Mitte 14. Jahrhundert) und der Algebra des arabischen Mathematikers al-Ḫwārizmī[133]. Zu den *Flores*, die von einer fremden Hand kopiert wurden, hat Regiomontan am Rande vor allem in den algebraischen Teilen öfters Bemerkungen hinzugefügt.

An diese Schrift schließt sich al-Ḫwārizmīs Algebra an. Diese Abhandlung, die in der 1. Hälfte des 9. Jahrhunderts entstand, ist neben einer erst kürzlich ausgewerteten algebraischen Schrift ibn Turks der älteste Traktat in arabischer Sprache, der sich mit diesem mathematischen Teilgebiet beschäftigt. Die Abhandlung hat auf die weitere Entwicklung der Mathematik einen großen Einfluß ausgeübt, und zwar nicht nur im arabischen Bereich, sondern auch in Westeuropa, da diese Schrift im 12. Jahrhundert gleich zweimal ins Lateinische übersetzt wurde[134]. Wir haben es in unserer Handschrift mit der Übersetzung durch Gerhard von Cremona zu tun. Es ist möglich, daß Regiomontan diesen Teil selbst kopierte; jedenfalls ähnelt die Schrift seiner Handschrift sehr, weicht allerdings etwas von Randbemerkungen und Ergänzungen ab, die mit Sicherheit von Regiomontan stammen. Für uns ist die Frage, wer diesen Traktat abschrieb, sekundär, da der Autor feststeht[135].

---

[132] Petz (A. 78) 258 = Zinner (A. 1) 358, Nr. 211. Im Katalog von 1512 gibt es keine vergleichbare Angabe. Vielleicht bezeichnet eine der Erwähnungen *Algebra* (Zinner A. 1, 358, Nr. 204, und 359, Nr. 239), *Opus in Arithmetica* (Zinner A. 1, 360, Nr. 259) oder *Aritmetica* (Zinner A. 1, 358, Nr. 211) diese Handschrift. Im Katalog von 1563 wird sie nicht mehr erwähnt.

[133] Eine weitere Handschrift dieses Textes ist mir nicht bekannt.

[134] 1145 durch Robert of Chester (siehe A. 131) und ferner durch Gerhard von Cremona. Diese Übersetzung hat G. Libri ediert (*Histoire des sciences mathématiques en Italie depuis la renaissance des lettres jusqu'à la fin du 17e siècle 1* (Paris 1838 = Hildesheim 1967) 253—289). Handschriften der Übersetzung Gerhards von Cremona sind etwa: Paris 9335, Paris 7377 B, Cambridge Mm. 2. 18, Madrid 9119 (= A a 30), Florenz, Conv. soppr. J. V. 18, Vat. lat. 5733.

[135] Zum Inhalt von al-Ḫwārizmīs Algebra siehe etwa Juschkewitsch (A. 41) 204—214.

Der al-Ḫwārizmī-Text endet in der Handschrift Plimpton 188 auf f. 82v. Es folgen auf f. 82v—96r noch Zusätze mit verschiedenartigen mathematischen Texten. Es bedarf noch sorgfältiger Analysen, um festzustellen, woher diese Teile stammen[136]. Zumindest einige davon sind von Regiomontan geschrieben. Ein paar Bemerkungen zum Inhalt: Man findet Verteilungsaufgaben[137], ferner Zahlenrätsel, die auf lineare Gleichungssysteme führen[138], und andere Aufgaben der Unterhaltungsmathematik. Es gibt neue Beispiele zu den sechs Gleichungstypen des al-Ḫwārizmī[139], einen längeren Abschnitt über das Ausziehen von Quadratwurzeln[140] und ein Resteproblem mit ausführlicher Lösungsangabe[141]. Das Verfahren beruht auf der chinesischen ta-yen-Methode, so daß wir hier einen relativ frühen Beleg für die Kenntnis dieser Methode in Westeuropa haben[142]. Gleichzeitig liefert diese Stelle einen Nachweis dafür, daß Regiomontan tatsächlich diese Art von Aufgaben lösen konnte. Immerhin formuliert er in seinem Briefwechsel mehr als einmal Resteaufgaben, ohne daß man bisher wußte, ob er die generelle Lösung kannte[143].

Es ist gegenwärtig noch zu früh, endgültige Aussagen über die Bedeutung dieser Texte für Regiomontans wissenschaftlichen Werde-

---

[136] Möglicherweise sind Teile davon mit Zusätzen identisch, die sich in der Handschrift Madrid 9119 auf f. 359r—363v im Anschluß an al-Ḫwārizmīs Algebra befinden.

[137] Auf f. 82v—84v.

[138] Auf f. 88v—89r.

[139] Auf f. 85r—88r.

[140] Auf f. 91v—92r.

[141] Auf f. 93r: $N \equiv 2 \pmod 3 \equiv 4 \pmod 5 \equiv 1 \pmod 7$.

[142] Siehe hierzu die hervorragende Monographie von U. Libbrecht in seinem Buch *Chinese mathematics in the thirteenth century: The Shu-shu chiu-chang of Ch'in Chiu-shao* (Cambridge / London 1973) Kapitel 14. Libbrecht geht ausführlich auf die Ansätze bei Leonardo von Pisa und Fridericus Gerhart ein, die sich von der (Libbrecht unbekannten) Aufgabe in Plimpton 188 in Zahlen und Begründung der Methode unterscheiden.

[143] Dies sind die Probleme $N \equiv 15 \pmod{17} \equiv 11 \pmod{13} \equiv 3 \pmod{10}$ (siehe Maximilian Curtze, Der Briefwechsel Regiomontan's mit Giovanni Bianchini, Jacob von Speier und Christian Roder. *Urkunden zur Geschichte der Mathematik im Mittelalter und der Renaissance 1* (Leipzig 1902 = New York / London 1968) 219, Nr. 8) und $N \equiv 12 \pmod{23} \equiv 7 \pmod{17} \equiv 3 \pmod{10}$ (siehe Curtze a. O. 295, Nr. 6). Der aufgrund der knappen Angaben im Briefwechsel berechtigte Schluß (Libbrecht a. O. 247—249), Regiomontan habe wohl kaum das allgemeine Lösungsverfahren gekannt, muß nach Auffinden der Stelle in Plimpton 188 neu überdacht werden.

gang zu machen; dazu müssen sie vollständig ausgewertet und mit anderen Traktaten insbesondere aus dem italienischen Bereich verglichen werden. Es ist auch unerläßlich, Klarheit darüber zu gewinnen, was Regiomontan selbst schrieb und was vielleicht von einer fremden Hand stammt[144]. Ich hoffe, in absehbarer Zeit an anderer Stelle eine ausführlichere Analyse dieser Handschrift vorlegen zu können. Sicher ist jedenfalls, daß Regiomontan den ganzen Abschnitt f. 1r—96r kannte und — wie seine zahlreichen Bemerkungen beweisen — intensiv studierte. Im folgenden möchte ich nur zwei Aspekte anschneiden: die algebraische Symbolik und Ansätze zur Behandlung von Gleichungen höheren Grades.

Regiomontans algebraische Symbolik besteht im wesentlichen darin, daß er Zeichen für die Unbekannte x (ꝛ), für x^2 (ꝗ, wie das Zeichen für x ebenfalls hochgestellt), für das Gleichheitszeichen (langgezogener Strich) und für die Wurzel (℞) benutzt; daneben kürzt er *minus* (ꝯ), nicht aber *plus*, ab[145]. All diese Eigentümlichkeiten finden sich auch schon in der sieben Jahre früheren Handschrift Plimpton an den Stellen, die mit Sicherheit von Regiomontan stammen[146]. Auf f. 90v gibt es ferner für x^4 die Abkürzung ꝗꝗ; dagegen wird x^3 (*cubus*) nicht abgekürzt[147]. Auch diejenigen Teile, die wahrscheinlich nicht von Regiomontan geschrieben wurden, verwenden eine Symbolik, allerdings in geringerem Umfang: Für die Unbekannte x finden wir in der Regel das Zeichen ℞ [148], selten ꝛ [149], wobei beide Symbole auf gleicher Höhe wie die Zahlenkoeffizienten stehen, und auch x^2 wird mindestens in einem

---

144 Die Hand, die den al-Ḫwārizmī-Text schrieb, verwendet durchgehend die alten Formen für die Ziffern 4 und 7. Dies könnte ein Indiz dafür sein, daß Regiomontan nicht der Schreiber ist, obwohl der Schriftduktus in diesem Teil demjenigen in Handschriften ähnelt, die sicher von Regiomontan geschrieben wurden. In Randbemerkungen des Codex Plimpton 188, die von Regiomontan stammen, benutzt dieser in der Regel, aber nicht ausschließlich die neuen Ziffernformen für 4 und 7.

145 Siehe Kaunzner A. 129 (Regensburg 1971) 178, und Kaunzner A. 129 (München 1971) 15 und Bild 1—4.

146 Sehr deutlich etwa auf f. 93v, wo die Gleichung $x^2+436-40x = 4x^2+144$ entsprechend geschrieben wird. Belege für das hochgestellte x und x^2 gibt es ferner etwa auf f. 51r oben, wo auch das Wurzelsymbol zu finden ist (℞ *de 8* $= \sqrt{8}$). Die Kubikwurzel wird mit ℞ *cubica de 7* ($= \sqrt[3]{7}$) bezeichnet: f. 44v, oben.

147 Dieser Teil wurde mit großer Wahrscheinlichkeit von Regiomontan geschrieben.

148 Z. B. auf f. 83r—84v.

149 Auf f. 85r—88r.

V

Text mit ꝗ abgekürzt[149]. *minus* und *plus* werden genauso geschrieben wie bei Regiomontan[148]. Sollte sich die Vermutung bestätigen, daß diese Abschnitte nicht von Regiomontan geschrieben wurden, so haben wir hiermit einen Vorläufer für Regiomontans Symbolik gefunden. Sicher ist schon jetzt, daß Regiomontan Gleichungen und Terme relativ stark abkürzt und daß er seine Symbolik schon Jahre eher als bisher angenommen besaß.

Regiomontans Anmerkungen in dieser Handschrift zeigen, daß er die verschiedenen Typen der quadratischen Gleichung, deren Auflösung al-Ḫwārizmī lehrt, vollkommen beherrscht. Darüber hinaus gibt es aber mehrere Stellen, die erkennen lassen, daß der Autor sich bemüht, die überlieferten Methoden auf Gleichungen 3. Grades anzuwenden. Dadurch verdient diese Handschrift höchstes Interesse, denn vor dem 16. Jahrhundert gibt es keinen ernsthaften Versuch, kubische Gleichungen im allgemeinen Fall algebraisch zu lösen.

Die fremden Texte in der New Yorker Handschrift beschränken sich auf die Behandlung linearer und quadratischer Gleichungen; die 3. Potenz der Unbekannten sowie die Bezeichnung *cubus* wird stets vermieden. Diese etwas künstlich wirkende Einschränkung hat für Regiomontan keine Gültigkeit, denn in seinen Zusätzen werden mehrfach Aufgaben formuliert, die auf kubische Gleichungen führen, wobei Regiomontan diesen Sachverhalt auch ausdrücklich hervorhebt. So schreibt er etwa auf f. 82r bei der Lösung der Gleichung $1x^2 \cdot 2x = 3x^2$: *est ergo cubus equalis* $^3/_2$ *census* bzw. *duc 2 res in censum fiunt 2 cubi*; ähnlich auf f. 84v bei der Aufgabe $(x - \frac{x}{3}) \cdot 2\sqrt{x} = x$. Wir können also feststellen, daß Regiomontan die 3. Potenz der Unbekannten genauso selbstverständlich benutzt wie die 2. und 1. und auch die Verbindung zwischen quadratischen und elementaren kubischen Gleichungen herzustellen versteht.

Dies wird noch deutlicher in einem etwas späteren Abschnitt (f. 90v), wo Regiomontan, offenbar angeregt durch die sechs Gleichungstypen des al-Ḫwārizmī, insgesamt 18 Formen 3. und 4. Grades aufstellt und dafür Lösungen angibt. Dabei handelt es sich meist um Gleichungen, die sich auf den Grad 2 reduzieren lassen; dreimal ergeben sich reine kubische Gleichungen[150]. Auch dieser Abschnitt bringt nichts grund-

[150] Es sind folgende Formen: $ax^4 = bx^3$, $ax^4 = bx^2$, $ax^4 = bx$, $ax^4 = b$, $ax^3 = bx^2$, $ax^3 = bx$, $ax^3 = b$, $ax^3 + bx^2 = cx$, $ax^3 = bx^2 + cx$, $ax^3 + bx = cx^2$, $ax^4 + bx^3 = cx^2$, $ax^4 + bx^2 = cx^3 + dx^2$, $ax^4 + bx^2 = cx^3$, $ax^2 = \sqrt{b}$, $x^2 = a\sqrt{x}$, $ax^4 + bx^2 = c$, $ax^4 + b = cx^2$, $ax^4 = bx^2 + c$.

legend Neues. Hervorzuheben ist die Selbstverständlichkeit, mit der auch von höheren Potenzen der Unbekannten geredet wird, und die Symbolik, die auch für x^4 eine Abkürzung aufweist[151].

Bekanntlich können quadratische Gleichungen auch geometrisch mit Hilfe von Flächenverwandlungen bei Quadraten gelöst werden. Eine entsprechende Idee führte im 16. Jahrhundert zur Lösung der Gleichung 3. Grades. In beiden Fällen benutzt man dabei einen Gnomon, d. h. die Restfigur, die man übrig behält, wenn man von einem Parallelogramm bzw. Quader eine dazu ähnliche kleinere Figur wegnimmt. Dieser Weg zur Lösung der kubischen Gleichung erfordert also die Beherrschung von Flächenverwandlungen und die Fähigkeit, algebraische Sachverhalte geometrisch zu deuten. Daß gerade Regiomontan hierzu in der Lage war, erkennt man an mehreren Stellen unserer Handschrift. Zwei Hinweise mögen genügen: Eine Verteilungsaufgabe[152] führt auf das Gleichungssystem $60 = xy = (x + 2) \cdot (y - 5)$. Im Text wird die algebraische Umformung in eine quadratische Gleichung und die sich daraus ergebende Lösung ausführlich dargestellt. Regiomontan liefert zusätzlich eine elegante geometrische Deutung und Lösung, die auf dem Satz vom Ergänzungsparallelogramm und auf Flächenverwandlungen beruht. Dabei konstruiert Regiomontan, ausgehend von einem gegebenen Rechteck 2×5, ein dazu ähnliches Rechteck dergestalt, daß das andere Rechteck um die Diagonale zusammen mit einem Ergänzungsparallelogramm die Fläche 60 hat:

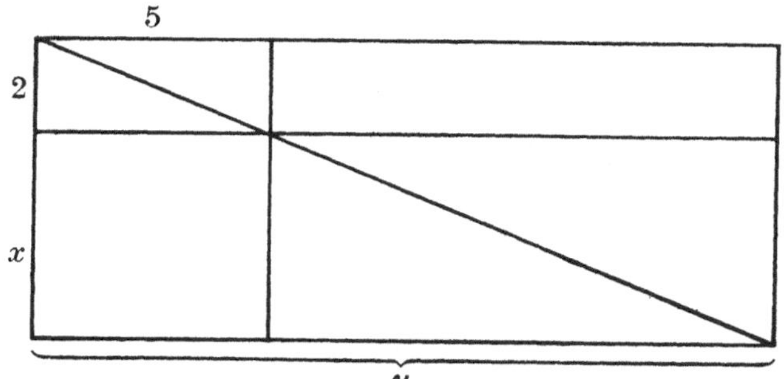

An einer anderen Stelle[153] gibt Regiomontan ein Verfahren an, um eine Summe von Quadratwurzeln durch eine einzige Wurzel auszu-

---

[151] Siehe oben S. 201.

[152] In den *Flores arithmeticae*, f. 54v.

[153] Auf f. 91r.

drücken: $\sqrt{a} + \sqrt{b} = \sqrt{2\sqrt{ab} + a + b} = \sqrt{c}$, wobei das Produkt ab eine Quadratzahl ist. Dieses Verfahren veranschaulicht Regiomontan mit Hilfe einer Gnomon-Figur. Diese erinnert an die geometrische Lösung der quadratischen Gleichung. Eine ähnliche Methode verwendet 100 Jahre später Michael Stifel in seiner *Arithmetica integra*, um kompliziertere Irrationalitäten auszudrücken[154].

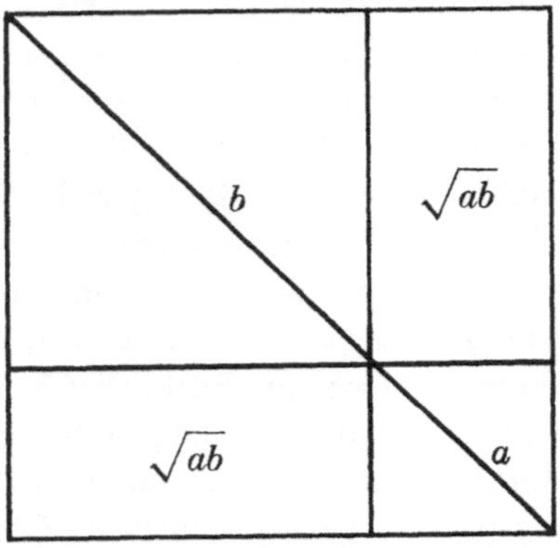

Die beiden genannten Stellen zeigen Regiomontans Vertrautheit mit der *geometrischen Algebra* nach Art der Griechen, die hier wie dort auf ebene Figuren beschränkt bleibt. Der entscheidende Schritt zu kubischen Identitäten in geometrischer Deutung, durch die erst die Gleichung 3. Grades lösbar wird, ist meines Erachtens in dieser Zeit noch nicht vollzogen worden. Regiomontan aber hat dies wenigstens im Prinzip versucht, wie man an einer anderen Stelle sieht. Er geht dort von der geometrischen Deutung der binomischen Formel $(a + b)^2 = a^2 + 2ab + b^2$ nach Euklid II 4 aus und erweitert diese durch Multiplikation mit der Strecke $(a + b)$ auf die kubische Identität $(a + b)^3 = a^3 + 3a^2b + 3ab^2 + b^3$. Regiomontan fügt beziehungsreich hinzu: *Auch andere Propositionen in Euklids Buch 2, die Flächen betreffen, kann man auf ähnliche Weise auf Würfel und Körper übertragen*[155].

---

[154] Vgl. HOFMANN (A. 34) 23f.

[155] f. 44r: *Si linea in duas partes dividitur, cubus tocius equalis est cubis ambarum parcium et ei quod fit ex una parte in triplum quadrati alterius cum eo quod fit ex altera in triplum quadrati illius. Ex quarta 2i Euclidis probabis si permiseris totam lineam multiplicare quadrata ambarum parcium et ad quod fit ex una in aliam bis per resolucionem sobriam. Alias propositiones 2i euclidis, quae de superficiebus sonant, ad cubos et corpora mutabis simili ingenio.*

Ob Regiomontan wußte, daß man kubische Identitäten zur Lösung kubischer Gleichungen heranziehen kann, war bisher unbekannt. Die New Yorker Handschrift ermöglicht es jetzt, diese Frage positiv zu beantworten: Am Schluß des Zusatzes nach al-Ḫwārizmī wird eine Zinsaufgabe gestellt: Wie hoch ist der Zinssatz, wenn das Kapital 100 Gulden nach 3 Jahren mit Zins und Zinseszins auf 265 Gulden anwächst? Das Problem führt zur kubischen Gleichung

$$x^3 + 300x^2 + 30000x = 1\,650\,000.$$

Der Autor gibt neben der richtigen Lösung $x = \sqrt[3]{2\,650\,000} - 100$ auch die Lösungsmethode an: Die Gleichung $x^3 + ax^2 + bx = c$ hat die Lösung

$$x = \sqrt[3]{(\frac{a}{3})^3 + c} - \frac{a}{3}.$$

Dabei wird ausdrücklich betont, daß diese Formel nur unter der Voraussetzung $(\frac{a}{3})^2 = \frac{b}{3}$ gilt[156]. Der Beweis ist exakt:

$$(x + \frac{a}{3})^3 = x^3 + (\frac{a}{3})^3 + ax^2 + 3(\frac{a}{3})^2 x$$

$$= x^3 + (\frac{a}{3})^3 + ax^2 + bx \qquad (1)$$

$$= c + (\frac{a}{3})^3, \quad \text{also}$$

$$x + \frac{a}{3} = \sqrt[3]{(\frac{a}{3})^3 + c}$$

$$x = \sqrt[3]{(\frac{a}{3})^3 + c} - \frac{a}{3}.$$

Nach den Worten des Autors ist $x^3 + ax^2 + bx$ ein körperlicher Gnomon, der den Würfel $(\frac{a}{3})^3$ umgibt. Das heißt, daß in (1) der Restwürfel $(x + \frac{a}{3})^3 - (\frac{a}{3})^3$ = körperlicher Gnomon $= x^3 + ax^2 + bx = c$ ist. Eine ähnliche Idee führte im 16. Jahrhundert zur endgültigen Lösung der kubischen Gleichung.

---

[156] *Hoc nequaquam procederet, nisi ita evenisset, quod tertia pars numeri censuum in se ducta quadrate produceret tertiam partem numeri rerum; alias enim gnomo solidus non resultasset integer.*

Der in Regiomontans Handschrift befindliche Lösungsansatz erinnert stark an die geometrische Deutung der quadratischen Gleichung $x^2 + px = q$ nach Art der Griechen: Es gilt die Identität

$$(x + \frac{p}{2})^2 = x^2 + px + (\frac{p}{2})^2$$

$$= q + (\frac{p}{2})^2, \quad \text{also}$$

$$x + \frac{p}{2} = \sqrt{(\frac{p}{2})^2 + q}$$

$$x = \sqrt{(\frac{p}{2})^2 + q} - \frac{p}{2}.$$

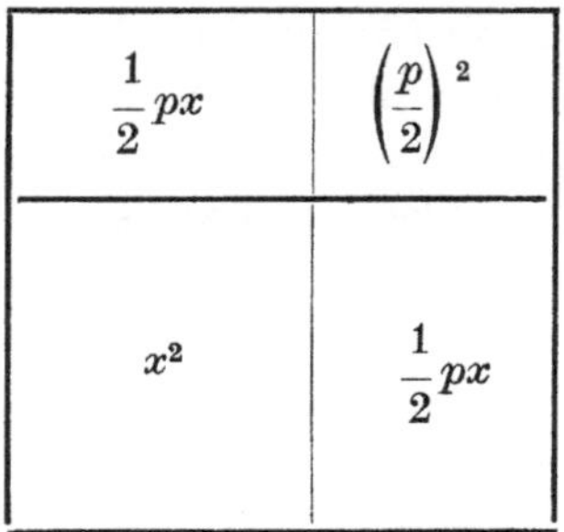

In der Handschrift Plimpton 188 übertrug der Autor offenbar das babylonisch-griechische Verfahren auf die kubische Gleichung und fand — wenigstens für den Spezialfall $(\frac{a}{2})^2 = \frac{b}{3}$ — die exakte Lösung. Er war sich der Beschränkung bewußt und ahnte wohl auch die Idee für die allgemeine Lösung, konnte sie aber nicht ausführen, weil er nicht das Verfahren kannte, um das quadratische Glied zu entfernen.

Es ist wahrscheinlich, wenn auch nicht mit letzter Klarheit beweisbar, daß der Text der kubischen Aufgabe in Plimpton 188 von Regiomontan selbst geschrieben wurde[157]. Doch auch wenn dies nicht der Fall sein sollte, ist aufgrund der Überlieferung des Codex sicher, daß Regiomontan diesen Text kannte. Unter diesen Umständen ist es nicht verwunderlich, daß Regiomontan acht Jahre später, also 1464, dieselbe Aufgabe seinem Korrespondenten Giovanni Bianchini stellt[158]. Bianchini gibt in seinem Antwortschreiben tatsächlich die richtige Lösung an[159] und verweist darauf, daß er in seinem *Liber florum* eine universelle Regel angegeben habe[160]. Bianchinis *Liber florum almagesti* ist entgegen CURTZES Ansicht[161] in mehreren Handschriften erhalten[162]. Diese Schrift,

---

[157] Die Ziffern 4 und 7, die für oder gegen Regiomontan sprechen könnten (s. A. 144), kommen in diesem Text nicht vor.

[158] CURTZE (A. 143) 219, Nr. 6.

[159] CURTZE (A. 143) 236 unten.

[160] *Sed multa similia per regulam argebre difficillime concluderentur, tamen in libro florum hec demonstravi et dedi regulam universalem ad omnem proportionem.*

[161] CURTZE (A. 143) 236, Anm. 3.

[162] Krakau 558; Bologna, UB 198 (293); Perugia 1004; Paris, BN lat. 10253; Vat. lat. 2228 und wohl auch Ferrara, Bibl. Comunale 775 (siehe P. O. KRISTELLER, *Iter italicum 1*, 53: s. XX).

die trotz der Vorarbeiten von BIRKENMAJER[163] und THORNDIKE[164] noch nicht näher untersucht ist, enthält in den drei ersten Büchern eine relativ ausführliche Darstellung der Arithmetik, Algebra und Trigonometrie. Die nach Bianchinis Worten naheliegende Vermutung, er habe wesentliche Beiträge zur Lösung von Gleichungen höheren Grades geliefert, scheint sich allerdings nicht zu bestätigen[165].

Nachdem Regiomontan sich wahrscheinlich schon in Wien mit algebraischen Fragen beschäftigte — erste Spuren haben wir in der Handschrift Plimpton 188 aus dem Jahre 1456 gefunden —, hat er offenbar an derartigen Problemen während seines Aufenthaltes in Italien intensiv gearbeitet. Dies ersieht man insbesondere aus seinem Briefwechsel mit Bianchini. Regiomontan besaß auch dessen *Flores almagesti*[166]; ob er diese Schrift, die vermutlich 1456 entstand[167], allerdings schon in der Wiener Zeit zu Gesicht bekam, ist fraglich. Auch Bianchini stellte übrigens dem Regiomontan ein Zinseszinsproblem. Dabei soll das Kapital 100 Dukaten nach 6 Jahren auf insgesamt 900 Dukaten anwachsen[168]. Regiomontan erkennt, daß dieses Problem auf eine Gleichung 6. Grades führt, findet aber trotzdem eine richtige Lösung

---

163 LOUIS BIRKENMAJER: Flores Almagesti. Ein angeblich verlorener Traktat Giovanni Bianchini's, Mathematikers und Astronomen von Ferrara aus dem XV. Jh. *Bulletin International de l'Académie des Sciences de Cracovie. Classe des sciences mathématiques et naturelles, Série A: Sciences mathématiques, Année 1911* (Krakau 1911).

164 LYNN THORNDIKE, Giovanni Bianchini in Paris manuscripts. *Scripta mathematica 16* (New York 1950) 5—12. 169—180, und DERSELBE, Giovanni Bianchini in Italian manuscripts. *Scripta mathematica 19* (1953) 5—17.

165 Ich kenne nur die Handschrift Krakau 558. Sie enthält keine derartigen Texte, jedoch fehlen hier das 3. und 4. Kapitel im *Tractatus secundus, Liber secundus* mit den Überschriften *Exempla regularum algebre de simplicibus* und *Exempla regularum algebre de compositis*.

166 Im Katalog von 1563 findet man folgenden Eintrag: *Alius fasciculus in quo insunt: Theonis traductio. Item distantiam Cometae a terra investigandi ratio. Item flores Almaiesti Joh. de Blanchi. Alles geschriben in 4°.* (ZINNER A. 1, 254, Nr. 45). Die Nennungen in den Katalogen von 1512 und 1522 sind zu unbestimmt. Regiomontans Codex ist nach ZINNER (A. 1) 109f. die Handschrift Krakau 558. Tatsächlich enthält dieser Codex einige wenige Randbemerkungen von Regiomontans Hand (z. B. f. 14v: Hinweis auf al-Ḫwārizmī; f. 23r: Erwähnung des *libellus trium fratrum*; f. 27r: Euklid; f. 33v: Theon).

167 Siehe BIRKENMAJER (A. 163) 273.

168 CURTZE (A. 143) 238, Nr. 4.

$\sqrt[6]{9\,000\,000\,000\,000} - 100$[169]; diese Lösung ist natürlich ohne Kenntnis eines Lösungsverfahrens für die Gleichung 6. Grades relativ elementar zu ermitteln. Regiomontan jedoch vermerkt, diese Aufgabe habe ihn an eine sehr große Klippe geführt[170], und es gebe noch keine Möglichkeit, alle Probleme der Algebra zu lösen. Dies sei ein *labyrinthus maximus*[171]. Er erwähnt nebenbei, daß er mit der Gleichung 4. Grades $5x^4 + 3x^3 + 8x^2 = 260x + 50$ noch nichts anfangen könne[172].

Erst jetzt, nachdem diese algebraischen Vorarbeiten Regiomontans bekannt sind, kann man jene berühmte Stelle aus seinem Brief an Christian Roder (1471) richtig einschätzen, die in allen einschlägigen mathematikhistorischen Arbeiten zitiert wird[173]:

Manche rühmen sich, die höhere Algebra schon in der zu besitzen, die in den so verbreiteten sechs Normalformen gelehrt wird. Sie übersehen vollkommen, daß sich diese Wissenschaft nicht auf dritte, vierte und höhere Potenzen erweitern läßt, wenn nicht die Geometrie volumengleicher Körper gefördert wird. Wie sich die drei zusammengesetzten Normalformen (der quadratischen Gleichung) auf Gleichheit von Flächen stützen, so wird notwendigerweise eine weitere Fortführung der Lehre von der Umwandlung von Körpern abhängen. Ich möchte daran deswegen erinnern, damit meine darauf gerichteten Bemühungen von anderer Seite unterstützt werden.

Zusammenfassend kann man also sagen, daß Regiomontan als einer der ersten intensiv versucht hat, die algebraische Lösung der kubischen Gleichung zu finden, und auch einige Fortschritte erzielte, ohne aber das Problem abschließend zu lösen. Seine Arbeiten berühren sich mit denjenigen Bianchinis, der sicherlich mit den Ansätzen anderer italienischer Algebraiker im 15. Jahrhundert vertraut war. Regiomontans Idee, nämlich die Verwandlung eines Würfels in eine andere volumengleiche Figur, wurde 50 Jahre später der Ausgangspunkt für die Lösung der allgemeinen kubischen Gleichung durch italienische Algebraiker. Es bedarf aber noch umfangreicher Untersuchungen, um festzustellen, wie weit Regiomontan dabei über Bianchini oder andere

---

169 Curtze (A. 143) 256 unten.

170 . . . *tamen in fine redigit me ad scopulum maximum.*

171 Curtze (A. 143) 280.

172 Curtze (A. 143) 256 unten. Zu den Aufgaben höheren Grades bei Regiomontan vergleiche man die lesenswerte Abhandlung von E. Bortolotti, Il carteggio matematico di Giovanni Regiomontano con Giovanni Bianchini, Giacomo Speier e Cristiano Roder. *Atti della accademia dalle scienze dell'Istituto di Bologna. Classe della scienze fisiche 9* (Bologna 1942) 81—90.

173 Curtze (A. 143) 335.

Vermittler von Bemühungen italienischer Algebraiker oder Rechenmeister profitieren konnte, die schon in dieser Zeit ähnliche Probleme behandelten.

Ich möchte an dieser Stelle abbrechen. Es dürfte klar geworden sein, daß es trotz Zinners grundlegender Arbeit möglich ist, präzisere Aussagen über Regiomontans Schaffen vor allem auf dem Gebiet der Mathematik zu gewinnen, wenn auch die Ergebnisse, die ich vorgelegt habe, in manchen Punkten noch unvollständig oder unsicher sind. Zwei Zugänge bieten sich an: Einmal gibt es noch unbekannte Handschriften aus dem Besitz Regiomontans. Ich habe einige davon genannt; weitere dürften sich finden lassen, wenn man die mittelalterlichen mathematischen Handschriften systematisch erforscht, dabei Regiomontans Bücherverzeichnisse heranzieht und Hinweise auf ihn (Eignerzeichen, eigenhändige Randbemerkungen u. ä.) beachtet. Andrerseits kann man durch eine inhaltliche Analyse der von Regiomontan geschriebenen Texte Auskunft über Originalität und Vorbilder erhalten. Hierfür müßte man freilich alle Arbeiten kennen, die Regiomontan hätten zur Verfügung stehen können; es liegt auf der Hand, daß bei unserem mangelhaften Wissen über die mittelalterlichen mathematischen Quellen hierin die eigentliche Schwierigkeit besteht. Wenn man aber so vorgeht, wird man in absehbarer Zeit eine Monographie über das *mathematische* Schaffen Regiomontans schreiben können, die weit über unser gegenwärtiges Wissen hinausgeht und viel klarer die hervorragenden Leistungen des Gelehrten auch auf diesem Gebiet würdigt.

Addenda to V

[p. 180] Regiomontanus' contributions to finding perfect numbers must today be seen in a rather different light: see item VIII, pp. 422f.

[pp. 180, note 33; 194; 200, note 142] The name "Fridericus Gerhart" must be replaced by "Fridericus Amann"; see item XII, p. 7.

[pp. 200–208] Regiomontanus' contributions to algebra must today be seen in a somewhat different light: see item XII, pp. 4–6, with references to modern literature on the collection of problems in manuscript Plimpton 188.

[p. 207] Giovanni Bianchini's *Flores almagesti* is still not yet edited. There is, however, a recent study of the algebraic parts of this treatise (Grażyna Rosińska, "Algebra w środowisku astronomów krakowskich w XV wieku. Traktat z *Flores almagesti* Jana Bianchiniego", *Kwartalnik Historii Nauki i Techniki* 39/2 (1994), 3–19. One of the manuscripts of Bianchini's *Flores*, Cracow, BJ 558, was in Regiomontanus' possession and contains marginalia in his hand.

VI

REGIOMONTANUS' ROLE IN THE TRANSMISSION AND TRANSFORMATION OF GREEK MATHEMATICS

Perhaps the most important figure in the mathematics of Central Europe in the fifteenth century was Johannes Regiomontanus (1436–1476). In the last 50 years his work has been studied in some detail. Starting with Zinner's monograph,[1] scholars have discovered a mass of new facts—albeit more about his astronomy than about his mathematics.[2] I will not assess here the mathematical achievements of Regiomontanus in detail, but will confine myself to his role in the transmission of Greek mathematics and to the question of how these works influenced him in his own mathematical work.

First a few words about his life. Even when he was still at Leipzig University (1447–1450), a part of his life about which we know little, Regiomontanus must have acquired sufficient knowledge to enable him to write an almanac for 1448. The next eleven years (1450–1461), which Regiomontanus spent in Vienna, were decisive for his development. In 1452 he became *baccalaureus* and in 1457 *magister* of the Faculty of Arts. Under the influence of John of Gmunden (before 1385–1442) and Georg Peurbach (1423–1461), Vienna became a center of astronomical and mathematical studies. After 1417 John of Gmunden treated only mathematical and astronomical subjects in his lectures. Although Regiomontanus did not meet him, he certainly was acquainted with his works, one of which was a trigonometrical treatise,[3] since Gmunden left his books and instruments to the Vienna library. The second important astronomer and mathematician in Vienna, Peurbach, who was active from 1448 as *baccalaureus* and from 1453 as *magister*, was respon-

---

[1] Ernst Zinner, *Leben und Wirken des Johannes Müller von Königsberg genannt Regiomontanus* (Munich, 1938); 2nd ed. (Osnabrück, 1968).

[2] The best modern secondary source for Regiomontanus' mathematics is Paul Lawrence Rose, *The Italian Renaissance of Mathematics*, chapter 4: "Regiomontanus in Italy" (Geneva, 1975), pp. 90–117.

[3] H. L. L. Busard, ed., *Der Traktat De sinibus, chordis et arcubus von Johannes von Gmunden* (Vienna, 1971).

sible—according to Doppelmayr's report of 1730[4]—for Regiomontanus' coming to Vienna; he became his teacher and friend. Peurbach was best known for his planetary theory, the topic of his lecture in 1454. He also wrote an elementary arithmetic[5] and a *Tractatus super propositiones Ptolemaei de sinibus et chordis*[6] which was heavily dependent on John of Gmunden's work of similar title. Regiomontanus studied these and other writings in Vienna and copied some of them; his autograph, the so-called Vienna *Rechenbuch*, from the years 1454–1462 (most of it was finished in 1458) is still extant.[7] This codex contains works on arithmetic, theory of proportions, calculations for polygons, circles, and spheres, but nothing on algebra.

In 1460, Cardinal Bessarion came to Vienna trying to persuade the *Kaiser* to support the struggle against the Turks. Regiomontanus' meeting with him changed his life. Bessarion planned to replace George of Trebizond's inferior translation of the *Almagest* by a better one, and he asked Peurbach and Regiomontanus to write an abridgment of the *Almagest*. He persuaded Peurbach to go to Italy with Regiomontanus in order to finish it. But since Peurbach died shortly before their departure, Regiomontanus himself in 1461 accompanied Bessarion to Italy. The next four years he spent partly in the cardinal's entourage and partly on his own travels. In both he studied mathematics and astronomy and learned Greek; furthermore he finished Peurbach's compendium of the *Almagest* and wrote the first four books of his *De triangulis*.

In 1467, at the latest, Regiomontanus left Italy for Hungary, where he was attached to the court of King Matthias Corvinus. In his years in Hungary he was occupied mainly with astronomical questions and drawing up trigonometrical tables. From 1471 to 1475 he lived in Nuremberg and *inter alia* prepared books for printing by his own press. In 1475 he went to Rome to lay the foundations of the calendar reform, but in 1476 he died suddenly.

Because of his early death, Regiomontanus could not carry out his plan to publish the most important mathematical and astronomical works in his own printing press. Among the books that did appear

---

[4] Johann Gabriel Doppelmayr, *Historische Nachricht von den Nürnbergischen Mathematicis und Künstlern* (Nuremberg, 1730), p. 1.

[5] *Algorismus*, printed several times from 1492 on.

[6] Printed in Nuremberg, 1541.

[7] Vienna, Österreichische Nationalbibliothek, MS 5203.

there is not a single mathematical work. But from a list printed on a single page—we will call it his *Programme*[8]—we know which texts he wanted to print, and indeed, although his *Nachlaß* was scattered in the sixteenth century, some of the manuscripts he evidently intended to use are still extant from his collection. Accordingly, we are able to make some deductions about his place in the transmission of Greek mathematics.

The mathematical works he intended to publish were:

1. Euclid's *Elementa* with the *Anaphorica* of Hypsicles, in Campanus' edition, but without his mistakes.[9]
2. A Latin translation of Archimedes' writings in the text of Jacobus Cremonensis.[10]
3. Menelaus' *Sphaerica.*[11]
4. Theodosius' *Sphaerica.*[12]
5. Apollonius' *Conica.*
6. Serenus' *Sectio cylindri.*[13]
7. Jordanus Nemorarius' *Arithmetica* and *De numeris datis.*[14]
8. Johannes de Muris' *Quadripartitum numerorum.*[15]
9. *Algorithmus demonstratus.*

To these he added some of his own works. Those that are mathematical in the strictest sense are:

10. Remarks about Campanus' edition of Euclid, in particular about Campanus' additions that should be left out.[16]

---

[8] Published in Zinner (see n. 1), Abb. 45.

[9] "Euclidis elementa cum anaphoricis Hypsiclis editione Campani evulsis tamen plerisque mendis: quae proprio etiam indicabuntur commentariolo."

[10] "Archimedis geometrae acutissimi opera de sphaera & cylindro. De circuli dimensione. De conalibus & sphaeralibus. De lineis spiralibus. De aequiponderantibus. De quadratura parabolae. De harenae numero. Cum commentariis Eutocii ascalonitae in tria opera ex praedictis: scilicet de sphaera & cylindro. de dimensione circuli. de aequiponderantibus. Traductio est Iacobi cremonensis: sed non nusquam emendata."

[11] "Menelai sphaerica nova editione."

[12] "Theodosii sphaerica. item de habitationibus. de diebus & noctibus nova traductione."

[13] "Apollonii pergensis conica. Item Sereni cylindrica."

[14] "Elementa arithmetica Iordani. Data eiusdem arithmetica."

[15] "Quadripartitum numerorum. Opus variis scatens argutiis."

[16] "Commentariolum quo commonstrantur placita Campani ex editione elementorum geometricorum reiicienda."

11. On the five regular bodies.[17]
12. Commentary on those treatises of Archimedes for which Eutocius did not provide a commentary.[18]
13. On the quadrature of the circle; against Nicholas of Cusa.[19]
14. On triangles, in five books.[20]
15. Geometrical problems of every sort. *Opus fructuosae iucunditatis*.[21]
16. On gauging rods of different kinds.[22]

From the *Programme* we learn that Regiomontanus knew the works of the most important Greek mathematicians (not only Euclid, Archimedes and Apollonius, but also Menelaus, Theodosius and Serenus) and that he intended to publish them in Latin translation, some with his own commentary. I shall later say something about the relevant manuscripts that he had collected. What Regiomontanus knew of the history of Greek mathematics we can see in his introductions to the astronomy of Alfraganus and Albategnius first published in 1537 from his *Nachlaß*.[23] Regiomontanus originally gave these introductions as lectures in Padua, probably in the spring of 1464.[24] He repeats the notion, common in his time, that geometry had been discovered by the Egyptians as a consequence of the division of the land after the inundation of the Nile; Euclid had collected the geometrical knowledge and written it down in 13 books; Hypsicles had added two more books. Regiomontanus further mentions

---

[17] "De quinque corporibus aequilateris quae vulgo regularia nuncupantur: quae videlicet eorum locum impleant corporalem & quae non. contra commentatorem Aristotelis Averroem."

[18] "Commentaria in libros Archimedis eos qui Eutocii expositione carent."

[19] "De quadratura circuli. contra Nicolaum Cusensem."

[20] "De triangulis omnimodis quinque volumina."

[21] "Problemata geometrica omnimoda. Opus fructuosae iucunditatis."

[22] "Radii visorii multorum generum cum usibus suis." To Regiomontanus' knowledge of the art of gauging (*Visierkunst*), see M. Folkerts, "Die mathematischen Studien Regiomontans in seiner Wiener Zeit," in *Regiomontanus-Studien*, ed. G. Hamann (Vienna, 1980), pp. 175–209 at 186–197.

[23] "Continentur in hoc libro: Rudimenta astronomica Alfragani. Item Albategnius astronomus peritissimus de motu stellarum ... Item oratio introductoria in omnes scientias Mathematicas Joannis de Regiomonte, Patavii habita, cum Alfraganum publice praelegeret. Eiusdem utilissima introductio in elementa Euclidis...." (Nuremberg, 1537). See Zinner (n. 1), pp. 321f.

[24] See Zinner (n. 1), pp. 110–118, and Rose (n. 2), pp. 95–97.

Boethius, Adelard, Alfred, and Campanus as redactors of the *Elements*. From Clagett's investigations in the fifties[25] we know that this Alfred is the Alfred given as author in manuscript Venice, B. Marciana f.a. 332—I shall come to this later. This manuscript was known to Regiomontanus through Cardinal Bessarion (about 1400–1472), who bought it in 1461 when he was in Vienna.[26] In his Padua lecture Regiomontanus particularly praises the writings of Archimedes; he also mentions the new translation of about 1450 by Jacobus Cremonensis. He says that Apollonius' *Conica*, when once translated into Latin, would excite the admiration of all mathematicians. As authors on spherics he mentions Theodosius and Menelaus. Under arithmetic—algebra is silently included—Jordanus Nemorarius is particularly praised. Of this author Regiomontanus cites here the *Arithmetica* in 10 books and the *De numeris datis*. Since, as he says, the 13 books of Diophantus have not yet been translated from Greek into Latin, the flowering of the higher arithmetic, the *ars rei et census*, still lies hidden. This subject is known by the Arabic name "algebra."[27] We know from Regiomontanus' correspondence that one year earlier, in 1463, he had discovered in Venice a Greek manuscript containing the first six books of Diophantus.[28] Before this discovery the work was apparently unknown in the West. Regiomontanus recognized immediately the importance of what he had found. In his Padua lecture he also mentioned Bianchini as learned in this art. Other important treatises on arithmetic are Johannes de Muris' *Quadripartitum numerorum*, the *Algorithmus demonstratus*, the *Arithmetica* of Boethius and the practical arithmetic of the monk Barlaam in Greek.

---

25 Marshall Clagett, "King Alfred and the *Elements* of Euclid," *Isis* 45 (1954): 269–277.

26 There is a note of ownership on f. 3v: "Est liber b. Car. niceni. emptus Wiene 1461."

27 "Diophanti autem tredecim libros subtilissimos nemo usque hac ex Graecis Latinos fecit, in quibus flos ipse totius Arithmeticae latet, ars videlicet rei et census quam hodie vocant Algebram arabico nomine."

28 "… me reperisse nunc Venetiis Diofantem, arithmeticum grecum nondum in latinum traductum.… Si liber hic … integer inveniretur, curarem eum latine facere … Interea tamen, si suadebitis, sex dictos libros traducere in latinum excipiam, quatenus latinitas hoc novo et pretiosissimo munere non careat." Edited by M. Curtze, *Urkunden zur Geschichte der Mathematik im Mittelalter und der Renaissance*, vol. 2: *Der Briefwechsel Regiomontan's mit Giovanni Bianchini, Jacob von Speier und Christian Roder* (Leipzig, 1902) at pp. 256f.

So much for the titles of the most important mathematical works. But how well was Regiomontanus acquainted with their contents?

First Euclid. It is clear[29] that Regiomontanus possessed manuscripts of at least two Latin versions of the *Elements*: the so-called "Adelard II" text, i.e. "version II," which, as it seems, in fact was compiled by Robert of Chester about 1140; and the Campanus redaction made shortly before 1260. The version II manuscript was copied by Regiomontanus himself from manuscript Venice, f.a. 332 already mentioned. Among the numerous Latin Euclid manuscripts, f.a. 332 is of particular importance for two reasons: (1) it contains two versions of Euclid—version II, which is ascribed at the beginning to "Adelhardus" and later to "Alfredus," and version III, ascribed to "Boethius"[30]—; (2) in the introduction to version III[31] is some historical information which is also found in Regiomontanus' Padua lecture. This introduction Regiomontanus copied from the Bessarion manuscript. The same text, under the title *Introductio in elementa Euclidis*, was published in the 1537 edition[32] and erroneously ascribed there to Regiomontanus, and this introduction turns up also in Regiomontanus' autograph copy of Euclid now in Nuremberg. We will come to it later.

In 1523 the famous painter Albrecht Dürer acquired Regiomontanus' copy of the Bessarion manuscript f.a. 332. This copy is lost. Another Euclid manuscript, now in Boston (Medical Library, MS 24), is apparently a copy of a Campanus codex in Regiomontanus' possession. With the help of these two manuscripts which we deduce to have been in Regiomontanus' possession—i.e. Campanus' Euclid and version II—, Regiomontanus compiled an improved text of the *Elements*, and the result is to be found in the Nuremberg manuscript Cent. VI. 13. The first part of this manuscript, up to III.8, is in Regiomontanus' own hand; the rest, which is the normal Campanus text, is in another hand, but has numerous interlinear and marginal glosses in Regiomontanus' writing. Some of these glosses contain

---

[29] See M. Folkerts, "Regiomontans Euklidhandschriften," *Sudhoffs Archiv* 58 (1974): 149–164.

[30] See Clagett (n. 25), pp. 271f.

[31] Edited by Clagett (n. 25), pp. 273–277. For the contents, see Marshall Clagett, "The Medieval Latin Translations from the Arabic of the *Elements* of Euclid, with Special Emphasis on the Versions of Adelard of Bath," *Isis* 44 (1953): 16–42.

[32] See n. 23.

alternative proofs. In the part written by Regiomontanus himself Campanus' proofs are radically shortened: among other things, he dispenses many of the formalities of the full-dress *Elements*, retaining only the essential mathematical content. In this manuscript he was almost certainly preparing the work mentioned in the *Programme*.

Later, Regiomontanus worked more intensively with the Campanus redaction, noticing some of the errors typical for the Arabic versions, particularly those found in the version II tradition. Criticism of these errors we find only in a letter in 1471.[33] Here he characterizes the parallel postulate as "obscure and remote from the understanding" (*obscurius et ab intellectu remotius*) and maintains that it is only introduced for the proof of proposition I.29; Campanus put it among the postulates (*petitiones*), while the Greeks counted it among the axioms (*communes scientias*). The Arabs, said Regiomontanus, did not need it at all, but defined parallels in another way.[34] Clearly, Regiomontanus knew a Greek manuscript in which this postulate is found amongst the axioms. For in some Greek manuscripts postulates 4 and 5 come at the end of the axioms (as nos. 10 and 11), and this is also true of the Greek *editio princeps*. From the modern point of view Regiomontanus' critique of the parallel postulate is not justified, but in his time his opinion was quite common. In fact, at least since the time of Proclus mathematicians had believed that this postulate could and should be proved.

But another of his criticisms is correct. He says that Campanus "tries in vain to establish the principles of Book V of the *Elements*." Tired out by work and his convolutions of words, Campanus brings the same material back reworked "by defining the same thing by itself."[35] In fact, Campanus' lucubrations at the beginning of Book V represent a failed attempt to restore a corrupt text: in the texts which

---

33 The addressee seems to have been Christian Roder in Erfurt. Edited by Curtze (n. 28), pp. 328f.

34 "Quem scrupulum et Campanus animadvertens hoc principium inter petitiones stolide collocavit, quamvis Graeci inter communes scientias ordinarint. Sed Arabes nonnulli a ministerio demonstrationis penitus reiecerunt hoc proloquium, aliter quidem equidistandes lineas diffiniendo."

35 "Pudet profecto recensere labores Campani, quibus frustra stabilire tentat principia quinti elementorum voluminis, quippe qui diuturno studio ac magno verborum involucro irretitus postremoque fessus atque desperabundus eadem semper recidiva laborat, idem scilicet per se ipsum diffiniendo."

are based on version II the Eudoxan-Archimedean axiom—that of two quantities of the same kind the less can always be multiplied so as to exceed the greater (V. def. 4)—is replaced by a definition of "continuous proportionality," which makes little sense.[36] The wording of the substituted definition 4 is similar to that of the genuine definition 5, so that the reader naturally took these definitions together. Here the Western commentators compounded the error by the assumption that equal multiples (*eque multiplicia*) *in the same way* (*eque*) exceed or fall short of one another. Campanus tried—in about ten pages—to explain this definition, naturally without success. Regiomontanus recognizes this and points out more of Campanus' mistakes. Unfortunately, his criticism is only known through this passage in the correspondence—the Nuremberg manuscript is only an autograph up to the beginning of Book III.

Another excursus by Campanus, on I.32 (on the sum of the angles of the triangle), led Regiomontanus to write a small treatise on star polygons, which is in the Nuremberg manuscript.[37] Already in the fourteenth century, Bradwardine in his *Geometria speculativa* had generalized to *n* sides the star pentagons of Campanus. Regiomontanus proved the theorem about the sum of the angles in a star *n*-gon, classified these figures and treated star polygons of a higher order.

Now a few remarks about Archimedes. Since Clagett leaves nothing out in the third volume of *Archimedes in the Middle Ages*,[38] we may simply repeat what he says.

Already in his Vienna period, Regiomontanus had fully assimilated those parts of the Archimedean tradition that were available at the time. In the *Rechenbuch* which he wrote mostly between 1454 and 1458, he copied a theorem from the redaction of the *Dimensio circuli* that a certain Gordanus had made before 1390.[39] There follow

---

36 "Quantitates que dicuntur continuam proportionalitatem habere, sunt quarum eque multiplicia aut equa sunt aut eque sibi sine interruptione addunt aut minuunt."

37 For this, see S. Günther, "Lo sviluppo storico della teoria dei poligoni stellati nell' antichità e nel medio evo," *Bullettino Boncompagni* 6 (1873): 313–340 at 332–335; and M. Cantor, *Vorlesungen über Geschichte der Mathematik*, vol. 2, 2nd ed. (Leipzig, 1900), pp. 277–279.

38 Marshall Clagett, *Archimedes in the Middle Ages*, vol. 3 (Philadelphia, 1978), pp. 342–383.

39 Edited by Marshall Clagett, *Archimedes in the Middle Ages*, vol. 1 (Madison, 1964), pp. 148–162. The text appears in Vienna 5203, f. 131v–133r. See also Clagett (n. 38), vol. 3, pp. 343–344.

some remarks, also traceable to Gordanus, about the relationships between circles and squares and between cubes and spheres. Here Regiomontanus made a number of changes and in many places shortened the text.[40] In a section entitled *Corpulenciam sive embadum spere comprehendere* there appears a proof of the theorem that the volume of the sphere is equal to the product of the radius and a third of the surface area.[41] The proof of the theorem is similar to that in the *Verba filiorum* of the Banū Mūsā,[42] but with an important modification: the Banū Mūsā, like Archimedes, inserted between two spheres a figure composed of conical segments and generated by the rotation of a regular polygon inscribed in a great circle of one of the spheres. Regiomontanus substituted this figure by a polyhedron composed of pyramids. This method is the same as that of Leonardo Fibonacci in his *Practica geometriae*. This polyhedron might be irregular. Regiomontanus masters the difficulties that may arise from this by an intuitive procedure to solve this problem.[43]

Archimedes' *Dimensio circuli* and parts of *De sphaera et cylindro* were known in the West in twelfth-century translations from the Arabic. From a list of Regiomontanus' *Nachlaß* we learn that he possessed a copy of the *Verba filiorum*.[44] In Italy he became acquainted with Archimedes' other writings in the translation which Jacobus Cremonensis had made, from the Greek manuscript A, about 1450 at the behest of Pope Nicholas V. Regiomontanus copied this translation during his stay in Italy, i.e. between 1461 and 1464. Both his copy and the manuscript he copied from are extant.[45] Regiomontanus' copy was used by Thomas Venatorius (Gechauff) when he published the *editio princeps* of the Greek text accompanied by

---

40 See Clagett (n. 38), vol. 3, pp. 344–346. The text is in Vienna 5203, f. 133r.

41 f. 134v: "Multiplica semidiametrum in terciam partem superficiei spere, et exibit ipsa corpulencia."

42 Proposition XV. Edited by Clagett (n. 39), vol. 1, pp. 332–334.

43 See Clagett (n. 38), vol. 3, pp. 348–351.

44 In the catalogue of 1522 the following manuscript is mentioned: "Liber in quo libellus trium fratrum Moisi; Comensurator; epistola Ahmeti de proporcione; Menelaus de spericis; de ponderibus; Tebith de Carascove (= carastone); planisperium Jordani; demonstracio circulorum azimuth dicte; isoperimetria Jordini; Teodosius de speris; et de speculis comburentibus." See Zinner (n. 1), p. 330.

45 The manuscript was copied from Bessarion's manuscript Venice, f.a. 327; see Clagett (n. 38), vol. 3, pp. 354f. Regiomontanus' autograph is Nuremberg, Cent. V. 15.

Jacobus' translation in 1544.[46] Since Regiomontanus intended to publish the translation of Archimedes by Jacobus Cremonensis, he compared the manuscript that he had used with all other manuscripts of the text available to him, noticing mistakes in the translation and correcting the Latin text. In this work he used a Greek codex owned by Bessarion, which is apparently the earliest Renaissance copy made from the old Byzantine manuscript **A**.[47] Regiomontanus remarks further that he had used a recent Greek manuscript[48] and an *exemplar vetus apud magistrum Paulum*. The latter manuscript is in all probability either the Greek manuscript **A** or the Latin manuscript Vat. Ottob. lat. 1850, which is the autograph of the translation from the Greek by William of Moerbeke in 1269.[49] We know that the Florentine astronomer Paolo dal Pozzo Toscanelli (d. 1482)—most famous because of his letters to Columbus and to Canon Martinez in which he encouraged the investigation of the westward route to India—owned a manuscript of Archimedes' works and that Regiomontanus at one time was in close contact with him. Whatever be the truth of this matter, Regiomontanus corrected and annotated his text of Jacobus' translation in many places with the help of at least four other manuscripts. Such a procedure in critical edition was very unusual in this time. Regiomontanus' remarks show that he read his texts critically and completely understood them;[50] as Clagett says, "Regiomontanus had a far greater understanding of the works of Archimedes and Eutocius than any other figure of the fifteenth century."[51]

One of the most striking emendations occurs in the book on conoids and spheroids. Even in the oldest manuscripts a passage on the segment of the paraboloid is spoiled by a scribal error. This led to

---

[46] Printed by Herwagen in Basel.

[47] This copy is manuscript **E** (Venice, Marc. gr. 305); see Clagett (n. 38), vol. 3, p. 355, n. 20.

[48] In MS Cent. V. 15, p. 119, he writes: "notos illos sumpsi ex greco novo exemplari 19." The manuscript mentioned is probably Nuremberg, Cent. V. 12; see Clagett (n. 38), vol. 3, p. 356. As Clagett has shown (n. 38, vol. 3, p. 356, n. 21), Heiberg's theory that the Nuremberg humanist Willibald Pirckheimer acquired this codex later in Rome seems unlikely.

[49] For this, see Clagett (n. 38), vol. 3, p. 357, n. 22.

[50] Some of the total of several hundred corrections and marginal additions are given by Clagett (n. 38), vol. 3, pp. 357–365.

[51] Clagett (n. 38), vol. 3, p. 357.

other errors and in the end to a faulty presentation of the proof. This was first pointed out in this century by Dijksterhuis[52] and Heller.[53] Regiomontanus was the first to emend the text in this passage.

Some time after this, Regiomontanus again studied the quadrature of the circle intensively. This was occasioned by the famous dispute with Cardinal Nicholas of Cusa. The latter, an amateur in mathematics, tried by mathematical and other means to reach the single real truth. *Inter alia*, he investigated the relationship between regular polygons and the circle of the same perimeter. To this end he wrote numerous treatises on the quadrature and rectification of the circle. Some of them were available to Regiomontanus through his friend Toscanelli, whom Nicholas was wont to consult on mathematical matters. When Cusanus sent Toscanelli a new tract on the quadrature of the circle, Regiomontanus saw it in 1462 and resolved to analyze the approximations used in it. The results are found in his *Dialogus de quadratura circuli secundum Nicolaum Cusensem*, which may be dated to 1464 or a little later. This pamphlet is a dialogue between Aristophilus, the proponent of Cusanus' view, and Critias, who represents Regiomontanus' own opinions.[54] In this dialogue and in his *De quadratura circuli*, which he dedicated to Toscanelli and pretty certainly wrote in 1464,[55] Regiomontanus appears as an intelligent disciple of Archimedes. He rejects Aristotle for not allowing the comparison of straight lines and curves and asserts that every curve may be rectified. Regiomontanus had a thorough knowledge and understanding of the methods used by Euclid and Archimedes in this and similar questions. He also used algebraic and trigonometrical methods, such as the rule of false position and calculations with chords, and was acquainted with the extraction of square roots. He

---

52 E. J. Dijksterhuis, *Archimedes* (Copenhagen, 1956), p. 245.

53 Siegfried Heller, *Ein Fehler in einer Archimedes-Ausgabe, seine Entstehung und seine Folgen* (Munich, 1954), especially pp. 24–27.

54 Regiomontanus' correspondence with Toscanelli was published by J. Schöner in 1533 as an appendix to *De triangulis*: "Ioannis de Regiomonte ... De quadratura circuli, dialogus, et rationes diuersae separatim aliquot libellis exquisitae: Ad ea de re Cardinalis Cusani tradita et inuenta." The matter is analyzed by J. E. Hofmann, "Über Regiomontans und Buteons Stellungnahme zu Kreisnäherungen des Nikolaus von Kues," *Mitteilungen und Forschungsbeiträge der Cusanus-Gesellschaft* 6 (1967): 124–154, and by Clagett (n. 38), vol. 3, pp. 366–381.

55 Edited in the appendix to *De triangulis omnimodis* (see n. 54), pp. 29–38. For discussion, see Hofmann and Clagett (n. 54).

accordingly showed that Nicholas' approximations to π were—except one—not even within the limits established by Archimedes. Later Regiomontanus characterized Cusanus as a "ridiculous geometer who tried to emulate Archimedes."[56]

In his various quadratures of the circle Nicholas of Cusa used the idea of isoperimetry, which is found, for instance, in the works of Jordanus Nemorarius (*Liber philotegni*) and Thomas Bradwardine (*Geometria speculativa*). Regiomontanus was well acquainted with this notion and knew the treatise on isoperimetric figures translated from the Greek in the twelfth century[57] and deriving ultimately from Zenodorus' treatment of the subject.[58] In fact, he had copied the text himself in Vienna.[59] We may note that the text he copied is not in the usual wording, but is apparently his own reworking of a late-medieval redaction.[60]

Regiomontanus also knew the work of the third great Greek mathematician, Apollonius. He possessed a Greek manuscript of the *Conica*, which incidentally also contained Serenus' work on the section of the cylinder.[61] This manuscript has no annotations by Regiomontanus. But in the lists of his *Nachlaß* a text *Inicium Apolonii ex Greco in Latinum traducti* and a Latin translation "per Bernardum" are mentioned. In his biography Zinner deduced that Regiomontanus had possibly begun the translation and Bernhard Walther (d. 1504) had finished it.[62] But it is also possible that Regiomontanus had in his possession a manuscript which contains the beginning of the *Conica* in Gerard of Cremona's translation. This

---

56 See Curtze (n. 28), p. 329: "geometra ridiculus Archimedisque emulus."

57 Edited by H. L. L. Busard, "Der Traktat *De isoperimetris*, der unmittelbar aus dem Griechischen ins Lateinische übersetzt worden ist," *Mediaeval Studies* 42 (1980): 61–88.

58 The Greek text was edited by F. Hultsch, *Pappi Alexandrini Collectionis quae supersunt*, vol. 3 (Berlin, 1878), pp. 1138–1165. A critical analysis of the text is in W. Knorr, *Textual Studies in Ancient an Medieval Geometry* (Boston, 1989), pp. 689–751.

59 Vienna 5203, f. 142r–146r.

60 Manuscripts of this later reworking are Florence, BN, cs. J. V. 18, f. 11ra–12va, a second copy on f. 96v; cs. J. V. 30, f. 12v (incomplete). On the authorship, see W. Knorr, "Paraphrase Editions of Latin Mathematical Texts: *De figuris ysoperimetris*," *Mediaeval Studies* 52 (1990): 132–189.

61 Now Nuremberg, Cent. V, app. 6. See also Zinner (n. 1), p. 328.

62 See Zinner (n. 1), p. 328.

text was regularly transmitted with Alhazen's *Liber de speculis comburentibus*.[63] At all events, it is clear that Regiomontanus' plans to publish Apollonius' work[64] did not get very far. The same goes for the works of Theodosius and Menelaus on spherics, of which he possessed a copy bound with several other texts, unfortunately no longer extant.[65] It is to be supposed that this collection was in the Arabic tradition.[66]

Besides the Greek texts, treatises from three medieval writers appear in the *Programme*: Jordanus Nemorarius, *Arithmetica* and *De numeris datis*; Johannes de Muris, *Quadripartitum numerorum*, and a work entitled *Algorithmus demonstratus*. This last work is probably the thirteenth-century treatise of this name in the *algorismus* tradition by a certain Gernardus, a treatise that was very popular in the later Middle Ages.[67] Regiomontanus copied this treatise in Vienna,[68] and on the basis of this copy J. Schöner published the work in 1534.[69]

The *Arithmetica* of Jordanus Nemorarius was one of the most important works on theoretical arithmetic in the West. As Dr. Busard, who recently edited this treatise,[70] has shown, Jordanus with this work intended to write the *Elements* of arithmetic, i.e. to elevate the subject to the level of Euclid. Jordanus' *Arithmetica* became the standard source for theoretical arithmetic in the Middle Ages; it was used by Campanus, Roger Bacon, Oresme and others. About one-

---

[63] Edited by J. L. Heiberg, *Apollonii Pergaei quae Graece exstant*, vol. 2 (Leipzig, 1893), pp. LXXV–LXXX; see Clagett: *Archimedes in the Middle Ages*, vol. 4 (Philadelphia, 1980), pp. 4–13.

[64] See *Programme*, no. 5.

[65] See n. 44.

[66] For Menelaus, see A. A. Björnbo, *Studien über Menelaos' Sphärik. Beiträge zur Geschichte der Sphärik und Trigonometrie der Griechen* (Leipzig, 1902), pp. 1-154. On the transmission of Theodosius' *Sphaerica*, see Richard Lorch, "The Transmission of Theodosius' *Sphaerica*," in *Mathematische Probleme im Mittelalter - der lateinische und arabische Sprachbereich*, ed. M. Folkerts [in press]. The Theodosius translation by Gerard of Cremona is also in MS Venice, f.a. 332, f. 261r–289r. This codex was in the possession of Bessarion and known to Regiomontanus (see above).

[67] See G. Eneström, "Der 'Algorismus de integris' des Meisters Gernardus," *Bibliotheca Mathematica*, Ser. 3,13 (1912–1913): 289–332; and idem, "Der 'Algorismus de minutiis' des Meisters Gernardus," ibid., 14 (1913–1914): 99–149. It is unknown whether this Gernardus in fact is the same as Jordanus Nemorarius.

[68] Vienna 5203, f. 148r–166v.

[69] *Algorithmus demonstratus* (Nuremberg, 1534).

[70] H. L. L. Busard, *Jordanus de Nemore, De Elementis Arithmetice Artis. A Medieval Treatise on Number Theory* (Stuttgart, 1991).

half of the work is devoted to the treatment of the arithmetical books VII–IX of Euclid's *Elements*. One of the many original propositions in the *Arithmetica*, III.31, is on solving in positive integers the indeterminate equation $ax - by = v$, for integral values of a, b, v (a, b prime to one another and $a > b$). The solution of Jordanus is in principle the same as that given by the Indian mathematicians. Book VIII treats the theory of figured numbers; some original conclusions can be found here, as elsewhere. In one manuscript[71] there is an additional proposition: the sum of the cube numbers beginning from the unit are square numbers. In his studies on the theory of numbers Regiomontanus may well have been influenced by this very important text. His manuscript of Jordanus' *Arithmetica*, which is mentioned in a list of his *Nachlaß* and which also contained Jordanus' *De numeris datis*,[72] seems to be lost, but it should be mentioned that the *Arithmetica* is part of manuscript Venice, f.a. 332, mentioned above.[73]

Regiomontanus knew Jordanus' treatise on proportions that is concerned with the different ways of presenting the compound ratios found in Menelaos' theorem.[74] As early as his Vienna period, Regiomontanus had written a short treatise based on this work of Jordanus. In it he showed how the three means are involved in this mathematical relation.[75]

Now to the *Quadripartitum numerorum* of Johannes de Muris. This treatise, written in 1343, was Johannes' principal work.[76] It contains, apart from arithmetic, music and mechanics, a detailed treatment of algebra, which is heavily dependent on al-Khwārizmī, though there are also practical problems in the style of Leonardo Fibonacci. Regiomontanus' manuscript of the *Quadripartitum nu-*

---

[71] Basel, F II 33.

[72] "Liber in quo Jordani arithmetica; Jordanus de datis; algorithmus demonstratus; tractatus de proporcionibus; propleumata arithmethicalia." The list is from 1522. See Zinner (n. 1), p. 318.

[73] f. 40r–85v.

[74] Edited by H. L. L. Busard, "Die Traktate *De Proportionibus* von Jordanus Nemorarius und Campanus," *Centaurus* 15 (1971): 193–227.

[75] This treatise was published by Schöner as an appendix to the *Algorithmus demonstratus* (see n. 69). New edition by Busard (n. 74), pp. 222–226.

[76] The *Quadripartitum numerorum* has now been published by G. l'Huillier, *Le* Quadripartitum numerorum *de Jean de Murs. Introduction et édition critique* (Geneva/Paris, 1990).

merorum is still extant.[77] It was not written by Regiomontanus, but there are corrections in his hand and also many annotations.[78] In several places he converts algebraic solutions into geometric solutions. Doubtless the annotations in the Plimpton manuscript would have become the basis of the commentary Regiomontanus hoped to publish on his own press.

In the last part of my paper I should like to mention some mathematical problems more or less in the Greek tradition, but not transmitted from Greek writers. Most of them are from three sources:

a. manuscript New York, Columbia University, Plimpton 188, 1456. It contains on f. 82v–96r a collection of arithmetic, algebraic und geometric problems written by Regiomontanus—as it seems, in 1456.[79]
b. the autograph of his *De triangulis*, now in St. Petersburg (first half of the 1460s).
c. correspondence with Giovanni Bianchini (1463–1464), Jacob of Speyer (1465) and Christian Roder (1471) (autograph now in Nuremberg).[80] In these letters Regiomontanus suggested a series of mathematical and astronomical problems and gave the solutions of some of them and hints for others.

I. Quadratic Equations

I.1. In *De triangulis* II.12 the base and height of a triangle and the ratio of the other two sides are given.[81] This is reduced to the quadratic equation $25x^2 + 3125 - 500x = 9x^2 + 180x + 1125$, which is then solved algebraically. In the autograph algebraic symbols are used, but in the printed edition of 1533 it is written out in words.[82]

---

[77] New York, Columbia Library, Plimpton 188, f. 1r–70v.

[78] l'Huillier describes their content: Ghislaine l'Huillier, "Regiomontanus et le *Quadripartitum numerorum* de Jean de Murs," *Révue d'histoire des sciences* 33 (1980): 193–214.

[79] The content of this part is given in M. Folkerts, "Regiomontanus als Vermittler algebraischen Wissens," in *Mathemata. Festschrift für Helmuth Gericke*, ed. M. Folkerts and U. Lindgren (Stuttgart, 1985), pp. 207–219. An edition is to be published in the series *Algorismus* (Institut für Geschichte der Naturwissenschaften, Universität München).

[80] Edited by Curtze (n. 28).

[81] Joannes de Regio Monte, *De triangulis omnimodis libri quinque* (Nuremberg, 1533), p. 51.

[82] See W. Kaunzner, "Über Regiomontanus als Mathematiker." in *Regiomon-*

I.2. In II.23 the height of a triangle, the difference between two of the sides and the difference between the parts of the base divided by the altitude are given.[83] In this case, the equation $\frac{1}{4}x^2 + 136 - 6x = 4x^2 + \frac{9}{4} - 6x$ must be solved.[84]

These problems—with a slight alteration in the second—also appear in manuscript Plimpton 188.[85]

Some quadratic problems are in the correspondence:

I.3. $x - y = \sqrt{5}$, $x^2 \cdot y^2 = 250$.[86] Regiomontanus refuses Bianchini's assertion that this is insoluble because 250 is not a perfect square,[87] but he gives no solution.[88]

I.4. A traditional purchase problem here leads to the quadratic equation $15 : x = (58 - x) : 27$.[89] Regiomontanus gives a detailed solution and finds correctly that $x = 29 - \sqrt{436}$.

I.5. Two numbers x, y must be found whose sum is 10 and such that $x/y + y/x = 25$.[90] This results in the quadratic equation $250x - 25x^2 = 2x^2 + 100 - 20x$.

I.6. A number x is sought for which $100/x + 100/(x+8) = 40$.[91] Here the quadratic is $40x^2 + 320x = 200x + 800$.

II. Cubic and Higher Equations

II.1. A problem of compound interest: to find the interest by which an amount of 100 *florins* comes up to 265 *florins* within three years.[92] This problem leads to the cubic equation $x^3+300x^2+30{,}000x$

---

tanus-Studien, ed. G. Hamann (Vienna, 1980), pp. 125–145 at p. 134f., with plate 2.

[83] *De triangulis*... (n. 81), pp. 55f.

[84] See Kaunzner (n. 82), pp. 134f., with n. 48 and plate 3.

[85] No. 61: Given are the base, height and the ratio of the other two sides; no. 62: Given are the base, height and difference of the other two sides.

[86] Curtze (n. 28), p. 219, no. 5, in a letter to Bianchini.

[87] Curtze (n. 28), p. 236.

[88] Curtze (n. 28), p. 253: "Quesivi duos numeros huiusmodi non in unitatibus, scilicet in actu tantum, sed in potentia etiam, si opus fuerit."

[89] Problem: Curtze (n. 28), p. 296, no. 9; answer by Jacob: p. 300; comment by Regiomontanus: pp. 304f., with calculations on pp. 317–319.

[90] Problem given by Bianchini: Curtze (n. 28), p. 209; answer by Regiomontanus: p. 216; calculations: pp. 232–234. See Kaunzner (n. 82), p. 136, with plate 4.

[91] Problem given by Bianchini: Curtze (n. 28), pp. 237f.; Regiomontanus' answer: p. 256; calculations: p. 278. See Kaunzner (n. 82), p. 136, with plate 5.

[92] Plimpton 188, no. 64. The same problem in the *Correspondence*: Curtze (n. 28), p. 219,6 and 236.

= 1,650,000. Regiomontanus solves this problem by using a three-dimensional *gnomon*; this may perhaps be seen as a step towards the solution of the general cubic found by the Italians in the first half of the sixteenth century. But in this case Regiomontanus was not original. His solution only works in a special case of the cubic equation $x^3 + ax^2 + bx = c$, namely if $a^2 = 3b$. In this case $x = \sqrt[3]{c + (b/a)^3} - b/a$. We know now that Master Dardi of Pisa had already solved this special problem in the second half of the fourteenth century,[93] and we find it somewhat later in anonymous texts[94] and in Piero della Francesca's *Trattato d'abbaco*.[95]

II.2. In 1464 Bianchini suggested a problem in compound interest: 100 *florins* become 900 *florins* in 6 years. Regiomontanus answered that the interest of the first year is $\sqrt[6]{9 \cdot 10^{12}} - 100$, but he added: *Labyrinthus maximus*.[96] Furthermore, he gave an example of a fourth-degree equation, $5x^4 + 3x^3 + 8x^2 = 260x + 5$, adding *quanta sit ipsa res, nondum habeo compertum*.[97]

II.3. In two cases Regiomontanus formulates problems on triangles which lead to cubic equations. One of them[98] is not solved by him. The other one is treated in more detail. The three sides are given. The problem is to find a division *BD* of the base *BC* so that $BD^2 + DA \cdot AB = AB^2$.[99] This leads to the cubic $x^3/324 + 540/29 = 3x$.[100] Regiomontanus gives no solution, but recognizes the similarity to the problem of finding the chord of 1° of arc when the chord of 3° is known (*Si dabitis lineam BD, dabo cordam unius gradus*). Thus he understood the correspondence of this problem with the trisection of the angle. It should be noted that Regiomontanus treated the Greek

---

[93] See W. Van Egmond, "The Algebra of Master Dardi of Pisa," *Historia Mathematica* 10 (1983): 399–421 at 417–418.

[94] For correct formulas to solve one particular class of cubic equations in Italian manuscripts of the fourteenth century, see R. Franci and L. Toti Rigatelli, "Fourteenth-century Italian algebra," in *Mathematics from Manuscript to Print, 1300–1600*, ed. Cynthia Hay (Oxford, 1988), pp. 11–29, especially pp. 19–20.

[95] Gino Arrighi, ed., *Piero della Francesca, Trattato d'abaco* (Pisa, 1970), p. 146.

[96] Bianchini's problem: Curtze (n. 28), p. 238,4; Regiomontanus' answer and comment: pp. 256, 280. See Kaunzner (n. 82), p. 138; Folkerts (n. 22), pp. 207f.

[97] Curtze (n. 28), p. 256; Kaunzner (n. 82), p. 138.

[98] Curtze (n. 28), p. 331, no. h.

[99] Figure in M. Folkerts, "Regiomontanus als Mathematiker," *Centaurus* 21 (1977): 214–245 at 228. The text of the problem is edited by Curtze (n. 28), p. 262, no. 17.

[100] See Cantor (n. 37), pp. 284f.

problem of the trisection of an angle in the same letter to Bianchini in 1464,[101] giving a proof which is similar to that in the *Verba filiorum* and in the *De triangulis* of Jordanus Nemorarius. As Clagett has shown, it seems more likely that Regiomontanus consulted Jordanus' treatise than that of the Banū Mūsā.[102]

II.4. Another problem in the correspondence is stated without further discussion: $x + y = 100$, $\sqrt{x} \cdot \sqrt[3]{y} = 25$.[103] We note that this is equivalent to a fifth-degree equation.

II.5. Although Regiomontanus did not possess the general solution of the cubic, he had an idea how one should solve it. A very similar idea was used in the first correct solution about 50 years later in Italy. In 1471 Regiomontanus wrote:

> Many flatter themselves that they understand the higher (*ampliorem*) algebra from the six standard forms. But they completely ignore the fact that this art cannot be extended to cubes or to fourth and higher powers, unless the geometry of solids of equal volume is first treated. Just as the three composed forms (of quadratic equations) are proved by means of figures of equal area, so the new extension of the art must be based upon the transformation of solids. I mention this to lighten, in part, my work for this task.[104]

III. Indeterminate Analysis

III.1. In the Plimpton collection, three problems (nos. 57–59) are on indeterminate analysis. Regiomontanus solves special cases of systems of simultaneous equations, two linear equations for three unknowns. He states, and proves geometrically, the conditions under which such problems are not soluble. For soluble problems he tries to give a general procedure to solve them.

III.2. In the correspondence he treats two indeterminate linear equations and one quadratic: $x + y + z = 240$, $97x + 56y + 3z =$

---

[101] Curtze (n. 28), pp. 258–259.

[102] Clagett (n. 38), vol. 3, pp. 352–353.

[103] Curtze (n. 28), p. 262, no. 20.

[104] Curtze (n. 28), p. 335: "Sunt enim, qui se iactant ampliorem habere artem algebricam, quam in sex capitulis vulgatissimis traditur. Sed ipsi profecto ignorant, hanc artem ad cubos, census censuum atque ulteriores potentias extendi non posse, nisi prius geometria solidorum equipollentium edatur. Quemadmodum enim tria capitula composita superficierum equipollentiis nituntur, ita novum artis additamentum ex commutatione solidorum hauriatur necesse est. Hoc ideo commemini, ut labor meus ad id negocium assumptus in parte levetur."

16047. This problem occurs twice in the correspondence.[105] The method of solution is not given, but when Regiomontanus mentions it first, he writes the solution (114, 87, 39) in the margin.[106]

III.3. $x + y + z = 214$, $x^2 - y^2 = y^2 - z^2$.[107] No solution is given.

III.4. $x + y + z = 116$, $x^2 + y^2 + z^2 = 4624$.[108] No solution is given.

III.5. Among the indeterminate problems the most important is the remainder problem which is treated as item 60 in the Plimpton collection. It gives the Chinese *ta yen* method in great detail. From this we see that Regiomontanus knew not only special solutions, but also the general method of solving the remainder problem. We know that Leonardo had already discussed a similar problem,[109] but later Fridericus Gerhart treats exactly the same problem with the same auxiliary numbers.[110] In this case—and it is true for others—Regiomontanus is the go-between in the transmission of the new mathematical knowledge from the Italians to the South German *cossists*. It should be mentioned here that Regiomontanus sets problems on the remainder problem in two places in his correspondence. The first runs: $N \equiv 15$ mod. $17 \equiv 11$ mod. $13 \equiv 3$ mod. 10.[111] Bianchini gives two special solutions,[112] but Regiomontanus is not content with them, pointing out that there is an infinite number of solutions and showing how to find them.[113] From here it is not clear that Regiomontanus had a general solution to such problems. The second problem of this kind is stated in the correspondence without proof: $N \equiv 12$ mod. $23 \equiv 7$ mod. $17 \equiv 3$ mod. 10.[114]

---

105 Curtze (n. 28), p. 262, no. 18, and p. 296, no. 8.

106 Curtze (n. 28), p. 262, no. 18. Jacob of Speyer finds the same solution: p. 300; Regiomontanus confirms its correctness: p. 305.

107 Curtze (n. 28), p. 262, no. 19.

108 Curtze (n. 28), p. 262, no. 21.

109 B. Boncompagni, ed. *Scritti di Leonardo Pisano*, vol. 1: *Il Liber abbaci* (Rome, 1857), p. 304. See U. Libbrecht, *Chinese Mathematics in the Thirteenth Century. The Shu-shu chiu-chang of Ch'in Chiu-shao* (Cambridge, Mass./London, 1973), pp. 213-413. Libbrecht did not know the Plimpton text.

110 Munich, Clm 14908, f. 124v–125r; edited by M. Curtze, "Ein Beitrag zur Geschichte der Algebra in Deutschland im fünfzehnten Jahrhundert," *Abhandlungen zur Geschichte der Mathematik* 7 (= *Supplement zur Zeitschrift für Mathematik und Physik* 40) (1895): 31–74 at 65–66.

111 Curtze (n. 28), p. 219, no. 8.

112 $N = 1103$ or 3313; Curtze (n. 28), p. 237.

113 $N_1 = 1103$; $N_{i+1} = N_i + 17 \cdot 10 \cdot 13$; Curtze (n. 28), p. 254.

114 Curtze (n. 28), p. 295, no. 6.

IV. Theory of Numbers

IV.1. Regiomontanus was also interested in the theory of numbers. In his Vienna *Rechenbuch* Regiomontanus calculates the fifth perfect number $(2^{13} - 1) \cdot 2^{12}$ and gives the formula $(2^{17} - 1) \cdot 2^{16}$ for the sixth perfect number—in the tradition of Euclid's IX.36—without calculating it.[115] Up to now it has been assumed that Regiomontanus was the first in the West to determine these numbers. But it is not so: in Bessarion's manuscript Venice, Biblioteca Nazionale Marciana, f.a. 332, of the thirteenth century, there is a short text on the first five perfect numbers.[116] The Regiomontanus text is almost identical with that in the Venice manuscript. Therefore we may assume that he made his copy from it.

In his correspondence he gives the following problems, without solving them:

IV.2. To find four squares whose sum is a square.[117]

IV.3. To find twenty squares whose sum is a square and of which the smallest is greater than 30,000.[118]

IV.4. To find three squares that are in harmonic proportion.[119]

IV.5. To find three numbers which are in harmonic progression and of which the smallest is greater than 500,000.[120]

IV.6. To find three squares which are in arithmetic progression and of which the smallest is greater than 20,000.[121]

It is not clear whether Regiomontanus could solve these problems. In three cases his correspondents gave a special solution, but Regiomontanus demanded general solutions.[122]

---

115 Vienna 5203, f. 167r. Edited by M. Curtze, "Eine Studienreise," *Centralblatt für Bibliothekswesen* 16 (1899): 257–306 at 288f.

116 On folios 36v–37v.

117 Curtze (n. 28), p. 296, no. 7. Answer by Jacob of Speyer: 1, 4, 16, 100 or 4, 16, 49, 100: Curtze, p. 300. Regiomontanus' comment: Curtze, p. 304.

118 Curtze (n. 28), p. 334, no. ε.

119 Curtze (n. 28), p. 296, no. 10.

120 Curtze (n. 28), p. 334, no. δ.

121 Curtze (n. 28), p. 334, no. γ.

122 Curtze (n. 28), p. 304: "Difficulter tamen quispiam decem huiusmodi societates numerorum quadratorum inveniet, ... nisi artem id parandi calleat, quam ego petivi."

V. Square Root Approximation

V.1. In the Vienna *Rechenbuch* there is a passage on square root approximation.[123] It gives the value common in Greek writings—which, indeed, goes back to the Babylonians—

$$\sqrt{n} = \sqrt{a^2 + r} = a + \frac{r}{2a} = n_1$$

and then, as a better approximation,

$$n_2 = \frac{n}{n_1} : 2 .$$

This method can be found in Hero's writings and later in Arabic texts.[124]

V.2. Somewhat different is the formula in the Plimpton collection (problem 56). It gives the common approximation

$$\sqrt{n} = \sqrt{a^2 + r} = a + \frac{r}{2a} = n_1$$

and also a better one, which can be expressed by

$$n_2 = a + \frac{4a^2 + 2r - 1}{(4a^2 + 2r) \cdot 2a} .$$

It is not clear where the second approximation comes from.

VI. Geometrical Problems

There are numerous geometrical problems in Plimpton 188 and in the correspondence. Sometimes Regiomontanus solves problems on triangles both geometrically and algebraically (cf. my remarks on quadratic and cubic equations). In the correspondence we also find problems about a circle quadrangle and two maximum problems:

---

[123] Vienna 5203, f. 167v–168r. See M. Curtze, "Die Quadratwurzelformel des Heron bei den Arabern und bei Regiomontan und damit Zusammenhängendes," *Zeitschrift für Mathematik und Physik, hist.-lit. Abt.* 42 (1897): 145–156.

[124] As to the history of this method, see J. Tropfke, *Geschichte der Elementar-Mathematik in systematischer Darstellung*, vol. 2, 3rd ed. (Berlin/Leipzig, 1933), pp. 170–176; 4th ed. (Berlin, 1980), pp. 263–268, 277.

VI.1. To determine the area of a circle quadrangle when the diameter and the ratio of the four sides to each other are given.[125] Regiomontanus rejects Bianchini's opinion that the problem is insoluble[126] and suggests a method of solution using Ptolemy's theorem.[127]

VI.2. Maximum problem: A 10-foot-long stick hangs vertically, so that it is 4 foot from the floor beneath it. To find the point from which the stick subtends the greatest angle.[128] No solution is given.[129]

VI.3. To inscribe the greatest possible square in a given triangle.[130]

How did it come about that Regiomontanus, who was one of the few humanists actively interested in mathematics and astronomy, could achieve so much in his short life? First, Vienna University, which became under the influence of John of Gmunden and Peurbach a center for the mathematical sciences, gave him as a young man the opportunity to broaden his knowledge of mathematical subjects. Secondly, through his contact with Bessarion he was able to improve his knowledge of Greek and had at his disposal the considerable library that the cardinal had accumulated. He was thereby introduced into the circle of Italian humanists with an interest in mathematics and astronomy.

Regiomontanus made full use of the chance presented to him: he used Latin translations from the Arabic and redactions of them together with Greek manuscripts to elucidate what he thought was the true meaning of Greek mathematical texts. Reconstruction of original texts as such was not his aim, but only the restoration of the authors' meaning. Further, on the basis of such texts he went on to

---

[125] The problem is given twice: Curtze (n. 28), p. 219, no. 3 (1464); and Curtze, p. 331, no. i (1471). Another problem given by Bianchini is Curtze, p. 236 (Regiomontanus' remarks on p. 253).

[126] Curtze (n. 28), pp. 235f.

[127] Curtze (n. 28), pp. 245–252.

[128] Curtze (n. 28), p. 333, no. α.

[129] Modern suggestions of how the problem could be solved are given by A. Lorsch, "Ueber eine Maximumaufgabe," *Zeitschrift für Mathematik und Physik, hist.-litter. Abth.* 23 (1878): 120 [also given by Cantor (n. 37), pp. 283f.]; and by S. Günther, *Geschichte der Mathematik*, 1. Teil (Leipzig, 1908), pp. 302f.

[130] Curtze (n. 28), p. 333, no. β.

derive his own results. For this purpose he made full use of the works of medieval and contemporary writers, such as Jordanus and Peurbach. On the other hand, he had no time for scholastic philosophy. Like a few other humanists, he made a determined effort to publish the ancient mathematical classics, but, unique among his contemporaries, he took the practical step of installing a printing press at his own expense in his house. Only his early death prevented his achieving his programme. But many sixteenth-century editions of mathematical and astronomical writings are based on his preliminary work.

His role in the transmission of mathematics is not confined to this editing activity. We have noted that he introduced the results of the Italian *maestri d'abbaco* to the German algebraists, and generally to the Latin-speaking world, and added results of his own. A similar claim could be made for his geometry. That he was one of the earlier transformers of the Greek heritage may be seen not only in his use of symbolism to explain arguments in the classical texts, but also in the more workaday business of calculating sine-tables, to say nothing of his monumental *De triangulis*.

Bibliography

Apollonius. See Heiberg.

Archimedes. *Opera quae quidem exstant omnia*. Basel: I. Herwagen, 1544.

Arrighi, Gino, ed. *Piero della Francesca, Trattato d'abaco*. Pisa: Domus Galilaeana, 1970.

al-Battānī. See *Rudimenta*.

Björnbo, A. A. *Studien über Menelaos' Sphärik. Beiträge zur Geschichte der Sphärik und Trigonometrie der Griechen*. Abhandlungen zur Geschichte der mathematischen Wissenschaften mit Einschluss ihrer Anwendungen 14. Leipzig: Teubner, 1902.

Boncompagni, B., ed. *Scritti di Leonardo Pisano*. Volume 1: *Il Liber abbaci*. Rome: Tipografia delle scienze matematiche e fisiche, 1857.

Busard, H. L. L., ed. *Der Traktat De sinibus, chordis et arcubus von Johannes von Gmunden*. Österreichische Akademie der Wissenschaften, math.-nat. Klasse, Denkschriften, 116 Bd., 3. Abh., pp. 73–113. Vienna: Springer, 1971.

——. "Die Traktate *De Proportionibus* von Jordanus Nemorarius und Campanus." *Centaurus* 15 (1971): 193–227.

——, ed. "Der Traktat *De isoperimetris*, der unmittelbar aus dem Griechischen ins Lateinische übersetzt worden ist." *Mediaeval Studies* 42 (1980): 61–88.

——. *Jordanus de Nemore, De Elementis Arithmetice Artis. A Medieval Treatise on Number Theory*. Stuttgart: Steiner, 1991.

Cantor, M. *Vorlesungen über Geschichte der Mathematik*. 4 vols. 2nd ed. Leipzig: Teubner, 1894–1908.

Clagett, Marshall. "The Medieval Latin Translations from the Arabic of the *Elements* of Euclid, with Special Emphasis on the Versions of Adelard of Bath." *Isis* 44 (1953): 16–42.

——. "King Alfred and the *Elements* of Euclid." *Isis* 45 (1954): 269–277.

——. *Archimedes in the Middle Ages.* Vol. 1. Madison: University of Wisconsin Press, 1964; vols. 2–5. Philadelphia: American Philosophical Society, 1976–1984.

Curtze, Maximilian. "Ein Beitrag zur Geschichte der Algebra in Deutschland im fünfzehnten Jahrhundert." *Abhandlungen zur Geschichte der Mathematik* 7 (= *Supplement zur Zeitschrift für Mathematik und Physik* 40) (1895): 31–74.

——. "Die Quadratwurzelformel des Heron bei den Arabern und bei Regiomontan und damit Zusammenhängendes." *Zeitschrift für Mathematik und Physik, hist.-lit. Abt.* 42 (1897): 145–156.

——. "Eine Studienreise." *Centralblatt für Bibliothekswesen* 16 (1899): 257–306.

——, ed. *Urkunden zur Geschichte der Mathematik im Mittelalter und der Renaissance.* Volume 2: *Der Briefwechsel Regiomontan's mit Giovanni Bianchini, Jacob von Speier und Christian Roder*, pp. 185–336. Abhandlungen zur Geschichte der mathematischen Wissenschaften 12. Leipzig: Teubner, 1902.

Dijksterhuis, E. J. *Archimedes.* Copenhagen: Ejnar Munksgaard, 1956.

Doppelmayr, Johann Gabriel. *Historische Nachricht von den Nürnbergischen Mathematicis und Künstlern.* Nuremberg: P. C. Monath, 1730.

Eneström, G. "Der 'Algorismus de integris' des Meisters Gernardus." *Bibliotheca Mathematica.* Ser. 3,13 (1912–1913): 289–332.

——. "Der 'Algorismus de minutiis' des Meisters Gernardus." *Bibliotheca Mathematica.* Ser. 3,14 (1913–1914): 99–149.

al-Farghānī. See *Rudimenta.*

Folkerts, Menso. "Regiomontans Euklidhandschriften." *Sudhoffs Archiv* 58 (1974): 149–164.

——. "Regiomontanus als Mathematiker." *Centaurus* 21 (1977): 214–245.

——. "Die mathematischen Studien Regiomontans in seiner Wiener Zeit." In *Regiomontanus-Studien*, edited by Günther Hamann, pp. 175–209. Österreichische Akademie der Wissenschaften, Phil.-Hist. Klasse, Sitzungsberichte 364. Vienna: Verlag der Österreichischen Akademie der Wissenschaften, 1980.

——. "Regiomontanus als Vermittler algebraischen Wissens." In *Mathemata. Festschrift für Helmuth Gericke*, edited by Menso Folkerts and Uta Lindgren, pp. 207–219. Stuttgart: Steiner, 1985.

Franci, R., and L. Toti Rigatelli. "Fourteenth-century Italian algebra." In *Mathematics from Manuscript to Print, 1300–1600*, edited by Cynthia Hay, pp. 11–29. Oxford: Clarendon Press, 1988.

Günther, Siegmund. "Lo sviluppo storico della teoria dei poligoni stellati nell' antichità e nel medio evo." *Bullettino Boncompagni* 6 (1873): 313–340.

——. *Geschichte der Mathematik.* 2 vols. in 3. Leipzig: G. J. Göschen, 1908–1921.

Heiberg, J. L., ed. *Apollonii Pergaei quae Graece exstant cum commentariis antiquis.* 2 vols. Leipzig: Teubner, 1891–1893.

Heller, Siegfried. *Ein Fehler in einer Archimedes-Ausgabe, seine Entstehung und seine Folgen.* Abhandlungen der Bayerischen Akademie der Wissenschaften, math.-nat. Klasse. N.F., Heft 63. Munich, Verlag der Bayerischen Akademie der Wissenschaften, 1954.

Hofmann, Joseph E. "Über Regiomontans und Buteons Stellungnahme zu Kreisnäherungen des Nikolaus von Kues." *Mitteilungen und Forschungsbeiträge der Cusanus-Gesellschaft* 6 (1967): 124–154.

l'Huillier, Ghislaine. "Regiomontanus et le *Quadripartitum numerorum* de Jean de Murs." *Révue d'histoire des sciences* 33 (1980): 193–214.

——. *Le* Quadripartitum numerorum *de Jean de Murs. Introduction et édition critique*. Geneva: Libraire Droz S. A., 1990.

Hultsch, F., ed. *Pappi Alexandrini Collectionis quae supersunt*. 3 vols. Berlin: Weidmann, 1876–1878.

Johannes de Muris. See l'Huillier.

Johannes de Regio Monte. *De triangulis omnimodis libri quinque*. Nuremberg: Johannes Petreius, 1533.

Kaunzner, Wolfgang. "Über Regiomontanus als Mathematiker." In *Regiomontanus-Studien*, edited by Günther Hamann, pp. 125–145. Österreichische Akademie der Wissenschaften, Phil.-Hist. Klasse, Sitzungsberichte 364. Vienna: Verlag der Österreichischen Akademie der Wissenschaften, 1980.

Knorr, Wilbur K. *Textual Studies in Ancient and Medieval Geometry*. Boston: Birkhäuser, 1989.

——. "Paraphrase Editions of Latin Mathematical Texts: *De figuris ysoperimetris*." *Mediaeval Studies* 52 (1990): 132–189.

Libbrecht, Ulrich. *Chinese Mathematics in the Thirteenth Century. The Shu-shu chiu-chang of Ch'in Chiu-shao*. Cambridge: MIT Press, 1973.

Lorch, Richard. "The Transmission of Theodosius' *Sphaerica*." In *Mathematische Probleme im Mittelalter - der lateinische und arabische Sprachbereich*, edited by M. Folkerts [in press].

Lorsch, A. "Ueber eine Maximumaufgabe." *Zeitschrift für Mathematik und Physik, hist.-litter. Abth.* 23 (1878): 120.

Peurbach, Georg von. *Algorismus*. 1492.

——. *Tractatus super propositiones Ptolemaei de sinibus et chordis*. Nuremberg: Johannes Petreius, 1541.

Rose, Paul Lawrence. *The Italian Renaissance of Mathematics: Studies on Humanists and Mathematicians from Petrarch to Galileo*. Geneva: Droz, 1975.

Rudimenta astronomica Alfragani. Item Albategnius astronomus peritissimus de motu stellarum.... Nuremberg: Johannes Petreius, 1537.

Schöner, Johannes, ed. *Algorithmus demonstratus*. Nuremberg: Johannes Petreius, 1534.

Tropfke, Johannes. *Geschichte der Elementar-Mathematik in systematischer Darstellung mit besonderer Berücksichtigung der Fachwörter*. Volume 2. 3rd ed. Berlin: W. De Gruyter, 1933; 4th ed. Berlin: W. De Gruyter, 1980.

Van Egmond, Warren. "The Algebra of Master Dardi of Pisa." *Historia Mathematica* 10 (1983): 399–421.

Zinner, Ernst. *Leben und Wirken des Johannes Müller von Königsberg genannt Regiomontanus*. Munich: C. H. Becksche Verlagsbuchhandlung, 1938. 2nd ed. Osnabrück: O. Zeller, 1968.

Addenda to VI

[pp. 94–96] Item VII of this volume is a more detailed study on Regiomontanus and Euclid.

[p. 107] The name "Fridericus Gerhart" must be replaced by "Fridericus Amann"; see item XII, p. 7.

VII

Regiomontanus' Approach to Euclid*

Although Johannes Regiomontanus was the most important European mathematician of the fifteenth century, his achievements have not yet been fully assessed. There are two reasons for this: his early death and the fate of his *Nachlaß*. Regiomontanus died at 1476 at the age of forty. He had founded a press in Nuremberg; and a one-page catalogue of the year 1474 has survived, containing a list of books, some of them his own, that he wished to print[1]. But his plan for printing these books was interrupted by his early death. Some of his own works were published posthumously; and some were never printed at all. His *Nachlaß* was scattered during the sixteenth century. We may guess what books had been in his possession from the rather inaccurate lists made some decades after his death. Ernst Zinner tried to identify the books in Regiomontanus' library with the entries in these lists[2]. Zinner's work shows that only a small part of Regiomontanus' *Nachlaß* has survived.

It is the purpose of this paper to try to determine Regiomontanus' knowledge of Euclid's *Elements* and how he built upon it. In particular, we shall gain some insight on the various versions of Euclid's text available to Regiomontanus, what manuscripts he had in his possession, and what he did with them.

1. Versions of Euclid in Regiomontanus' possession

Regiomontanus' first contact with Euclid was probably during his study in Vienna, before his baccalaureate in 1452, for this required knowledge of the first book of the *Elements*. In his time in Vienna Regiomontanus must have occupied himself more intensively with Euclid: the topic of one of his first lectures at Vienna university (winter semester 1460/61) was the first book of Euclid; also, in the so-called "Wiener Rechenbuch", which he wrote between 1454 and 1462, he mentions Euclid[3]. His knowledge of the *Elements* is clear in the *De triangulis*, which was essentially the work of

*Completely revised translation of the article "Regiomontans Euklidhandschriften", *Sudhoffs Archiv* 58 (1974), 149–164.

[1] Regiomontanus, 'Verlagsanzeige'.

[2] Zinner, 'Leben und Wirken', pp. 297–363.

[3] E.g. on f. 131v of this "Rechenbuch" (now Vienna, ÖNB, Ms. 5203).

years 1462–1464[4], in his correspondence, especially in a letter to Giovanni Bianchini at the beginning of 1464[5], and in the above-mentioned list of intended publications. For in this list we find

> *Euclidis elementa cum anaphoricis*[6] *Hypsiclis editione Campani evulsis tamen plerisque mendis, quae proprio etiam indicabuntur commentariolo.*

and

> *Commentariolum quo commonstrantur placita Campani ex editione elementorum geometricorum reiicienda.*

Thus Regiomontanus intended to establish a purified Euclidean text in Campanus' version and also to print a commentary to show that Campanus' additions should be removed. He died before he could achieve this.

Further information about the Euclid texts in Regiomontanus' possession may be gained from the lists of his *Nachlaß*. For several decades after his death his library was kept together, though individual items disappeared and books were sometimes added to the collection. At first the library went *en bloc* into the possession of the Nuremberg patrician Bernhard Walther, who died in 1504. When Walther's books had to be sold, a list of them was drawn up, in 6 Dec. 1512, comprising 199 items[7]. This list mentions Euclid four times[8]:

> *Euclides elementa*
> *Euclidis*
> *Euclides*
> *Alardus In elementa Euclid.*

But the sale did not take place. On 1 Oct. 1522 another attempt was made to sell the library and another list was drawn up, this time with 215 items, among them 11 printed books[9]. In general, the second list is more detailed than the first. The following editions of Euclid were mentioned[10]:

> *Euklides. Scriptus in bapiro et parva forma*
> *Euklides alius. Scriptus in bapiro et parva forma*

---

[4] *De triangulis omnimodis libri quinque*, edited by J. Schöner in Nuremberg in 1533.

[5] This letter was edited by Curtze, 'Briefwechsel', pp. 242–266.

[6] Almost certainly the so-called books 14 and 15 of the *Elements* are meant rather than Hypsicles' astronomical work *Anaphorikos*.

[7] See Zinner, 'Leben und Wirken', pp. 248 and 261.

[8] Zinner, 'Leben und Wirken', p. 316.

[9] See Zinner, 'Leben und Wirken', pp. 250 and 261. The list is published in Petz, 'Urkundliche Nachrichten', pp. 247–262.

[10] Zinner, 'Leben und Wirken', p. 316.

Euclides. Impressus
Alhardus super auxiomata Euclidis. In pergameno scriptus
Campanus in elementa Euclidis.

The third list itemizes the books that were in the library in 1 June 1563[11]. It contains only 148 items of which 30 were printed books. Almost always there is exact information about the format and binding of the books and indication of whether they are printed books or manuscripts. There are five editions of Euclid in this list:

21. *Elementa Euclidis geschriben Vnnd gepünden in 4°.*
22. *Euclides latinus geschriben Vnnd gepünden in 4°.*
23. *Euclides latine gedrückt in folio. cum theoricis Purbachij.*
24. *Euclides latine gedrückt in folio. Vngepünden.*
26. *In elementa Euclidis Campanus geschriben Vnnd in perment gehefft in 4°.*

The printed editions of Euclid cannot have come from Regiomontanus' library, since they were printed in 1482 or later. If these are left aside, there remain four manuscripts – which Bernhard Walther, and presumably Regiomontanus before him, had possessed. One of these is further characterized by the name *Alardus.* The other three, of two of which it is explicitly mentioned that they were paper manuscripts, were very probably of the Campanus text[12].

Another piece of information may be gained from the *Nachlaß* of the Altdorf professor Petrus Saxonius (d. 1625). According to a list of this *Nachlaß* Saxonius possessed a manuscript of Euclid once in the possession of Regiomontanus, characterized as follows[13]:

Euclidis Elementorum libri XV in membrana latine manu Johannis Regiomontani. Est antiquissima translatio succinctis constans demonstrationibus, Authoris Adelhardi aut Alfredi quorum meminit Regiomont. in Oratione sua habita Patavij, praefixa Alfragano. Diss ist ein schene Antiquitet[14]*, fuit etiam Alberti Düreri, ad quem ex bibliotheca Regiomontani devenit.*

---

[11] See Zinner, 'Leben und Wirken', pp. 250f.; the list is published ibid., pp. 252–258.

[12] In the fifteenth century the Campanus text was the dominant version of Euclid's *Elements.* A Latin manuscript described as being of Euclid without closer specification would normally mean this version. For the transmission of Euclid in the Middle Ages, see Folkerts, 'Euclid'. Zinner's assertion that Regiomontanus possessed six Euclid manuscripts (Zinner, 'Leben und Wirken', p. 115) is unfounded.

[13] See Zinner, 'Leben und Wirken', p. 316.

[14] "This is a fine old piece".

Accordingly, it is a copy of Euclid made by Regiomontanus and labelled *authoris Adelhardi aut Alfredi*; the famous German artist Albrecht Dürer bought it some decades after Regiomontanus' death and it was later in the possession of Saxonius.

In the city archive of Nuremberg there is documentary evidence that Dürer bought ten books from Walther's *Nachlaß* at the beginning of 1523[15]. Clearly, one of these was the above-mentioned Euclid manuscript. This accords well with the fact that in the lists of 1512 and 1522 *Al(h)ardus* is mentioned, but in the list of 1563 it is not there.

This manuscript, here designated as **A**, has been lost. However, we can deduce which tradition it belonged to: the two names *Adelhardus aut Alfredus* immediately connect Dürer's manuscript with the manuscript Venice, Bibl. Marciana, fondo antico 332 (**V**) of the thirteenth century. **V** contains two Latin versions of Euclid, the so-called "Version II" (earlier known as 'Adelard II'), and the so-called "Version III" (earlier known as 'Adelard III')[16], and the proofs of Version II are attributed in the beginning to *Alfredus* and from book 4 they were ascribed to *Adelhardus.* Manuscript **V** belonged to Cardinal Bessarion with whom Regiomontanus was in close contact since 1460 – he followed Bessarion to Italy in 1461. It is clear that manuscript **A** is a copy of **V**, probably made by Regiomontanus himself.

As mentioned above, the other three Euclid manuscripts from Regiomontanus' library contained most probably the Campanus text. This text is known in well over 100 manuscripts[17]. Two of them (Nuremberg, Stadtbibliothek, Cent. VI. 13, and Boston, Medical Library, Ms. 24) are connected with Regiomontanus.

2. The manuscripts N (Nuremberg, Cent. VI. 13) and B (Boston, Medical Library, Ms. 24)

Manuscript Nuremberg, Stadtbibliothek, Cent. VI. 13 (**N**) has been known for some time, but has been misinterpreted[18]. It comprises 162 paper folios and is bound in a brown Renaissance binding. At the top of f. 1r is *IHC*, the sign for Jesus which Regiomontanus characteristically wrote on his books as his owner's mark. Underneath this mark is,

[15] Zinner, 'Leben und Wirken', p. 259.

[16] For these versions see Folkerts, 'Euclid', pp. 14–18 and 19–21.

[17] See the list in Folkerts, 'Euclid', pp. 48–52.

[18] Cantor, 'Vorlesungen', p. 277, and Heiberg, 'Euclidis opera omnia', p. CI, call it an Adelard manuscript and Zinner ascribes it in one place to Gerard of Cremona and in another to Adelard (Zinner, 'Leben und Wirken', pp. 75 and 316).

in the hand of J. Schöner, *In Elementa Euclidis praefatio Joh. de Regiomonte* αὐτου. The manuscript contains all 15 books of the *Elements*. The greater part (from f. 14v, the end of Euclid III.8) was copied in a humanistic hand by an unknown copyist. The beginning of the codex to f. 14r was written by Regiomontanus himself. In the latter part to the end of book 7 (f. 64v) there are numerous marginal notes and interlinear glosses in Regiomontanus' writing.

The part not written by Regiomontanus contains the normal Campanus text. Thus **N** is one of the three Campanus manuscripts that Regiomontanus possessed. Much more interesting is the beginning of the codex written by Regiomontanus himself: it is not a copy of any other known manuscript. In Section 3 we give some details of it.

The codex begins (f. 1r–3r) with a chapter on the importance of mathematics and its various parts. This chapter was published by J. Schöner in 1537 as a work of Regiomontanus[19]. Parts of it were taken over by Regiomontanus in his introduction to his lecture on Alfraganus held at the University of Padua in 1464. This introduction described the development of mathematical sciences and has been considered as one of the first essays in the history of mathematics[20]. In fact the chapter appears in some manuscripts as an introduction to Version III of the *Elements*[21] – of which manuscript **V** is one. A comparison of **N**, **V** and the edition of 1537 (**Ed**) shows first that **Ed** is immediately dependent on **N**[22]. This is no surprise, since Schöner had access to Regiomontanus' manuscripts. Secondly there are many mistakes common to **NEd** that are not in **V**, but no mistakes in **V** not in **NEd**. Hence either **N** was copied from **V** or from a copy of **V**. There are reasons to suppose the second possibility and that the copy of **V** is the lost manuscript **A**, which Regiomontanus himself copied from **V**.

Directly after this introduction in manuscript **N** comes a short text giving information on the content of the individual books of the *Elements*. In content, though not *verbatim*, this section corresponds to a scholium in **V** (f. 86v, bottom).

The main section of manuscript **N** contains a text of Euclid that is either the Campanus text or based on it. **V** cannot be the source for this

[19]Regiomontanus, 'Continentur', last item.

[20]Edited in Regiomontanus, 'Continentur', penultimate item.

[21]Edited by Busard, 'Johannes de Tinemue', pp. 31–33.

[22]Two mistakes in **Ed** may be explained by readings in **N**: *practicam* <u>*liberalem*</u> (i.e. *liberalem* written and deleted) **N**, *practicam liberalem* **Ed** (correct: *practicam*) and *cā* **N**, *causa* **Ed** (correct: *cetera*).

text, since **V** does not contain Campanus' text, but Versions II and III. We know that Regiomontanus possessed three Campanus manuscripts (presumably **N** and two others): no doubt one of these is the source for **N**. Unfortunately, no extant Campanus manuscript (except **N**) clearly shows that it belonged to Regiomontanus. However, there is a manuscript which must be closely related to a Campanus manuscript in Regiomontanus' possession: Boston Medical Library, Ms. 24 (**B**). This manuscript is in two parts: Johannes Merbolt de Varnpach's lecture on Aristotle's *De generatione et corruptione* and then Campanus' version of Euclid, books 1–9. The catalogue says that the manuscript was written in 1466[23].

On f. 110v–111r of **B** is the introduction sometimes found in Version III. As in **N** this is followed, on f. 111rv, by the short text describing the content of the books of the *Elements*. On f. 111v–147r follows the text of books 1–9 of Campanus' Euclid. **B** was not written by Regiomontanus and has no indication of his ownership.

The introduction and the short notice about the contents of the *Elements* in **B** agree fully with manuscript **N** up to unimportant orthographical variants. They are obviously very closely related – more closely than **N** and **V**.

After the introduction and contents summary the similarity ends: in **B** we have Campanus, but in **N** we have, first, Regiomontanus' own redaction and only then Campanus. In general **B** follows Campanus, albeit with variations[24], but we should note that in I.19 the text in **B** is different from Campanus and is very similar to the Regiomontanus text in **N**[25]. This example may indicate that Regiomontanus used for his reworking of Euclid in **N** either **B** or the manuscript from which it was copied. Perhaps we may speculate that Regiomontanus himself copied the manuscript from which **B** was copied, and in doing this he took over the introduction that he found in **V**, adding an expanded version of the gloss he found concerning the contents of the *Elements*. If this is so, he copied the "normal" Euclid text, i.e. that of Campanus; when he was writing **N**, he started replacing Campanus' text by his own.

---

[23] De Ricci, 'Census', p. 916 (no. 25); Ballard, 'Catalogue', p. 17 (no. 24). It is not clear where the date 1466 comes from. The date does not appear in the Euclid section of the codex. We note that the copy of Johannes Merbolt's lecture is also dated in Ballard's catalogue to 1466 (p. 22, no. 36).

[24] By "Campanus" I mean the text edited by Busard from early manuscripts and from the edition of 1482 (Busard, 'Campanus').

[25] **B** ends with: ... *contra ypothesym vel erit minor C. tunc per premissam infertur BA esse maius. quod est contra ypothesym quare habetur propositum.*

Accordingly, we may make some strong suggestions about when **N** was written: Regiomontanus was introduced to Bessarion when the latter visited Vienna in 1460; earlier access to manuscript **V** is highly improbable. On the other hand, Regiomontanus was certainly acquainted with the introduction found in **V** at the latest in 1464, as the two names "Adelard" and "Alfred" in the lecture on Alfraganus held in Padua in this year prove. He therefore seems to have written manuscript **N** (or at least the manuscript from which **B** was copied) in his years in Italy when he was in Bessarion's entourage and was able to use his library.

3. Regiomontanus' attempt to establish a correct text of Euclid

What Regiomontanus did in order to establish a better text of the *Elements* we know principally from the part of manuscript **N** written by Regiomontanus himself, i.e. from the text up to III.8, and from his additions in the later books in his own hand.

In the text from the beginning to III.8 the definitions, axioms, postulates and enunciations correspond *verbatim* to the Campanus text. But the proofs differ radically from Campanus; in particular, they are much shorter. They embody the same mathematical content, but sometimes consider more cases. For instance, in I.2 Euclid shows how to construct a line equal to a given line from a given point. Euclid, followed by Campanus, gives a construction and a detailed proof; Regiomontanus does the same, only much more concisely, for the case that the point is not on the line (silently assumed by Euclid), and then adds two special cases (in which the point is on the line, either an endpoint or not). These special cases are considered by neither Euclid nor Campanus. Proclus considers three special cases, but it is unlikely that he influenced Regiomontanus. Regiomontanus' proof of the case considered by Euclid runs[26]:

> Datus punctus sit A, linea data $B\ C$. Aut punctus est in linea $B\ C$ aut extra. Sit primo extra. Coniungam A cum C per primam peticionem, et per premissam super $A\ C$ constituatur triangulus yssopleurus $A\ C\ D$, et per 2^{am} peticionem super C fiat circulus occupando $C\ B$, et $D\ C$ continuetur in E occursum circuli, et secundum spacium $D\ E$ super D fiat circulus, in cuius occursum $D\ A$ continuetur in F. Dico $A\ F$ equalem $B\ C$. Patet ex diffinicione circuli et prima et tercia concepcionibus.

The reader may compare this text with that of Campanus[27]. We note that Regiomontanus refers explicitly to the postulates, axioms (both by

[26]Manuscript **N**, f. 4v.

[27]See Busard, 'Campanus', pp. 60f.

number) and the definition of the circle as well as the previous proposition, whereas Campanus mentions only the proposition. Later proofs are similarly treated.

It should be noted that the extra section on star polygons introduced by Campanus after I.32 is in Regiomontanus' text somewhat differently treated and is on the whole shorter[28].

In his reworking of the *Elements* Regiomontanus made use not only of Campanus' text, but also of Version II. There are traces of this version *inter alia* in some proofs of book 1. Thus, the proof for I.1 is identical to that in Version II with the addition of references to the relevant postulates and axioms; also I.32–38, particularly I.33, show great similarities to this text. Nonetheless the reworking was based on Campanus. The manuscript of Version II that Regiomontanus used was probably codex **A**.

On the last two folios (f. 13r–14r) Regiomontanus suddenly changes his approach: from III.5 on, his text agrees *verbatim* with Campanus. At the bottom of f. 14r is the end of the text from Regiomontanus' hand, in the middle of proposition III.8; the rest of this proposition and the continuation of the Campanus text is in another hand, which starts at the top of f. 14v. Since the new hand begins in the middle of a theorem, Regiomontanus cannot have added the beginning of the Euclid text later. Rather in writing the first 12 folios he brought the text of the *Elements* to a form that he wanted; but after this he simply copied Campanus, a task that was later taken over by another hand. The reason for this change of purpose we can only speculate about. Perhaps Regiomontanus had at the beginning the intention of writing an improved text of Euclid on the basis of Campanus' version, a text that was later to be printed. For some reason, perhaps lack of time, he changed his plan when writing book 3.

Regiomontanus clearly read through the text written by the second hand, for there are alterations and marginal notes from f. 14v up to the end of book 7 on f. 64v. He inserted all the figures and supplied the titles of the books. But even here he mostly limited himself to correcting faultily written words (sometimes *esset* into *erit*) and, in the wording of Campanus, to filling lacunae left by the scribe. Most of the rest are short remarks like that to I.2 (f. 4v) or the words *tunica francisci* over the figure for Pythagoras' theorem I.46 C[ampanus] = I.47 H[eiberg]. At the end of the proof by *reductio ad absurdum* in III.18 C. = III.19 H. he added:

---

[28]Regiomontanus' text was edited by Siegmund Günther (Günther, 'Sviluppo', pp. 332–338) and paraphrased by Moritz Cantor (Cantor, 'Vorlesungen', pp. 277–279).

fieret enim angulus rectus minor recto (f. 16v). Campanus' conclusion of the side of the square from its area in III.35 C. = III.36 H. gave rise to Regiomontanus' marginal note *Quarum quadrata sunt equalia, lineas rectas equales esse concipio* [?] (f. 22r). In VI.1 Regiomontanus explains the term used for parallelogram (*rectilinee superficies equedistancium laterum*) by *que sunt inter lineas equedistantes, cum conveniunt* [?] *supra eandem lineam quantamlibet etc.* (f. 44v). In seven places Regiomontanus comments on Campanus' text, with one exception (f. 44r, see below) very briefly. Thus we find *hic vide, an demonstrat* against V.25 (f. 42v) and *non demonstrat hic campanus* against V.34 C. (not in H.) (f. 44v). Regiomontanus remarks that the nine propositions after V.25 are additions by Campanus with the words *hee sequentes sunt addite* (f. 42v); no doubt he deduced this from a comparison with another text, probably his Version II manuscript, which did not contain these additions. Beside VII.3 Regiomontanus wrote in the margin *premittit notabile* (f. 55r), beside VII.11 *Addita a Campano* (f. 57r) and beside VII.19b *Campani est hec* (f. 59v)[29]. Even Nicholas of Cusa is mentioned in one place: next to the note *Sed infinitae sunt proportiones irrationales, quarum denominatio scibilis non est* in Campanus' explanation of V. def. 16 C. = 17 H. Regiomontanus adds, perhaps ironically: *ex cusano pulcher*[30].

In the part of the manuscript that Regiomontanus looked through but did not write (i.e., f. 14v–64v) he made eight substantial additions of his own[31]. The first two refer to III.12 C. = III.13 H.: no. 1 contains a new, elegant proof of the first part of the theorem, namely that it is impossible that two circles touch each other internally at more than one point; no. 2 shows that they also cannot touch externally more than once. Text no. 3 is concerned with Euclid III.30 C. = III.31 H.; it shows how the theorem about the angle in a semicircle makes it possible to draw a perpendicular from a given point to a given line[32]. The fourth addition is written at the bottom of the page on which V. def. 9–11 C. = 8–10 H. are written, although there is no recognizable connection with this part of Euclid's text. It is a *verbatim* citation of a passage of a theological work by Boethius, his *Quomodo substantiae in eo quod sint bonae sint cum non sint substantialia bona*[33]. Boethius here gives

[29]The last two propositions are missing in the Greek text and in Version II.

[30]We know that Regiomontanus criticised in detail the attempts by Cusanus to square the circle; see Hofmann, 'Stellungnahme'.

[31]These additions are edited in the appendix.

[32]Regiomontanus' procedure is mentioned in Cantor, 'Vorlesungen', p. 280.

[33]Edited in Peiper, 'Philosophiae consolationis libri quinque', p. 169, 17–25.

an explanation of the term *communes animi conceptiones* and its two possible meanings. No. 5 belongs to V. def. 10/11 C. = 9/10 H. It is an attempt to explain the expression *proporcio duplicata* or *triplicata*. It seems that Regiomontanus' text is based upon a passage in manuscript **V**[34]. No. 6 appears to be an explanation of a part of the procedure in the proof of V.4. In no. 7 Regiomontanus points to a mistake in the text of V.34 in Campanus' numbering (this is one of the propositions that Campanus added to Euclid): from $a : d > b : e > c : f$ it follows that $a : d > (a + b + c) : (d + e + f) > (b + c) : (e + f)$, but not $(a + b + c) : (d + e + f) > b : e$. Regiomontanus demonstrates this by means of the counter example $(a, b, c) = (8, 4, 18)$; $(d, e, f) = (3, 2, 15)$. He accordingly changed the text, saying that he did not believe that Campanus had written the words *quam alicuius subsequencium ad suam comparem*, as can be understood from his method of proof. Clearly, Regiomontanus had been at some pains to understand this rather difficult part of the text. No. 8 is attached to VI.17 C. = VI.19 H. In the proof of the theorem that similar triangles are in duplicate proportion to their sides Campanus mentions the trivial case of two congruent triangles and remarks *equalis autem proporcio componitur ex equali duplicata vel triplicata vel quotienscumque sumpta*. Regiomontanus takes this as a reason for the remark that in this case the ratio is equal to its duplicate.

There are no further remarks or improvements worth recording in Regiomontanus' hand. Shortly before the end of book 7, at the end of f. 64v, Regiomontanus' notes stop suddenly. In the rest of the codex there are no traces of his reading the text; even the diagrams and book titles are missing.

To sum up, we can say that the beginning of manuscript **N**, written by Regiomontanus, was probably preliminary work for his planned edition of a purified text of Euclid on the basis of Campanus' version. Besides his copy of Campanus, Regiomontanus also used the text of the so-called Version II which was available to him in Bessarion's manuscript **V**. This procedure, with the help of various texts to establish a mathematically correct text (not to be understood in modern text-critical sense of a reconstruction of the original ancient text) is typical for Regiomontanus[35]. However, there are only a few attempts in **N** at a critique of the texts of the manuscripts he used. It appears that Regiomontanus at this time,

---

[34] In **V**, f. 292–293, there is an anonymous text *De proportione* beginning with *Volo ostendere quod omnium trium linearum proportionalium.* This text was edited from other manuscripts by Clagett, 'Archimedes', pp. 601–603.

[35] See Folkerts, 'Transmission', p. 98.

i.e. the beginning of the 1460s, had not yet studied Campanus' reworking at all intensively and that he in general agreed with it. He had not yet noticed the fundamental misunderstandings evident in Campanus' version of the definitions in book 5; only in one place did he criticise Campanus, and that very mildly. Even in his lecture given in Padua in 1464 Regiomontanus said that Campanus had reproduced the 15 books of Euclid very lucidly[36]. Later, apparently at the end of the 60s or the beginning of the 70s, Regiomontanus must have been seriously occupied with Euclid's text. Whether he used for this purpose Greek manuscripts, which were available to him in Italy[37], is not known. However this may be, he radically changed his originally positive opinion of the Campanus text and made severe criticisms of it. His critical remarks on Campanus, which he intended for publication, are known today only from a letter to Christian Roder dated 4 July, 1471. In this letter he expressed himself very clearly[38]:

> Et ut ab ipso quasi vestibulo geometrie quedam promantur libamenta, quid dices de principio, quod de recta quavis linea duabus incidente assumitur? Nonne illud longe obscurius et ab intellectu remotius est quam vigesima nona primi elementorum Euclidis conclusio, cuius gratia principium illud premittitur? Quem scrupulum et Campanus animadvertens hoc principium inter petitiones stolide collocavit, quamvis Greci inter communes scientias ordinarint. Sed Arabes nonnulli a ministerio demonstrationis penitus reiecerunt hoc proloquium, aliter quidem equidistantes lineas diffiniendo. Pudet profecto recensere labores Campani, quibus frustra stabilire tentat principia quinti elementorum voluminis, quippe qui diuturno studio ac magno verborum involucro irretitus postremoque fessus atque desperabundus eadem semper recidiva laborat, idem scilicet per seipsum diffiniendo. Nam quid tam [?] aliud pretendit diffinitio sexta secundum mentem eius, quam quod quantitates incontinue proportionales sunt, quarum prime et tertie equemultiplices itemque secunde et quarte equemultiplices, si conferantur, eodem

---

36 *Hinc Adelhardus et Alfredus, ac tandem Campanus eos quindecim libros tanquam solo Euclide editos retractare ausi sunt, illi quidem eleganter ac brevissime, iste vero lucidissime.*

37 We know that Bessarion had at least five Greek manuscripts of Euclid in his library in Italy; see Labowsky, 'Bessarion's Library', pp. 197, 222, 242 (items 98, 107, 109, 569, 1004).

38 The text is based on Curtze's edition (Curtze, 'Briefwechsel', pp. 328f.), but mistakes have been corrected by reference to the autograph (Nuremberg, Stadtbibliothek, Cent. V, app. 56c).

ordine, quo submultiplices proponuntur, incontinue proportionales sunt? Illud enim significat similitudo additionis aut minuitionis vel equationis, quam ipse insinuat. Undecime autem diffinitionis expositio quam ridicula sit, facile quispiam intelliget, qui proportionalitates irrationalis denominationis penitus exortas duplicare, triplicare aut quantumlibet multiplicare didicit. Porro principium decimi voluminis quomodo constabit, si sesquitertia proportio infinities etiam coacervata sesquialteram neque superare potest, neque etiam attingere? Quod si proportiones minoris equalitatis non sunt de eodem genere quantitatis, quomodo altere ex alteris componi poterunt? Si autem a genere quantitatis omnes prorsus excludende sunt proportiones, nulla omnino erit proportionum proportio, et ideo fabulas narrabunt, quicumque septimi physicorum regulas de velocitate motuum traditas subtiliter exponere conantur. Omnis etiam ferme calculationum in Italia celebratarum corruet argutia. Nugari videbuntur postremo quicumque de proportionum proportionibus disserere quidpiam tentant. Audire velim, qui primam undecimi demonstret, postquam secundam cum tercia decima primi, cuius succursu prima undecimi fulcire debet, lineas suas in eadem planicie supponat, tales autem consideratas lineas, que in figura prime sunt, in eodem esse plano non nisi per secundam constare potest.

His criticism was thus directed especially against the parallel postulate (wrongly), the definitions of book 5 and the beginning of book 11. His objections to the definition of the equality of two ratios (V. def. 6 C. = 5 H.) are justified, since Campanus had completely misunderstood the matter[39].

Regiomontanus died before the first publication (1482) of the Campanus version. A general criticism of Campanus' text began only after Zamberti's translation from the Greek (1505), but not before the second half of the sixteenth century did Campanus loose his importance.

Appendix: Regiomontanus' longer additions to the Campanus text in manuscript N

1) f. 15r, on III.12 C. = III.13 H., after *contrarium premisse*:

Si vero dices quod circuli sese contingant in K (ne cogeris dicere contrarium premisse), quia centra horum circulorum per 6^{tam} *huius sunt diversa, sint ipsa G et H. Ductis itaque lineis AG et AH, erit HK equalis*

[39]For a more detailed account of these criticisms, see Folkerts, 'Transmission', pp. 95f.

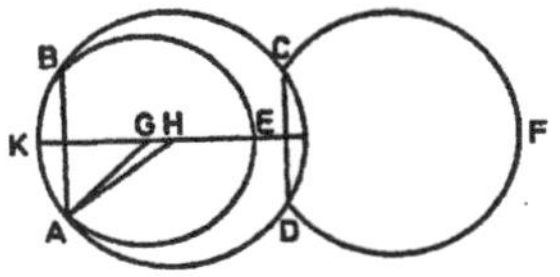

HA, quia ambe exeunt a centro circuli ad circumferenciam. Item GK equalis GA. Addita igitur eorum GH erunt due AG et GH equales HK. Quare eciam equales AH, quod est contra 20^{am} *primi. Ergo etc.*

2) f. 15r, on III.12 C. = III.13 H., after *contrarium 2*^{e} *huius*:

quia ipsa cadet extra circulum et tamen ducitur ab uno puncto circumferencie ad alium punctum eiusdem. Si autem diceres, quod caderet intra utrumque circulum, ergo utriusque circuli esset pars, quare sequitur, quod secabunt se et ideo non contingent se circuli, quod est contrarium ypothesi.

3) f. 19v, on III.30 C. = III.31 H.:

Hac propositione faciliter duces perpendicularem supra quamcumque lineam voles hoc modo: Signa punctum, a quo vis educere perpendicularem. Deinde posito pede circini extra lineam illam extende unum pedem ad punctum in linea signatum et describe circulum occultum, qui a proposita linea secabitur in duobus punctis. Ab uno illorum, a quo non vis educere perpendicularem, duc lineam rectam per centrum circuli, donec secabit circumferentiam, et punctum sectionis nota. Deinde ab eodem ad punctum primo signatum duc lineam rectam, que erit perpendicularis. Racio patet per hanc 30^{am}.

4) f. 31v, beneath V. def. 9–11 C. = 8–10 H.:

Boecius in libro de trinitate: Communis animi concepcio est enunciacio, quam quis probet auditam. Huius duplex est modus. Nam una ita communis est, ut omnium hominum sit, veluti si hanc proponas: Si de equalibus quantitatibus equalia auferas, que relinquuntur equalia esse. Nullus intelligens id negabit. Alia vero est doctorum tantum, que cum enunciabilibus communibus animi concepcionibus venit, ut est: que incorporalia sunt, in loco non esse, et omnia que vulgus non comprobat.

5) f. 32r, on V. def. 10–11 C. = 9–10 H.:

In proporcione continua proporcio primi ad 3^{m} *est proporcio primi ad 2*^{m} *duplicata. Ipsa enim, scilicet primi ad tercium, proporcio denominatur a quadrato denominatoris*[40] *proporcionis primi ad 2*^{m}. *Quadratum autem*

[40]Perhaps *denominacionis* is intended.

duabus lineis equalibus continetur. Sic in 4 terminis proporcio primi ad 4^m^ est proporcio primi ad 2^m^ triplicata. Denominatur enim a cubo denominatoris[41]*. Cubus autem tribus lineis equalibus continetur ad signandum* [?] *istum, quod fit ex ductu linee in se et iterum* [?] *in productum.*

6) f. 36r, on V.4:

Primo oportet argui per locum a ductu [?] *incontinue proporcionalitatis, 2° per locum a ducendo* [?], *ex quo 4 quantitates A B C D sunt incontinue proporcionales, et K L M N sunt ipsarum equemultiplices.*

7) f. 44r, on V.34 C. (missing in H.):

Si quotlibet quantitates ad totidem alias comparentur fueritque cuiuslibet precedentis ad suam relativam maior proporcio quam subsequentis ad suam, erit omnium harum pariter acceptarum ad omnes illas pariter acceptas maior proporcio quam omnium subsequencium pariter acceptarum ad omnes subsequentes pariter acceptas, minor autem quam prime precedencium ad 2^am^. Non credo quod Campanus addiderit illam particulam 'quam alicuius subsequencium ad suam comparem', ut potest elici ex modo demonstrandi. Ad hoc enim non tendit sua demonstratio, sed nititur ostendere, quod omnium harum simul acceptarum ad omnes illas pariter acceptas maior sit proporcio quam omnium subsequencium simul sumptarum ad omnes subsequentes pariter acceptas etc.

In the margin is added:

8	3	
4	2	*hic habes instanciam*
18	15	

8) f. 48v, on VI.17 C. = VI.19 H.:

Dicere poteris ex hoc infero proporcionem esse duplam ad seipsam. Signo enim duos triangulos similes et equales; horum bases necesse est esse equales. Itemque reliquum latus unius reliquo alterius equale. Inde [?] *ergo trianguli ad triangulum sit proporcio equalitatis, et ipsa sit proporcio lateris unius trianguli ad suum relatum alterius trianguli, que etiam equalitatis est proporcio. Dico duplicata erit proporcio equalitatis, duplicata vel dupla, ad proporcionem equalitatis, quod factum* [?] *apparet.*

---

[41] Perhaps *denominacionis* is intended.

Bibliography

J. F. Ballard, *A Catalogue of the Medieval and Renaissance Manuscripts and Incunabula in the Boston Medical Library*, Boston 1944.

H. L. L. Busard (ed.), *Johannes de Tinemue's Redaction of Euclid's Elements, the so-called Adelard III Version*, Stuttgart 2001.

H. L. L. Busard (ed.), *Campanus of Novara's Edition of Euclid's* Elements, Stuttgart 2005.

M. Cantor, *Vorlesungen über Geschichte der Mathematik*, vol. II, 2nd ed., Leipzig 1900.

M. Clagett, *Archimedes in the Middle Ages*, vol. 5, Philadelphia 1984.

M. Curtze, "Der Briefwechsel Regiomontan's mit Giovanni Bianchini, Jacob von Speier und Christian Roder", *Abhandlungen zur Geschichte der mathematischen Wissenschaften mit Einschluß ihrer Anwendungen* 12 (1902), 185–336.

S. De Ricci, *Census of Medieval and Renaissance Manuscripts in the United States and Canada*, vol. 1, New York 1935.

M. Folkerts, "Regiomontanus' role in the transmission and transformation of Greek mathematics", in F. J. Ragep, S. P. Ragep (ed.), *Tradition, Transmission, Transformation. Proceedings of two conferences on pre-modern science held at the University of Oklahoma*, Leiden / New York / Cologne 1996, pp. 89–113. (Reprinted in this volume, item VI.)

M. Folkerts, "Euclid in Medieval Europe", this volume, item III.

S. Günther, "Lo sviluppo storico della teoria dei poligoni stellati nell' antichita' e nel medio evo", *Bullettino di bibliografia e di storia delle scienze matematiche e fisiche* 6 (1873), 313–340.

I. L. Heiberg, H. Menge (ed.), *Euclidis opera omnia*, vol. 5, Leipzig 1888.

J. E. Hofmann, "Über Regiomontans und Buteons Stellungnahme zu Kreisnäherungen des Nikolaus von Kues", *Mitteilungen und Forschungsbeiträge der Cusanus-Gesellschaft* 6 (1967), 124–154.

L. Labowsky, *Bessarion's Library and the Biblioteca Marciana. Six Early Inventories*, Rome 1979.

[Johannes Regiomontanus], *Continentur in hoc libro: Rudimenta Astronomica Alfragani. Item Albategnius astronomus peritissimus de motu stellarum, ex observationibus tum propriis, tum Ptolemaei, omnia cum demonstrationibus Geometricis et Additionibus Joannis de Regiomonte. Item Oratio introductoria in omnes scientias Mathematicas Joannis de Regiomonte, Patauij habita, cum Alfraganum publice praelegeret. Eiusdem utilissima introductio in elementa Euclidis*, Norimbergae 1537.

[Johannes Regiomontanus, *Verlagsanzeige*]. Faksimile in Zinner, 'Leben und Wirken', Tafel 26, Abbildung 45.

R. Peiper (ed.), *Anicii Manlii Severini Boetii philosophiae consolationis libri quinque. Accedunt eiusdem atque incertorum opuscula sacra*, Leipzig 1871.

H. Petz, "Urkundliche Nachrichten über den literarischen Nachlaß Regiomontans und B. Walters 1478–1522", *Mitteilungen des Vereins für Geschichte der Stadt Nürnberg* 7 (1888), 237–262.

E. Zinner, *Leben und Wirken des Joh. Müller von Königsberg, genannt Regiomontanus*, 2nd ed., Osnabrück 1968.

VIII

Regiomontanus' Role in the Transmission of Mathematical Problems

Johannes Regiomontanus (1436–1476) played an important role in the transmission of mathematical knowledge from Italy to Central Europe. Even in Vienna, as a student of, and collaborator with, Georg Peurbach, and especially during his stay in Italy (1461–1467) — where he was in contact with Bessarion, Bianchini and other humanists — Regiomontanus became acquainted with theoretical texts and with the mathematical knowledge of the Italian practitioners. There are three manuscripts in Regiomontanus' own handwriting containing mathematical problems. The article lists the groups of problems treated by Regiomontanus. Most of them go back to Italian sources (e. g. Leonardo Fibonacci, Paolo Gerardi). Many of these problems became known to Fridericus Amann (d. 1464 or somewhat later), who lived in a monastery near Regensburg, and to Johannes Widmann, who taught algebra at Leipzig University in the 1480s, and were later to be found in "Rechenbücher" in Germany and elsewhere.

One of the central figures in 15th-century mathematics was Johannes Regiomontanus (1436–1476). He is well known for his contributions to astronomy (including astronomical instruments), algebra and trigonometry: his *De triangulis omnimodis libri quinque* (Five Books on Triangles of Every Kind), which was published posthumously in 1533 in Nuremberg, marks the beginning of a new phase in trigonometry in the West. But it is not so generally known that Regiomontanus also played an important role in the transmission of mathematical problems from Italy to Central Europe. This is the topic of the present paper. But first a few facts about his life.[1]

Regiomontanus was born in Königsberg (Franconia). At the early age of 11 he became a student at Leipzig university. There he acquired sufficient knowledge to write an almanac for 1448. From 1450 to 1461 he lived in Vienna. These years were decisive for his intellectual development. In 1452 he became *baccalaureus* and in 1457 *magister* of the faculty of arts. At that time Vienna was a centre of astronomical and mathematical studies, mainly under the influence of John of Gmunden (before 1385–1442) and Georg Peurbach (1423–1461). Although John of Gmunden had died before Regiomontanus went to Vienna, Regiomontanus was certainly acquainted with his works on astronomy and trigonometry, since John had left his books and instruments to the Vienna library. The other important astronomer and mathematician in Vienna, Peurbach, became Regiomontanus' teacher and friend.

1 The most detailed biography is [Zinner 1968]. A survey of Regiomontanus' mathematical contributions is given in [Folkerts 1977]. Some aspects of his role in the transmission of mathematical knowledge are discussed in [Folkerts 1985, 1986].

Peurbach was best known for his planetary theory, but he also wrote a treatise on sines and chords, which was heavily dependent on John of Gmunden's work, and a treatise on elementary arithmetic, *Algorismus*. Regiomontanus studied these and other writings in Vienna and copied some of them; the so-called Vienna *Rechenbuch*, in his hand, from the years 1454–1462 (most of it was finished in 1458) is still extant.[2] This manuscript contains works on arithmetic, theory of proportions, calculations involving polygons, circles and spheres, but no recreational mathematics or special mathematical problems.

When Cardinal Bessarion came to Vienna in 1460, he met Peurbach and Regiomontanus and persuaded them to go to Italy in order to finish an abridgment of the *Almagest* to replace George of Trebizond's inferior translation of it. Since Peurbach died shortly before their departure, only Regiomontanus accompanied Bessarion to Italy in 1461. The next four years he spent partly in the cardinal's entourage and partly traveling on his own. During both phases of this period he studied mathematics and astronomy and learned Greek; furthermore he finished Peurbach's compendium of the *Almagest* and wrote the first four books of his *De triangulis*.

In 1467, at the latest, Regiomontanus left Italy for Hungary, where he was attached to the court of King Matthias Corvinus. During his years in Hungary he was occupied mainly with astronomical questions and the drawing up of trigonometrical tables. From 1471 to 1475 he lived in Nuremberg and *inter alia* prepared books for printing on his own press. In 1475 he went to Rome to lay the foundations of the currently planned reform of the calendar, but in 1476 he died suddenly.

Regiomontanus was one of the few humanists actively interested in mathematics and astronomy, but not in philosophy. As a young man, he had the chance to broaden his knowledge of mathematical subjects at the university of Vienna. Later, through his contact with Bessarion he had at his disposal the considerable library that the cardinal had accumulated. He was thereby introduced into the circle of Italian humanists with an interest in mathematics and astronomy. One of them was Giovanni Bianchini (after 1400–after 1469).

Regiomontanus used Latin translations from the Arabic and redactions of them together with Greek manuscripts to elucidate what he thought was the true meaning of Greek mathematical texts. Reconstruction of original texts as such was not his aim, but only the restoration of the authors' meaning. Further, on the basis of such texts he went on to derive his own results. For this purpose he made full use of the works of mediaeval and contemporary writers, such as Jordanus Nemorarius and Peurbach.

Although Regiomontanus' *Nachlaß* was scattered in the 16th century, we have three catalogues of books that belonged to his library, and some of the manuscripts from his collection are still extant. In addition, there is a list printed on a single page — we will call it his *Programme*[3] — containing the titles of texts that Regiomontanus wanted to publish on his printing press. He had the idea of publishing the

2 Wien, Österreichische Nationalbibliothek, MS 5203.

3 Published in [Zinner 1968, Abb. 45].

mathematical and astronomical works that seemed to him to be the most important, but because of his early death, he could not carry out this plan. From all of these sources we are able to make some deductions about his place in the transmission of mathematics in general and of mathematical problems in particular.

We know that Regiomontanus was very interested in algebra. In 1463 he discovered in Venice a Greek manuscript containing the first six books of Diophantus. Before that, the work was apparently unknown in the West. In his lectures, which Regiomontanus gave at the university of Padua in 1464, he praised among the works on arithmetic (which tacitly includes algebra) not only Diophantus' treatise, but also two works by Jordanus Nemorarius: *Arithmetica* and *De numeris datis*. These two treatises are also mentioned in his *Programme*. There are only two other mediaeval writers whose treatises appear in the *Programme*: Johannes de Muris' *Quadripartitum numerorum* and a work entitled *Algorithmus demonstratus*. This last is probably the 13th-century treatise by a certain Gernardus, which was very popular in the later Middle Ages and which was copied by Regiomontanus in Vienna; on the basis of his copy it was published by Johannes Schöner in 1534.

Because the *Algorithmus demonstratus* belongs to the *algorismus* tradition and does not deal with special mathematical problems, we need not treat it here, but I should like to make some remarks on the other texts mentioned.

The *Arithmetica* of Jordanus Nemorarius was one of the most important works on arithmetic in the West.[4] It became the standard source for theoretical arithmetic in the Middle Ages and was used by Campanus, Roger Bacon, Oresme and others. About one half of the work is devoted to the treatment of the arithmetical books VII–IX of Euclid's *Elements*. One of the many original propositions in the *Arithmetica*, III.31, is on solving in positive integers the indeterminate equation $ax - by = v$, for integral values of a, b, v (a, b prime to one another and $a > b$). Jordanus' solution is in principle the same as that given by the Indian mathematicians. Book VIII treats the theory of figured numbers; some original conclusions can be found here, as elsewhere. In one manuscript[5] there is an additional proposition: the sum of the cube numbers beginning from the unit is a square number. Because Regiomontanus' manuscript of Jordanus' *Arithmetica* seems to have been lost, we do not know how deeply he studied the work. The same is true of Jordanus' *De numeris datis*, which was probably used as a university textbook for algebra. In it Jordanus gave the problems he treated an abstract form and he used a special system of symbols, which is not very suitable and does not make the understanding easier.

In the 14th century, Johannes de Muris treated algebra in great detail in his *Quadripartitum numerorum*. This treatise, written in 1343, was Johannes' principal work.[6] It contains, apart from arithmetic, music and mechanics, a detailed treatment of algebra, which is heavily dependent on al-Khwārizmī, though there are also practical problems in the style of Leonardo Fibonacci. Regiomontanus' manuscript

4 Edited in [Busard 1991].

5 Basel, Universitätsbibliothek, F II 33.

6 Edited in [L'Huillier 1990].

of the *Quadripartitum numerorum* is still extant.[7] It was not written by Regiomontanus, but there are corrections in his hand and also many annotations. In several places he converts algebraic solutions into geometric solutions. I shall not go into details here, because the content of his annotations has been described by [L'Huillier 1980].

Now I come to mathematical problems that exist in Regiomontanus' own handwriting and that do not belong to a foreign text by a known author. They are from three sources:

a. Manuscript New York, Columbia University, Plimpton 188. This manuscript starts with Johannes de Muris' *Quadripartitum numerorum* (see above). There follows, on fol. 73r–82v, a copy of al-Khwārizmī's algebra, in the translation by Gerard of Cremona, apparently in Regiomontanus' hand. On fol. 82v–96r it contains a collection of 64 arithmetic, algebraic and geometric problems, most of which were written by Regiomontanus, apparently in 1456. These problems have not yet been edited.
b. The autograph of his *De triangulis*, now in St. Petersburg (early 1460s).[8]
c. Correspondence with Giovanni Bianchini (1463–64), Jacob von Speyer (1465) and Christian Roder (1471) (autograph now in Nuremberg).[9] In these letters Regiomontanus suggested a series of mathematical and astronomical problems and gave the solutions of some of them and hints for others.

Of special interest are the problems in the Plimpton manuscript.[10] Some correspond to problems found partly in the older and partly in the more recent algebraic texts of the thirteenth to the fifteenth century. The older writings belong to the tradition of the *libri d'abbaco*, which are based upon Leonardo Fibonacci's *Liber abbaci*. There are some hundreds of texts that were written in Italian from the 13th to the 15th centuries; they were first investigated systematically in the last few decades, especially by Warren Van Egmond and the Siena group of historians of mathematics (R. Franci, L. Toti Rigatelli and their students). A book of this type, the *Libro di ragioni*, written in Montpellier in 1328 by Paolo Gerardi of Florence, seems to have been the first algebra in the vernacular.[11] By "the more recent algebraic texts" of the period I am referring to the development in Southern Germany. Though algebra enjoyed a secure place in Italian mathematical teaching from the 13th century, it did not come to Germany before the 15th. The decisive period was the second half of this century, and the most important places were Vienna, Regensburg and Leipzig. Of special importance was Fridericus Amann (d. 1464/5), a monk from the Emmeram monastery near Regensburg, who in 1461 copied, besides other mathematical texts,

7 New York, Columbia Library, Plimpton 188, fol. 1r–70v.

8 The work was not printed until 1533 [Regiomontanus 1533].

9 Published by Curtze [1902].

10 Information about the content is given in [Folkerts 1985].

11 Edited in [Arrighi 1987]; edition of the algebraic part with English translation in [Van Egmond 1978].

an algebraic treatise in the Italian tradition and many problems in arithmetic, algebra and geometry.[12]

The problems of the Plimpton manuscript constitute an important testimony to the transmission of mathematical problems from Italy to Central Europe. Even if it is not clear which problems were Regiomontanus' own work and which he had simply copied from older sources, his importance for the transmission of algebraic and geometrical texts must now be seen in a new light.

I should like to take a brief look at the most interesting groups of problems, most of which belong to the text in the Plimpton collection. I start with problems that might be derived from the practical arithmetic of the merchants. Then I proceed to problems of theoretical arithmetic and geometry.

1. Exchange of Goods (Barter)

Since the time of Leonardo Fibonacci the custom had been established among merchants that two prices were settled for goods to be exchanged, one for cash-payment (b) and the other for barter (s).[13] In the simplest case the exchange was acceptable if the ratio of $b : s$ was the same for both articles. In several problems a merchant demands cash for some fraction $\frac{1}{n}$ of his goods. In this case the formula is

$$\left(b_1 - \frac{s_1}{n}\right) : \left(s_1 - \frac{s_1}{n}\right) = b_2 : s_2 \, .$$

One such problem (Plimpton collection, no. 24) leads to the equation

$$\left(9 - \frac{x}{4}\right) : \left(12 - \frac{x}{4}\right) = 10 : x \, ,$$

so that a quadratic must be solved.[14]

2. Compound Interest

2.1. No. 22 in the Plimpton collection is a problem in compound interest: if 20 dinars increase to 30 dinars in two years, how much is the rate of interest? This problem is treated twice by Fridericus,[15] and it is also to be found in Gerardi [Van Egmond 1978: 167–168, no. 4].

2.2. Another problem of compound interest occurs in the Plimpton collection and in the correspondence: to find the interest by which an amount of 100 *florins* comes

12 Edited in [Curtze 1895]. It should be noted that the name "Fridericus Gerhart", as given by Curtze and all later historians of mathematics, must be corrected to Fridericus Amann; see [Gerl 1999: 1–2].

13 On the barter problems see [Tropfke 1980: 519–527 (esp. p. 522); Vogel 1978: 68–73].

14 This problem corresponds with München, Bayerische Staatsbibliothek, Clm 14908, fol. 141v–142v: [Curtze 1895: 54–55].

15 On fol. 140v–141r and fol. 149r–150v; see [Curtze 1895: 54, 61–62].

up to 265 *florins* within three years.[16] This problem leads to the cubic equation $x^3+300x^2+30{,}000x = 1{,}650{,}000$. Regiomontanus solves this problem by using a three-dimensional *gnomon*;[17] this may perhaps be seen as a step towards the solution of the general cubic found by the Italians in the first half of the 16th century. But in this case Regiomontanus was not original. His solution only works in a special case of the cubic equation $x^3 + ax^2 + bx = c$, namely if $a^2 = 3b$. In this case we find $x = \sqrt[3]{c + (b/a)^3} - b/a$. We know now that Master Dardi of Pisa had already solved this special problem in the second half of the 14th century [Van Egmond 1983: 417–418], and we find it somewhat later in anonymous texts.[18]

2.3. In 1464 Bianchini suggested another problem in compound interest: 100 *florins* become 900 *florins* in 6 years. Regiomontanus answered that the interest of the first year is $\sqrt[6]{9 \cdot 10^{12}} - 100$, but he added: *Labyrinthus maximus.*[19] Furthermore, he gave an example of a fourth-degree equation, $5x^4 + 3x^3 + 8x^2 = 260x + 5$, adding *quanta sit ipsa res, nondum habeo compertum.*[20]

3. People Buying a Horse

3.1. In problem 27 of the Plimpton collection three people buy a horse. None of them can pay for the horse by himself, but each can do so with help from one of the others. The way they can do it may be expressed in the equations

$$x + \frac{y}{2} = 100\,, \quad y + \frac{z}{3} = 100\,, \quad z + \frac{x}{4} = 100\,.$$

3.2. A similar problem is no. 33 of the Plimpton collection, which results in the following system of equations:

$$x + \frac{y}{3} = 100\,, \quad y + \frac{z}{4} = 100\,, \quad z + \frac{x}{5} = 100\,.$$

This system is solved with the help of the rule of double false position.

These problems had been, since the time of Leonardo Fibonacci, a standard item in recreational mathematics.[21]

16 Plimpton 188, no. 64. The same problem in the *Correspondence*: [Curtze 1902: 219, no. 6; 236].

17 See [Folkerts 1980: 205; Tropfke 1980: 443–444].

18 For correct formulas to solve one particular class of cubic equations in Italian manuscripts of the 14th century, see [Franci & Toti Rigatelli 1988], especially pp. 19–20.

19 Bianchini's problem: [Curtze 1902: 238, no. 4]; Regiomontanus' answer and comment: [Curtze 1902: 256, 280]. See [Kaunzner 1980: 138; Folkerts 1980: 207–208].

20 [Curtze 1902: 256; Kaunzner 1980: 138].

21 See [Tropfke 1980: 608–609]. The same problem is in München, Clm 14908, fol. 155rv: [Curtze 1895: 70–71].

4. "Give and Take"

From the 64 propositions in the Plimpton collection, about 15 are of the "give and take" type.[22] I present here no. 3:

4.1. There are 3 *socii*. The first says: If I had 10 from your (money), I should have as much as both of you and even 2 more. The second says: If I had 8 from you, I should have as much as both of you, half of it and even 4 more. The third says: If I had 6 from you, I should have the double of both of you minus 3.
This leads to the system of equations

$$\begin{aligned} x+10 &= (y+z-10)+2 \\ y+8 &= (x+z-8)\cdot\tfrac{3}{2}+4 \\ z+6 &= (x+y-6)\cdot 2-3\,. \end{aligned}$$

The solution is original. First, one puts $x+y+z=R$. Then another unknown, a, is introduced. If the first *socius* has $\frac{1}{2}R+a$, then the other two have together $\frac{1}{2}R-a$. From the first equation, which becomes $\frac{R}{2}+a=\frac{R}{2}-a+2$, one finds that $a=1$. Hence the first *socius* has $\frac{R}{2}+1$ after the exchange, or $\frac{R}{2}-9$ before the exchange. A similar treatment of the second and third equations yields $\frac{3}{5}R-\frac{32}{5}$ and $\frac{2}{3}R-7$. From the three individual results and the sum R the three values x, y, z, i. e. the amounts of money in possession of the three *socii*, may be determined. This remarkable type of algebraic reasoning is otherwise unknown to me, but it might exist in some Italian text. It may be noted that the idea of considering the sum of the constituents is to be found in Leonardo Fibonacci's treatise [Boncompagni 1857: 190–203].

4.2. There is a similar problem (Plimpton collection, no. 30) which leads to the linear system

$$\begin{aligned} x+7 &= (y+z-7)\cdot 3 \\ y+9 &= (x+z-9)\cdot 4 \\ z+11 &= (x+y-11)\cdot 5\,. \end{aligned}$$

The same problem is also given by Fridericus Amann[23] and in the 1480s (with the same method of solution) by Johannes Widmann in Leipzig.[24] This seems to be an indication that this problem was brought to Germany by Regiomontanus.

4.3. There are two problems in the Plimpton collection (nos. 11, 12) where the first *socius* has no money, but debts; these debts he can only repay when he gets his share from the other two. In no. 11 we have the following system:

$$\begin{aligned} -x+132 &= y+z-132 \\ y+66 &= (-x+z-66)\cdot 2 \\ z+22 &= (-x+y-22)\cdot 3\,. \end{aligned}$$

22 For this type, see [Tropfke 1980: 609–611].

23 München, Clm 14908, fol. 146rv (without proof) and fol. 155v–156r: [Curtze 1895: 58, 71–72].

24 As part of the so-called *Latin Algebra*; see [Wappler 1899: 543–544].

5. To Find a Number

In the Plimpton collection there are some problems in which a number is to be found. Most of them serve as examples for the six types of equations given by al-Khwārizmī[25] (nos. 16–32). It is worth mentioning that the Latin text in the Plimpton manuscript, which describes the six forms of equations, agrees word-for-word with the German translation that Fridericus Amann wrote five years later,[26] and that all problems in this group are also to be found in this manuscript.[27] This part of MS Plimpton 188 may thus be the source of the examples that Fridericus took over in 1461 to illustrate the six types of equations.

5.1. In no. 17 of the Plimpton collection two numbers are to be found whose sum is 10 and whose quotient is 5, i. e. to solve

$$\frac{10-x}{x}=5\,.$$

This problem had already been given by Gerardi [Van Egmond 1978: 166] and turns up later in the collection of mathematical texts in a manuscript which originated in Leipzig.[28]

5.2. In no. 19 a problem is given that leads to the equation

$$\left(x-\frac{x}{3}-\frac{x}{4}\right)^2=16\,.$$

This, too, is treated by Fridericus[29] and is also to be found in some form in the 14th-century Italian *Columbia Algorismus*[30] and in the so-called *Deutsche Algebra* written in 1481.[31] Interestingly enough, Gerardi had treated the same problem, with 12 instead of 16 [Van Egmond 1978: 166, no. 2].

5.3. Again, problem no. 20 is given in similar form by Gerardi [Van Egmond 1978: 167, no. 3] and by Fridericus.[32] It leads to the system of equations

$$x+y=x\cdot y\,,\qquad x:y=2:3\,.$$

There are other problems of finding numbers in Regiomontanus' correspondence:

5.4. $x-y=\sqrt{5}$, $x^2\cdot y^2=250$.[33] Regiomontanus rejects Bianchini's assertion that this is insoluble because 250 is not a perfect square [Curtze 1902: 236], but he gives no solution.[34]

25 $ax=b, ax^2=b, ax=bx^2, ax^2+bx=c, ax^2+b=cx, ax+b=cx^2$.

26 In a manuscript now in Munich (Clm 14908). Edited in [Curtze 1895: 50].

27 Edited in [Curtze 1895: 50–73].

28 Now Dresden, C 80, fol. 352v.

29 On fol. 139rv. Edited in [Curtze 1895: 53].

30 $(x-\frac{x}{3}-\frac{x}{4})^2=x$. See [Vogel 1977: 31–32].

31 $(x-\frac{x}{3}-\frac{x}{4})^2=2x$. See [Vogel 1981: 36–37].

32 München, Clm 14908, fol. 139v–140v: [Curtze 1895: 53–54].

33 [Curtze 1902: 219, no. 5], in a letter to Bianchini.

34 [Curtze 1902: 253]: *Quesivi duos numeros huiusmodi non in unitatibus, scilicet in actu tantum, sed in potentia etiam, si opus fuerit.*

5.5. Another problem of finding two numbers is disguised in a traditional purchase problem.[35] The problem is: x *brachia* cost 15 *ducati*, 27 *brachia* cost y *ducati*, and $x + y = 100$. This leads to the quadratic equation

$$15 : x = (58 - x) : 27 .$$

Regiomontanus gives a detailed solution and finds, correctly, that $x = 29 - \sqrt{436}$.
5.6. Two numbers x, y must be found whose sum is 10 and such that $\frac{x}{y} + \frac{y}{x} = 25$.[36] This results in the quadratic equation

$$250x - 25x^2 = 2x^2 + 100 - 20x .$$

5.7. A number x is sought for which $\frac{100}{x} + \frac{100}{x+8} = 40$.[37] Here the quadratic equation is

$$40x^2 + 320x = 200x + 800 .$$

5.8. Another problem in the correspondence is stated without further discussion: $x + y = 100$, $\sqrt{x} \cdot \sqrt[3]{y} = 25$ [Curtze 1902: 262, no. 20]. We note that this is equivalent to a fifth-degree equation.

6. A Special Arithmetical Problem

In problem 15 in the Plimpton collection somebody wants to go as many miles as he has dinars. After every mile the dinars he possesses are doubled, but he loses 4 dinars. At the end he has 10 dinars.

After 1 mile he has $2x - 4$ *denarii.*
After 2 miles he has $2 \cdot (2x - 4) - 4 = (2^2 \cdot x - 2 \cdot 4) - 4$.
After 3 miles he has $(2^3 \cdot x - 2^2 \cdot 4 - 2 \cdot 4) - 4$.
After n miles he has $2^n \cdot x - 4(2^{n-1} + 2^{n-2} \ldots + 2 + 1) = 2^n \cdot x - 4(2^n - 1)$.

When $n = x$, the equation to be solved is $2^x \cdot x - 4(2^x - 1) = x$. This implies $x \cdot (2^x - 1) = 4 \cdot (2^x - 1)$, so that x is 4. In the margin is the remark *prepostero solvitur* ("it is solved in a reversed order"). Perhaps this means "assuming the answer and trying it in the problem" — i. e. trial and error.

7. Water and Wine

Problem 34 of the Plimpton collection is about a vessel containing 12 measures of wine. By removing one measure of the vessel's contents and adding one measure of

35 Problem: [Curtze 1902: 296, no. 9]; answer by his correspondent Jacob von Speyer: p. 300; comment by Regiomontanus: pp. 304–305, with calculations on pp. 317–319.

36 Problem given by Bianchini: [Curtze 1902: 209]; answer by Regiomontanus: p. 216; calculations: pp. 232–234. See [Kaunzner 1980: 136 (with Plate 4)].

37 Problem given by Bianchini: [Curtze 1902: 237–238]; Regiomontanus' answer: p. 256; calculations: p. 278. See [Kaunzner 1980: 136 (with Plate 5)].

water, and repeating the procedure, the wine is diluted. The question arises: when does the vessel contain only water? The reasoning given for its never coming about is correct.

8. Weight Problem

No. 40 of the Plimpton collection concerns the so-called Bachet's weight problem, in which every weight up to 2560 units must be represented with 10 standard weights, each of which is an integral number of units. As usual, the first seven powers of three, $3^0 = 1$, $3^1 = 3$, ... , $3^6 = 729$ are chosen, and then — surprisingly — the extension to 2560 is achieved by adding three weights of 489 units each.

This problem was first given by the Persian mathematician Abū Jaʿfar Muḥammad ibn Ayyūb Ṭabarī (11th c.)[38] and is found later in Italian texts (Leonardo Fibonacci, *Columbia Algorismus*). As far as I know, it does not appear in Central Europe before manuscript Plimpton 188.[39]

9. Indeterminate Analysis

9.1. In the Plimpton collection there are three problems which belong to indeterminate analysis. In no. 57 the system of equations

$$\begin{aligned} x + 13y + 14z &= 391 \\ x + y + z &= 40 \end{aligned}$$

is solved by subtracting the equations. This gives

$$12y + 13z = 351 .$$

From 351 the various multiples of 12 — i.e. 12, 24, ... —, are subtracted, and at each stage the difference is tested to see if it is a multiple of 13. This happens with 195, since $351 - \mathit{13} \cdot 12 = \mathit{15} \cdot 13$. Hence $y = 13$, $z = 15$, $x = 12$.

9.2. This method does not work with no. 58:

$$\begin{aligned} x + 4y + 10z &= 389 \\ x + y + z &= 100 . \end{aligned}$$

For the result of subtracting the equations is $3(y + 3z) = 289$, and 3 does not divide 289. Therefore there is no solution in integers.

9.3. In the next problem (no. 59) the author tries to find a general method for solving such problems when they are soluble.

38 See [Hermelink 1975].

39 See [Tropfke 1980: 633–636].

9.4. In the correspondence Regiomontanus treats three indeterminate problems. One is linear, two are quadratic. The linear system is

$$x + y + z = 240\,, \quad 97x + 56y + 3z = 16047\,.$$

This problem comes twice in the correspondence.[40] The method of solution is not given, but when Regiomontanus mentions it first, he writes the solution (114, 87, 39) in the margin.[41]

9.5. The first quadratic system is

$$x + y + z = 214\,, \quad x^2 - y^2 = y^2 - z^2\,.$$

No solution is given [Curtze 1902: 262, no. 19].

9.6. The second quadratic system is

$$x + y + z = 116\,, \quad x^2 + y^2 + z^2 = 4624\,.$$

No solution is given [Curtze 1902: 262, no. 21].

10. Remainder Problem

10.1. Among the indeterminate problems the most important is the remainder problem which is treated as no. 60 in the Plimpton collection. It is equivalent to

$$N \equiv 2 \bmod 3 \equiv 4 \bmod 5 \equiv 1 \bmod 7\,.$$

The solution follows the Chinese *ta yen* (*da yan*) method[42]: the auxiliary numbers $70 = 2 \cdot 5 \cdot 7$ ($\equiv 1 \bmod 3$), $21 = 3 \cdot 7$ ($\equiv 1 \bmod 5$) and $15 = 3 \cdot 5$ ($\equiv 1 \bmod 7$) are formed. From this we see that Regiomontanus knew not only special solutions, but also the general method of solving the remainder problem. We know that Leonardo had already discussed a similar problem[43] and later Fridericus Amann treats exactly the same problem with the same auxiliary numbers.[44] Once again we find Regiomontanus as the intermediary in the transmission of the new mathematical knowledge from the Italians to the South German *cossists*.

10.2. It should be mentioned here that Regiomontanus states examples of the remainder problem in two places in his correspondence. The first runs [Curtze 1902: 219, no. 8]:

$$N \equiv 15 \bmod 17 \equiv 11 \bmod 13 \equiv 3 \bmod 10\,.$$

40 [Curtze 1902: 262, no. 18], and [Curtze 1902: 296, no. 8].

41 [Curtze 1902: 262, no. 18]. His correspondent, Jacob von Speyer, finds the same solution: p. 300; Regiomontanus confirms its correctness: p. 305.

42 For this method and its history see [Libbrecht 1973: 213–413].

43 [Boncompagni 1857: 304]. Libbrecht [1973: 213–413] did not know the Plimpton text.

44 München, Clm 14908, fol. 124v–125r; ed. in [Curtze 1895: 65–66].

Bianchini gives two special solutions,[45] but Regiomontanus is not content with them, pointing out that there is an infinite number of solutions and showing how to find them.[46] From this example alone it would not be clear that Regiomontanus had a general solution to such problems.

10.3. The second problem of this kind is stated in the correspondence without proof [Curtze 1902: 295, no. 6]:

$$N \equiv 12 \bmod 23 \equiv 7 \bmod 17 \equiv 3 \bmod 10\,.$$

11. Square Root Approximation

11.1. In the Vienna *Rechenbuch* there is a passage on square root approximation.[47] It gives the procedure common in Greek writings — which, indeed, goes back to the Babylonians:

$$\sqrt{n} = \sqrt{a^2 + r} \approx a + \frac{r}{2a} = n_1$$

and then, as a better approximation,

$$n_2 = \left(\frac{n}{n_1} + n_1\right) : 2\,.$$

This method can be found in Hero's writings and later in Arabic texts.[48]

11.2. Another square root approximation is in the Plimpton collection (no. 56), $\sqrt{10}$ being taken as example. The method may be expressed by the formula

$$\sqrt{n} = \sqrt{a^2 + r} \approx a + \frac{r}{2a} \approx a + \frac{4a^2 + 2r - 1}{(4a^2 + 2r) \cdot 2a}\,.$$

The author considers the formula to be not particularly useful, for he writes: *Hec pro exercicio ingenii scripsi. Non enim est in his utilitas digna tantis laboribus, cum neque in omnibus numeris locum habeat illa invencio, neque quantum licet iri possit.*

12. Theory of Numbers

12.1. Regiomontanus was also interested in the theory of numbers. In his Vienna *Rechenbuch* he calculates the fifth perfect number $(2^{13} - 1) \cdot 2^{12}$ and gives the formula $(2^{17} - 1) \cdot 2^{16}$ for the sixth perfect number — in the tradition of Euclid's IX.36 — without calculating it.[49] Up to now it has been assumed that Regiomontanus was

45 $N = 1103$ or 3313: [Curtze 1902: 237].

46 $N_1 = 1103$; $N_{i+1} = N_i + 17 \cdot 10 \cdot 13$: [Curtze 1902: 254].

47 Wien 5203, fol. 167v–168r. See [Curtze 1897].

48 For the history of this method, see [Tropfke 1980: 263–268, 277].

49 Wien 5203, fol. 167r. Edited in [Curtze 1899: 288–289].

the first in the West to determine these numbers. But it is not so: in Bessarion's 13th-century manuscript (Venice, Biblioteca Nazionale Marciana, f. a. 332), there is a short text on the first five perfect numbers.[50] The Regiomontanus text is almost identical with that in the Venice manuscript. Therefore we may assume that he made his copy from it.

In his correspondence he gives the following problems, without solving them:

12.2. To find four squares whose sum is a square.[51]

12.3. To find twenty squares whose sum is a square and of which the smallest is greater than 30,000 [Curtze 1902: 334, no. ϵ].

12.4. To find three squares that are in harmonic progression [Curtze 1902: 296, no. 10].

12.5. To find three numbers that are in harmonic progression and of which the smallest is greater than 500,000 [Curtze 1902: 334, no. δ].

12.6. To find three squares that are in arithmetic progression and of which the smallest is greater than 20,000 [Curtze 1902: 334, no. γ].

It is not clear whether Regiomontanus could solve these problems. In three cases his correspondents gave a special solution, but Regiomontanus demanded general solutions.[52]

13. Geometrical Problems

13.1. In the Plimpton collection there are three geometrical problems about solving triangles. In the first case (no. 61) c, one side, is 20, the corresponding height, h_c, is 6 and one of the remaining sides (b) is double the third side (a). Two applications of Pythagoras' theorem lead to a quadratic equation, from which the part of c that is cut off by the height h_c may be determined. Regiomontanus then tries to find a non-algebraic solution by using Euclid II.13, the equivalent of the cosine-theorem. He thereby finds the same quadratic equation, but needs to perform only a few algebraic operations. The same problem can be found in *De triangulis* II.12, where the ratio of the two sides is 3 : 5 [Regiomontanus 1533: 51]. This is reduced to the quadratic equation $25x^2+3125-500x = 9x^2+180x+1125$, which is then solved algebraically. In the autograph algebraic symbols are used, but in the printed edition of 1533 they are written out in words.[53]

13.2. Problem 62 of the Plimpton collection differs from problem 61 only in that b is put equal to $a+4$ and not to $2a$. It is solved by the theorem in Euclid on intersecting chords. Here again there is a second solution.

50 On folios 36v–37v.

51 [Curtze 1902: 296, no. 7]. Answer by Jacob von Speyer: 1, 4, 16, 100 or 4, 16, 49, 100: [Curtze 1902: 300]. Regiomontanus' comment: [Curtze 1902: 304].

52 [Curtze 1902: 304]: *Difficulter tamen quispiam decem huiusmodi societates numerorum quadratorum inveniet, ...nisi artem id parandi calleat, quam ego petivi.*

53 See [Kaunzner 1980: 134–135 (with Plate 2)].

13.3. A slightly altered problem is in *De triangulis*, II.23: the height of a triangle (h_c), the difference between two of the sides ($b - a$) and the difference between the parts of the base divided by the altitude ($q - p$) are given [Regiomontanus 1533: 55–56]. In this case, the equation $\frac{1}{4}x^2 + 136 - 6x = 4x^2 + \frac{9}{4} - 6x$ must be solved.[54]

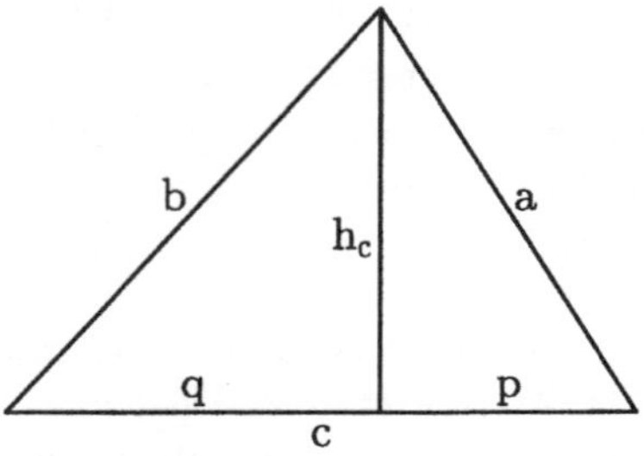

13.4. In the Plimpton collection another geometrical problem is proposed (no. 63). Of three mutually touching circles, the radii of two are given, and the area of the triangle whose vertices are the centres of the circles is also given: to find the radius of the third circle. The solution is based upon Hero's formula on the area of a triangle, which is explicitly taken from the *libellus trium fratrum*.[55] This procedure leads to a cubic equation.

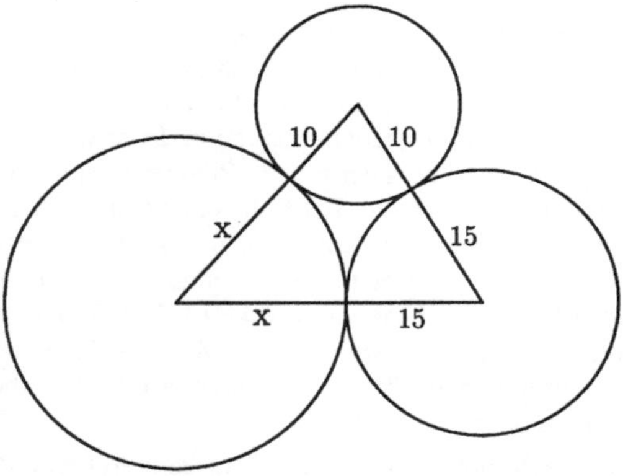

13.5. In two other cases Regiomontanus formulates problems on triangles which lead to cubic equations. One of them [Curtze 1902: 331, no. h] is not solved by him. The other one is treated in more detail. The three sides are given. The problem is to

54 See [Kaunzner 1980: 134–135 (with note 48 and Plate 3)].

55 The *Verba filiorum* of the Banū Mūsā; ed. in [Clagett 1964: 223–367].

find a division BD of the base BC so that $BD^2 + DA \cdot AB = AB^2$.[56] This leads to the cubic equation $x^3/324 + 540/29 = 3x$.[57]

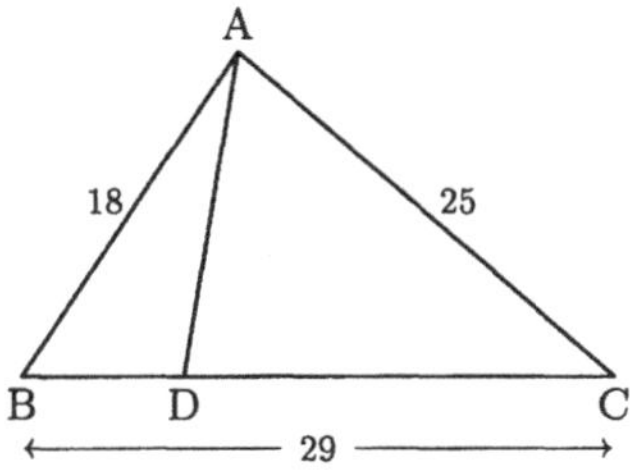

Regiomontanus gives no solution, but recognizes the similarity to the problem of finding the chord of 1° when the chord of 3° is known.[58] Thus he understood the correspondence of this problem with the trisection of the angle. It should be noted that Regiomontanus treated the Greek problem of the trisection of an angle in the same letter to Bianchini in 1464 [Curtze 1902: 258–259], giving a proof that is similar to that in the *Verba filiorum* and in the *De triangulis* of Jordanus Nemorarius. It seems more likely that Regiomontanus consulted Jordanus' treatise than that of the Banū Mūsā.[59]

In the correspondence we also find problems about a cyclic quadrilateral, a tetrahedron and two maximum problems:

13.6. To determine the area of a cyclic quadrilateral when the diameter and the ratio of the four sides to each other are given.[60] Regiomontanus rejects Bianchini's opinion that the problem is insoluble [Curtze 1902: 235–236] and suggests a method of solution using Ptolemy's theorem [Curtze 1902: 245–252]. By solving this problem Regiomontanus finds the relation between the diagonals e, f of the cyclic quadrilateral and its sides a, b, c, d:

$$\frac{e}{f} = \frac{ad + bc}{ab + cd}.$$

This formula was already known by Brahmagupta (598–after 665), but there is no reason to assume that Regiomontanus was acquainted with Brahmagupta's solution. Regiomontanus' results may have influenced the German *Rechenmeister* Simon Jacob (1510–1564) and the professor of the University of Altdorf, Johannes Praetorius (1537–1616), who dealt with the construction of the cyclic quadrilateral.[61]

56 See [Folkerts 1977: 228]. The text of the problem is edited in [Curtze 1902: 262, no. 17].

57 See [Cantor 1900: 284–285].

58 *Si dabitis lineam BD, dabo cordam unius gradus.*

59 See [Clagett 1978: 352–353].

60 The problem is given twice: [Curtze 1902: 219, no. 3] (1464); and [Curtze 1902: 331, no. i] (1471). Another problem given by Bianchini is [Curtze 1902: 236] (Regiomontanus' remarks on p. 253).

61 The history of this problem is presented in [Tropfke 1940: 150–169].

13.7. An extension of the previous problem to three dimensions is the problem of determining the volume of a tetrahedron from the ratios of its six sides and the diameter of the circumscribed sphere [Curtze 1902: 219, no. 4]. Bianchini could not solve this problem, either [Curtze 1902: 236]. Regiomontanus knew the solution, but in his correspondence he only gives hints about the way to solve the problem, with the remark that he has written down all theorems necessary for it elsewhere.[62] The history of this problem has not yet been fully investigated. J. Tropfke erroneously wrote that N. Tartaglia (1499?–1557) was the first to determine the volume of a general tetrahedron from the lengths of the sides [Tropfke 1924: 26].

13.8. Maximum problem: A 10-foot-long stick hangs vertically, so that it is 4 feet above the floor. To find the point from which the stick subtends the greatest angle [Curtze 1902: 333, no. α]. No solution is given.[63]

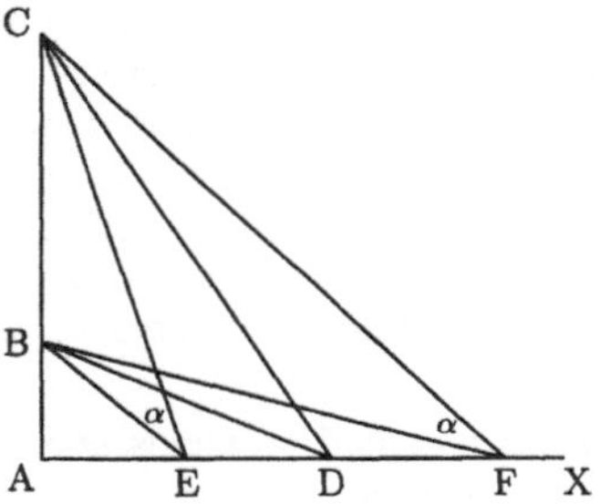

13.9. To inscribe the greatest possible square in a given triangle [Curtze 1902: 333, no. β]. Only the problem is given.

I now come to the end. We have seen that in his Vienna period, when he (partially) wrote the text in the Plimpton collection, Regiomontanus already had a good knowledge of algebra and geometry. Almost all the problems of the collection go back to Italian sources. I have shown that some of them were given by Leonardo Fibonacci, others by Paolo Gerardi, and there might also be other sources for them. Fridericus Amann must have learnt something of the contents of MS Plimpton 188 soon after it was finished — he was at least acquainted with the treatment of the six al-Khwārizmī types of equation. The same can be said of Johannes Widmann, who taught algebra at Leipzig University in the 1480s — as is witnessed by manuscript Dresden C 80. After writing this part of MS Plimpton 188, in the 1460s, when Regiomontanus was in Italy, he continued stating and solving mathematical problems, as can be seen from his correspondence. Thus Regiomontanus played a crucial role in transmitting mathematical knowledge from Italy to Central Europe in the 15th century.

62 *Hec autem omnia alibi demonstrata dedi*: [Curtze 1902: 252]. These texts do not appear to be extant.

63 Modern suggestions of how the problem could be solved are given in [Lorsch 1878] (also given in [Cantor 1900: 283–284]), and in [Günther 1908: 302–303].

Bibliography

Arrighi, Gino 1987. Paolo Gherardi, *Opera matematica. Libro di ragioni — Liber habaci. Codici Magliabechiani Classe XI, nn. 87 e 88 (sec. XIV) della Biblioteca Nazionale di Firenze*. Lucca: Maria Pacini Fazzi.

Boncompagni, Baldassarre, Ed. 1857. *Il Liber abbaci di Leonardo Pisano*. Rome: Tipografia delle scienze matematiche e fisiche.

Busard, Hubertus L. L. 1991. Jordanus de Nemore, *De Elementis Arithmetice Artis. A Medieval Treatise on Number Theory*. Stuttgart: Steiner.

Cantor, Moritz 1900. *Vorlesungen über Geschichte der Mathematik*, 2nd ed., Vol. 2. Leipzig: Teubner.

Clagett, Marshall 1964. *Archimedes in the Middle Ages*, Vol. 1: *The Arabo-Latin Tradition*. Madison: The University of Wisconsin Press.

—— 1978. *Archimedes in the Middle Ages*, Vol. 3: *The Fate of the Medieval Archimedes*. Philadelphia: The American Philosophical Society.

Curtze, Maximilian 1895. Ein Beitrag zur Geschichte der Algebra in Deutschland im 15. Jahrhundert. *Abhandlungen zur Geschichte der Mathematik* **7** (= *Supplement zur Zeitschrift für Mathematik und Physik* **40**): 31–74.

—— 1897. Die Quadratwurzelformel des Heron bei den Arabern und bei Regiomontan und damit Zusammenhängendes. *Zeitschrift für Mathematik und Physik, Historisch-literarische Abtheilung* **42**: 145–156.

—— 1899. Eine Studienreise. *Centralblatt für Bibliothekswesen* **16**: 257–306.

—— 1902. *Urkunden zur Geschichte der Mathematik im Mittelalter und der Renaissance*, Vol. II: *Der Briefwechsel Regiomontan's mit Giovanni Bianchini, Jacob von Speier und Christian Roder*. Leipzig: Teubner.

Folkerts, Menso 1977. Regiomontanus als Mathematiker. *Centaurus* **21**: 214–245.

—— 1980. Die mathematischen Studien Regiomontans in seiner Wiener Zeit. In *Regiomontanus-Studien*, Günther Hamann, Ed., pp. 175–209. Vienna: Verlag der Österreichischen Akademie der Wissenschaften.

—— 1985. Regiomontanus als Vermittler algebraischen Wissens. In *Mathemata. Festschrift für Helmuth Gericke*, Menso Folkerts & Uta Lindgren, Eds., pp. 207–219. Stuttgart: Steiner.

—— 1996. Regiomontanus' Role in the Transmission and Transformation of Greek Mathematics. In *Tradition, Transmission, Transformation. Proceedings of Two Conferences on Pre-Modern Science Held at the University of Oklahoma*, F. Jamil Ragep & Sally P. Ragep, with Steven Livesey, Eds., pp. 89–113. Leiden: Brill.

Franci, Raffaella, & Toti Rigatelli, Laura 1988. Fourteenth-Century Italian Algebra. In *Mathematics from Manuscript to Print 1300–1600*, Cynthia Hay, Ed., pp. 11–28. Oxford: Clarendon Press.

Gerl, Armin 1999. Fridericus Amann. In *Rechenbücher und mathematische Texte der frühen Neuzeit*, Rainer Gebhardt, Ed., pp. 1–12. Annaberg-Buchholz: Adam-Ries-Bund e.V.

Günther, Siegmund 1908. *Geschichte der Mathematik*, 1. Teil. Leipzig: G. J. Göschen.

Hermelink, Heinrich 1975. The Earliest Reckoning Books Existing in the Persian Language. *Historia Mathematica* **2**: 299–303.

Kaunzner, Wolfgang 1980. Über Regiomontanus als Mathematiker. In *Regiomontanus-Studien*, Günther Hamann, Ed., pp. 125–145. Vienna: Verlag der Österreichischen Akademie der Wissenschaften.

L'Huillier, Ghislaine 1980. Regiomontanus et le Quadripartitum numerorum de Jean de Murs. *Révue d'histoire des sciences* **33**: 193–214.

—— 1990. *Le* Quadripartitum numerorum *de Jean de Murs. Introduction et édition critique*. Genève / Paris: Droz.

Libbrecht, Ulrich 1973. *Chinese Mathematics in the Thirteenth Century. The Shu-shu chiu-chang of Ch'in Chiu-shao*. Cambridge, MA: MIT Press.

Lorsch, Ad. 1878. Ueber eine Maximumaufgabe. *Zeitschrift für Mathematik und Physik, Historisch-literarische Abtheilung* **23**: 120.

Regiomontanus, Johannes 1533. Joannes de Regio Monte, *De triangulis omnimodis libri quinque*. Nuremberg: Io. Petreius.

Tropfke, Johannes 1924. *Geschichte der Elementar-Mathematik in systematischer Darstellung mit besonderer Berücksichtigung der Fachwörter*, Siebenter Band: *Stereometrie, Verzeichnisse*, 2nd ed. Berlin / Leipzig: de Gruyter.

—— 1940. *Geschichte der Elementar-Mathematik in systematischer Darstellung mit besonderer Berücksichtigung der Fachwörter*, Vierter Band: *Ebene Geometrie*, 3th ed. Berlin: de Gruyter.

—— 1980. *Geschichte der Elementar-Mathematik in systematischer Darstellung*, Band 1: *Arithmetik und Algebra*, 4th ed. Berlin / New York: de Gruyter.

Van Egmond, Warren 1978. The Earliest Vernacular Treatment of Algebra: The *Libro di Ragioni* of Paolo Gerardi (1328). *Physis* **20**: 155–189.

—— 1983. The Algebra of Master Dardi of Pisa. *Historia Mathematica* **10**: 399–421.

Vogel, Kurt 1977. *Ein italienisches Rechenbuch aus dem 14. Jahrhundert (Columbia X511 A13)*. Munich: Deutsches Museum.

—— 1978. *Beiträge zur Geschichte der Arithmetik*. Munich: Minerva.

—— 1981. *Die erste deutsche Algebra aus dem Jahre 1481, nach einer Handschrift aus C 80 Dresdensis herausgegeben und erläutert*. Bayerische Akademie der Wissenschaften, Mathematisch-Naturwissenschaftliche Klasse, Abhandlungen, Neue Folge, Heft 160. Munich: Beck.

Wappler, Emil 1899. Zur Geschichte der deutschen Algebra. *Abhandlungen zur Geschichte der Mathematik* **9**: 539–554.

Zinner, Ernst 1968. *Leben und Wirken des Joh. Müller von Königsberg genannt Regiomontanus*, 2nd ed. Osnabrück: Otto Zeller.

IX

Leonardo Fibonacci's Knowledge of Euclid's *Elements* and of Other Mathematical Texts*

This paper is concerned with the sources which Leonardo Fibonacci knew and used for his mathematical works. I have limited the field to those writings which were available in Latin in Leonardo's time and which Leonardo explicitly cited. Accordingly I shall not discuss possible Arabic sources and even in Latin I shall not in general speculate about possible sources that Leonardo does not expressly cite. Since Euclid's most important writing, the *Elements*, is cited particularly often, I shall ask the question which version or versions of the *Elements* Leonardo knew. In the following the content of the chapters in Leonardo's works will be sketched and for each chapter the sources cited by Leonardo will be given.

1. *Liber abbaci*

Let us begin with Leonardo's principal work, the *Liber abbaci*. It is well known that in this book Leonardo expounded the arithmetic of his time and that he included algebra and geometry. The principal matters discussed are: practical calculations, commercial problems and recreational mathematics.

The work is divided into 15 chapters. In the first five[1] Leonardo treats calculation with integers using the new, i.e. the Hindu-Arabic, number system. The individual chapters are on: how to write the numbers, multiplication, addition, subtraction, division. In chapter 5 (and later, too) he writes fractions with multiple numerators and denominators ("aufsteigende Kettenbrüche") – a notation which is also found in Abū Kāmil.

The only author mentioned by Leonardo in these chapters is Euclid. He cites him twice. The first citation is from Book 7 (though this is not explicitly mentioned). Leonardo is discussing the multiplication of

*To appear in the proceedings of the conference "Leonardo Fibonacci. Matematica e società nel Mediterraneo del secolo XIII. Convegno internazionale di studi. Pisa–Firenze, 20–23 novembre 2002".

[1]Boncompagni, 'Scritti 1', pp. 2–47.

numbers of more than one decimal place. It is necessary to know in which places to put the result of multiplying the various digits together. This may be derived from the law of multiplying the powers of 10: $10^m \times 10^n = 10^{m+n}$. Leonardo proves this rule by citing Euclid explicitly for the law of three or four proportional numbers: from $a : b = b : c$ follows $a \times c = b^2$ and from $a : b = c : d$ follows $a \times d = b \times c$ [2]. The rule for four numbers is proved in Euclid VII.19; the corresponding rule for three appears in some Greek manuscripts after VII.19, though Heiberg does not think the passage is genuine[3].

The second citation is also from Book 7, this time one of the definitions (VII. def. 16). It occurs in an excursus in the section on division. Here Leonardo defines prime numbers (*incompositi* or *primi numeri*) and composite numbers. Leonardo points out that Euclid calls composite numbers "plane numbers"[4].

* * *

In chapters 6 and 7 (Boncompagni, 'Scritti 1', pp. 47–83) Leonardo explains calculation (multiplication, addition, subtraction, division) with fractions and with numbers comprising integers and fractions. Leonardo places the fraction to the left of the integer, in the Arabic way.

In these chapters Euclid is again the only author cited. There are four citations: in the transformation of fractions Leonardo uses the theory of proportion and cites Euclid for the statement that the whole to the whole is as part to the part[5]. This is the proposition V.15. A little later in the text Leonardo gives a detailed procedure for finding the highest common

---

[2] *Nam cum tres numeri proportionales sunt, ita quod sicut primus est ad secundum, ita secundum sit ad tertium; tunc multiplicatio primi in tertium equatur multiplicationi secundi in se. Et cum quattuor numeri sunt proportionales, fueritque sicut primus ad secundum, ita tertius ad quartum; tunc multiplicatio primi in quartum equa est multiplicationi secundi in tertium, ut in euclide reperitur.* (Boncompagni, 'Scritti 1', p. 15, lines 6–10.)

[3] See the remarks in Heath, 'Euclid's Elements, 2', p. 320. The Greek text of this rule is edited in Heiberg, 'Euclidis Elementa, 2', pp. 428,22–430,17. The rule is also extant in the translation made direct from the Greek (Busard, 'Mediaeval Latin Translation', p. 165, lines 1–3).

[4] *Reliqui vero compositi, vel epipedi, idest superficiales, a peritissimo geometriae Euclide appellantur* (Boncompagni, 'Scritti 1', p. 30, lines 2–1 from the bottom). In the Greek-Latin translation the text of Euclid's definition of "composite number" (VII. def. 17) is: *Quando vero duo numeri multiplicantes se invicem faciunt aliquem, factus epipedos appellatur* (Busard, 'Mediaeval Latin Translation', p. 154, lines 25f.).

[5] Leonardo justifies the proportion $12 : 72 = 12/12 : 72/12$ with the words: *sicut totum ad totum, ita pars est ad partem* (Boncompagni, 'Scritti 1', p. 51, lines 9f.).

factor of two numbers by using the so-called Euclidean algorithm. Here he also cites Euclid[6]. Thus he knows the procedure given in Euclid VII.2. When he is performing a division Leonardo refers to Euclid's statement (V.15) that the ratio between two numbers is the same as the ratio of equimultiples of these numbers[7]. At the end of chapter 7 Leonardo explains how a fraction can be expressed as a sum of unit fractions[8]. Here he again cites Euclid's proposition about the equivalence of $a : b = c : d$ and $a \times d = b \times c$, i.e. VII.19[9].

* * *

After treating the elementary arithmetic operations with integers and fractions Leonardo turns to their application to practical questions: commercial problems and recreational mathematics. In chapters 8 to 10 (Boncompagni, 'Scritti 1', pp. 83–143) problems involving the theory of proportion are addressed. In chapter 8 Leonardo treats the "rule of three". In chapter 9 he gives a rule, known in the eighteenth and nineteenth centuries as the "chain rule", according to which a ratio between two numbers may be determined when their ratios to intermediate quantities are known. The short chapter 10 is about commercial partnerships, i.e. sharing a profit amongst partners. In chapter 11 (Boncompagni, 'Scritti 1', pp. 143–166) mixture problems are treated.

In the section on the chain rule Leonardo refers to Ptolemy's *Almagest* and to the sector-figure (*figura cata*), and he mentions the book on proportions of "Ametus filius" (i.e. Aḥmad b. Yūsuf) in which 18 forms of the compound proportion derivable from one basic form are proved geometrically. Leonardo writes that Ptolemy used the sector-figure to prove a theorem about the declination of points on a circle (i.e. the ecliptic cir-

[6]Boncompagni, 'Scritti 1', p. 51, penultimate paragraph. He ends with *ut in Euclide apertis demonstrationibus declaratur*. Leonardo's quotation of Euclid VII.2 is not word-for-word.

[7]*et hoc est quod Euclides peritissimus geometra in suo libro declarat: quod quam proportionem habet quilibet numerus ad quemlibet numerum, eandem proportionem habent equa quelibet multiplicia illorum* (Boncompagni, 'Scritti 1', 69 middle). It arises in the justification of the proportion $83 : 5\frac{2}{3} = 249 : 17$.

[8]For this, see Lüneburg, 'Lesevergnügen', pp. 80f.

[9]*quia cum IIIIor numeri sunt proportionales, est multiplicatio secundi in tertium equa multiplicationi primi in quartum, ut ab Euclide demonstratum est* (Boncompagni, 'Scritti 1', p. 82, third to last paragraph). See note 2.

cle) from the equator[10]. He refers to Ptolemy, *Almagest*, Book I, sections 13–14. Euclid is not quoted in chapters 8–11.

* * *

The long chapter 12, which comprises about a third of the whole work (Boncompagni, 'Scritti 1', pp. 166–318), contains a large number of different problems. Among these are: the summation of arithmetical and geometrical progression; the formula for the sum of the squares of the natural numbers; systems of linear equations; problems of simple and compound interest; the so-called remainder problem; perfect numbers. But the most famous of these problems is the "rabbit problem" in which the sum of a special recursively defined series is to be found (Boncompagni, 'Scritti 1', pp. 283f.). Many of the problems in this chapter Leonardo solves by means of the "rule of false position". The algebraic method is used by Leonardo for a group of problems which were very popular in the Middle Ages: e.g. several men wish to buy a horse, but none can do this alone; one borrows a specified part of the available money from the others; the problem is to find how much each man has at the beginning[11]. In these problems it is remarkable that in some cases there is a negative solution[12].

The question which sources Leonardo used for this chapter cannot be answered here. Many of the problems in it can be found, sometimes with the same numerical values, in Arabic texts. We may assume that Leonardo is here taking advantage of what he learned in his journeys in the Mediterranean area. He mentions a problem which was propounded to him in Constantinople by a very learned "magister Muscus"[13]. In chapter 12 there is also a short section on perfect numbers[14]. Here Leonardo gives the procedure formulated by Euclid (IX.36) to find perfect numbers and he proves that the first three numbers found in this way (6, 28,

[10] *est enim hec talis propositio proportionum ea que ostenditur in figura cata, scilicet sectoris, per quam Tholomeus docuit in almagesti reperire demonstrationem* (lege: *declinationem*) *circulorum a circulo recto, et multa alia; et Ametus filius ponat* (lege: *posuit*) *decem et octo combinationes ex ea in libro, quem de proportionibus composuit* (Boncompagni, 'Scritti 1', p. 119, first paragraph). For the *figura cata* and its history see Lorch, 'Thābit ibn Qurra'. The Latin text of Aḥmad b. Yūsuf is edited in Schrader, 'Epistola'.

[11] See Vogel, 'Geschichte'.

[12] See Sesiano, 'Appearance', pp. 116–133.

[13] *Questio nobis proposita a peritissimo magistro musco constantinopolitano in constantinopoli* (Boncompagni, 'Scritti 1', p. 249). In this problem five people together buy a ship.

[14] *De inventione perfectorum numerorum* (Boncompagni, 'Scritti 1', p. 283).

496) are indeed perfect. His presentation ends with "By always doing this, you can find perfect numbers *in infinitum*" (*Et sic semper faciendo, poteris in infinitum perfectos numeros reperire*). Even if Leonardo does not mention Euclid explicitly here, we may assume that his knowledge of perfect numbers came from Euclid, Book 9.

* * *

In chapter 13 (Boncompagni, 'Scritti 1', pp. 318–352) problems which lead to linear equations are solved by means of the "rule of double false position". The name "Regula elchatayn", which Leonardo uses, shows that he here used Arabic or Arabic-based material. In this chapter Leonardo does not mention any sources.

* * *

In chapter 14 (Boncompagni, 'Scritti 1', pp. 352–387) Leonardo explains by numerical examples the method of approximate calculation of square and cubic roots. He further considers numerous problems which display calculations with various kinds of irrational numbers.

The calculation of square roots is based on the binominal theorem, $(a+b)^2 = a^2 + b^2 + 2ab$. It is well known that this theorem is proved in geometrical form in Book 2 of the *Elements*. It is no wonder that Leonardo at the beginning of this chapter quotes word for word six of the first seven enunciations of Book 2 with express citation of Euclid. Since he does not need, as Euclid did, the propositions for general quantities, he substitutes everywere the term "number" for "line"[15]. – In the second part of chapter 14 calculation with expressions in which there are square roots is shown. In this part Leonardo bases himself on Book 10 of Euclid's *Elements* – as he explicitly confirms[16]. In the discussion on the various roots[17] Leonardo makes use of the following notions which occur in Book 10 of the *Elements*: *riti* (= rhete) for "rational", and *medialis, bimedialis, maior, recisum, apothami* (= apotome) for the various irrational quantities. In

[15]*Liceat mihi in hoc de radicum capitulo quedam necessaria, que claves dicuntur, inserere; que cum sint in secundo euclidis apertis demonstrationibus demonstrata, sufficit super diffinitiones earum hoc tantum secundum numerum procedere* (Boncompagni, 'Scritti 1', p. 352). There follow Propositions II.2, 1, 4, 7, 5, 6, each illustrated by numerical examples (Boncompagni, 'Scritti 1', p. 352). It may be mentioned here that in his *Practica geometrie*, too, Leonardo cites some propositions of Book 2 (see note 27).

[16]*diffinitiones duarum linearum ratiocinatarum, super quas decimus euclidis liber geometrie tractat, assignare disposui* (Boncompagni, 'Scritti 1', p. 356, last paragraph).

[17]For the content, see Lüneburg, 'Lesevergnügen', pp. 254–257.

his description of multiplication of irrationals Leonardo quotes Euclid explicitly, who – according to Leonardo – has said that between two similar numbers there is always another number[18].

* * *

The final, fifteenth, chapter (Boncompagni, 'Scritti 1', pp. 387–459) contains, besides a series of geometrical problems involving the application of Pythagoras' theorem, numerous problems that may be reduced to quadratic equations, which are solved by means of "algebra and almuchabala". Leonardo assumes the same six types of equation as al-Khwārizmī, though the fifth and sixth cases are interchanged. He uses the same terms for the unknown and its powers as are found in the translations of al-Khwārizmī's *Algebra*: *res* or *radix* for the unknown, *census* or *quadratus* for its square, *numerus* or *denarius* for a given number.

Many problems in this chapter are taken from Abū Kāmil's "Algebra", several with exactly the same numbers[19], though Leonardo adds problems of his own. Here again Euclid is mentioned, in three places:

At the beginning of the chapter Leonardo presents (on ten pages) proportions among three and four numbers – proportions that he needs later to solve quadratic equations. In his treatment of the proportion $c : b = (c - b) : (b - a)$ (with $a < b < c$) Leonardo used the fact that this proportion implies $c : b = b : a$ and refers to Euclid's Book 5 for the proof[20]. At the beginning of the geometric section Leonardo presents the well-known ladder problem: a 20-foot rod rests in its whole length against the wall of a tower. How many feet does the top of the rod move when the lower end moves 12 feet? Of course, this is easy to solve by applying Pythagoras' theorem, which is proved – as Leonardo points out – as penultimate proposition of Book 1 of the *Elements*[21]. In the solution

---

[18] *et hoc est quod euclides ostendit, cum dixit: inter duos numeros similes unum intercidere numerum* (Boncompagni, 'Scritti 1', p. 359, lines 8f.). The definition of similar numbers is in Euclid VII. def. 21: "Similar plane and solid numbers are those which have their sides proportional" (Heath, 'Euclid's Elements, 2', p. 278). In VIII.18 Euclid proves that between two similar plane numbers there is one mean proportional number; the wording in the direct translation from the Greek is: *Duobus similibus epipedis numeris unus medius proportionalis incidit numerus* (Busard, 'Mediaeval Latin Translation', p. 188).

[19] For a list of the same or similar problems, see Levey, 'Abū Kāmil', pp. 217–220. See, too, Sesiano, 'Abū Kāmil'.

[20] *ut in quinto euclidis ostenditur* (Boncompagni, 'Scritti 1', p. 391, last line). The proposition to which Leonardo refers is V.19; see Heath, 'Euclid's Elements, 2', p. 174.

[21] *et quoniam ut heuclides testatur in penultima sui primi libri, quod in trigonis rectiangulis quadratus lateris subtendentis angulum rectum equatur quadratis duobus*

of a group-purchase problem Leonardo refers to the fact established by Euclid that between two cubic numbers two mean proportions can be found[22].

2. *Practica geometrie*

Now for the *Practica geometrie.* This work, which is subdivided into eight sections, is, according to Leonardo's own words[23], a perfect handbook (*perfectum documentum*) for those who want to work in geometry, both those who do this with proofs and those who are laymen in this matter[24].

The work begins with the introduction of fundamental geometrical terms and the explanation of measures of length, area and volume (Boncompagni, 'Scritti 2', pp. 1–5). The first part of the introduction is mainly occupied with citations of Book 1 of Euclid's *Elements*: the definitions, axioms, postulates and some propositions from this book, but differently arranged[25].

Section 1 (Boncompagni, 'Scritti 2', pp. 5–13) contains the calculation of the area of a rectangle when the sides are known in integers, in fractions or in various units of measure. At the end of this section (Boncompagni, 'Scritti 2', pp. 13–18) Leonardo gives some general theorems which he says are necessary for the understanding of the second section[26]. Among these theorems is the greater part of the propositions of Euclid, Book

---

lateri<bu>s reliquorum duorum laterum angulum rectum continentium (Boncompagni, 'Scritti 1', p. 397, middle).

[22] *Et quoniam ut euclides dicit, inter duos cubos numeros duo medi<i> intercidunt numeri continuati cum ipsis in proportione continua; ideo cubicetur 100, erunt 1000000, quorum proportio est ad cubum denariorum primi fori, sicut primus numerus ad quartum, ut euclides ostendit* (Boncompagni, 'Scritti 1', pp. 399f.). Here Leonardo refers to Euclid VIII.19: "Between two similar solid numbers there fall two mean proportional numbers", but he replaces "similar numbers" by "cube numbers" and thus restricts the statement, because all cube numbers are similar numbers, but not all similar numbers are cube numbers.

[23] Boncompagni, 'Scritti 2', p. 1.

[24] *hi qui secundum demonstrationes geometricas: et hi qui secundum vulgarem consuetudinem, quasi laicali more, in dimensionibus voluerint operari.*

[25] Boncompagni, 'Scritti 2', pp. 1 middle – 3 beginning. To the contents see p. 19 below. Leonardo does not write at the beginning, but only in the middle of this section that his text comes from Euclid: *Multa enim sunt que oportet scire eos, qui in mensuratione, et divisione corporum, secundum subtilitatem geometricam procedere volunt, que in euclide aperte monstrantur* (Boncompagni, 'Scritti 2', p. 2, lines 15–17).

[26] *Et ut ea, que in hac secunda distinctione promisimus plenarie demonstrentur, quedam huic operi necessaria dignum duximus preponenda* (Boncompagni, 'Scritti 2', p. 13, last paragraph).

2[27], the equivalence of proportions of three and four numbers and the corresponding equality of products[28], and the theorem on intersecting chords in a circle[29]. Leonardo does not mention Euclid in the beginning of this passage, but later refers to him twice[30].

* * *

In section 2 (Boncompagni, 'Scritti 2', pp. 18–30) Leonardo shows how to determine square roots, both arithmetically and geometrically. He also treats calculation with expressions involving roots. Here he mentions no sources.

* * *

Section 3 (Boncompagni, 'Scritti 2', pp. 30–110) is on the measure of "fields" (*campi*) of various shapes, i.e. to find the area and associated quantities of plane figures (triangles, various quadrilaterals, polygons, circles). In the part on quadrilaterals he also applies quadratic equations. This section is similar in content and in some details to Savasorda's *Liber embadorum.* Some examples remind us of the *Liber mensurationum* of Abū Bakr. But Leonardo does not mention these authors.

Among those cited by name Euclid is in this section, too, the most frequently quoted. Leonardo cites passages from Books 1–3, 6, 12–14 of the *Elements.* These passages are[31]:

Book 1:
I.32 (sum of angles in a triangle): p. 31 middle (ECBP)
I.43: p. 69, line 9 (E)
I.47 (Pythagoras' theorem): p. 32, beginning of the second paragraph (C)

---

[27] Boncompagni, 'Scritti 2', pp. 13 bottom – 17, second third. Leonardo cites propositions II.2, 1, 3, 4, 7, 5, 6, 9 and 10. At all places he brings the enunciation, the proof and a numerical example.

[28] This is Prop. VII.19 with its addition; see above. The text is on Boncompagni, 'Scritti 2', pp. 17 bottom – 18, second third. Leonardo gives the text of the enunciations with numerical examples.

[29] Euclid III.35. Leonardo gives only the text of the enunciation (Boncompagni, 'Scritti 2', p. 18, end of the penultimate paragraph).

[30] *Quoniam superfluum videtur nostris demonstrationibus demonstrare omnia, que Euclides super ostensionibus assignavit* (Boncompagni, 'Scritti 2', p. 17, lines 17–16 from the bottom); *in Euclide ostenditur* (Boncompagni, 'Scritti 2', p. 18, end of the first paragraph).

[31] The letters in brackets have the following meanings: E = Euclid, cited by name; C = text given as citation; B = book number cited; P = proposition number cited. All page numbers are from Boncompagni, 'Scritti 2'.

Book 2:
II.11 (the calculation of the greater part of a number divided according to the "golden section"): p. 73 middle (E)
II.12: p. 38, line 3 from the bottom (EB)
II.13: p. 36, end of the first paragraph (EB)

Book 3:
III.3: p. 102 middle (EB)
III.27: p. 97 middle (ECB)
III.31 (angle in a semicircle): p. 32 middle (ECB); p. 90, lines 10–9 from the bottom (ECB)

Book 6:
VI.1: p. 110, line 6 from the bottom (CB)
VI.2: p. 44, beginning of the third paragraph (EB)
VI.5: p. 36, line 13 (EC)[32]
VI.8: p. 32, line 13 from the bottom (ECB); p. 95, lines 6f. (EBP)
VI.19: p. 43, beginning of the second section (EB)

Book 12:
XII.2 (the areas of circles are proportional to the squares of the diameters): p. 88, line 3 (ECBP)

Book 13:
XIII.9: p. 105 middle (EB)
XIII.10: p. 105 middle (EB)

Book 14:
XIV.3[33]: p. 106, line 9 from the bottom (EB).

Besides his borrowing from Euclid, Leonardo also uses in this chapter material from the *agrimensores*, i.e., the Roman surveyors. He cites them explicitly twice: first, he details the "common method" used by the *agrimensores*, sufficient for the measure of all triangles[34]. Secondly, in the section on calculating a circular arc he says that his procedure differs

[32]Boncompagni's text *ostenditur, quod omnes figure que habent latera opposita, equalia habebunt similiter et angulos equales* is corrupt: *opposita, equalia* should be replaced by *proportionalia*.

[33]In Heiberg's edition the propositions of Book 14 are not numbered; the numbers given here and later are taken from Busard's edition of the Greek-Latin translation (Busard, 'Mediaeval Latin Translation', pp. 399–407).

[34]*Modus vulgaris quo uti debent agrimensores, et est sufficiens in mensuratione omnium trigonorum* (Boncompagni, 'Scritti 2', p. 43 bottom). The problem is the calculation of the area of a triangular field with the help of a unit of length (*mensuret cum pertica cathetum*) and a simple instrument (*lensa fili*).

from that of the *agrimensores*, who measure a curved line with the help of a rope which can be bent[35].

Other authors mentioned by Leonardo in section 3 are:

In the determination of the area of the circle Leonardo expressly names Archimedes for proving that the circumference of the circle is $3\frac{1}{7}$ times the diameter. Leonardo describes in detail how the approximate value of $1440 : 458\frac{1}{3}$ ($= 864 : 275$) for π can be deduced by Archimedes' method[36].

Leonardo draws up a table of chords, which, according his own words, he has put together himself[37]. His procedure is based on the method that Ptolemy described in the *Almagest.* Leonardo mentions Ptolemy expressly and sketches his procedure[38]. A little later he mentions Ptolemy again in connection with his diagram of a quadrilateral in a circle from which his calculation of chords depends. Leonardo then states and proves Ptolemy's theorem[39]. He also mentions the 18 combinations of the ratios which can be obtained from the *figura cata* (that is Menelaus' theorem) and refers to his own proof in his *regula baracti* – as it seems, his *Liber Abbaci*, chapter 9[40].

---

[35] *Sed hec talis investigatio non est operanda ab agri mensoribus, qui secundum vulgarem modum procedere volunt …* (Boncompagni, 'Scritti 2', p. 95, second paragraph).

[36] Boncompagni, 'Scritti 2', pp. 88 bottom – 90. The text starts with: *Ostendendum est etiam quomodo inventum fuit, lineam circunferentem omnis circuli esse triplam et septimam sui dyametri ab Archimedine phylosopho; et fuit illa inventio pulcra et subtilis valde: quam etiam reiterabo non cum suis numeris, quibus ipse usus fuit demonstrare; cum possibile sit cum parvis numeris ea que ipse cum magnis ostendit plenissime demonstrare* (Boncompagni, 'Scritti 2', p. 88 bottom).

[37] *sequentes tabulas composui* (Boncompagni, 'Scritti 2', p. 95 bottom). The table itself is on page 96.

[38] *Est enim alius modus reperiendi cordas semiarcuum, ex quibus arcubus corde sunt note; et quem Tholomeus posuit in almagesto* (Boncompagni, 'Scritti 2', p. 94 bottom). Ptolemy's procedure is described directly afterwards (up to Boncompagni, 'Scritti 2', p. 95 middle). Ptolemy has this in *Almagest* I.10: see Ptolemäus, 'Handbuch', pp. 30f.

[39] *et hoc demonstravit Tholomeus in compositione tabule arcuum, et cordarum in almagesti per alium modum. Demonstravit quidem Tholomeus per consimilem figuram, quod omnis quadrilateri qualitercumque cadentis in circulo, si duo dyametri protrahantur in ipso, multiplicatio unius dyametri in alium equatur duabus multiplicationibus oppositorum laterum; quod sic probatur* (Boncompagni, 'Scritti 2', p. 104, first paragraph). See Ptolemäus, 'Handbuch', pp. 28f.

[40] *in hac figura cata, et quas etiam in libro meo in regula baracti demonstravi* (Boncompagni, 'Scritti 2', p. 54 bottom).

Of considerable interest is Leonardo's proof of "Hero's formula" of the area of a triangle. It is similar to the proof of the Banū Mūsā, but Leonardo does not mention any names[41].

Another solution of a problem which Leonardo presents in this chapter was given to him by a man living in Verona[42].

* * *

In section 4 (Boncompagni, 'Scritti 2', pp. 110–148) Leonardo treats the division of rectilinear or circular fields by lines so that the areas of the parts are in a given proportion. He studies in detail the problem: to divide a triangle by a straight line, which is drawn through a given inner point of the triangle, into two equal parts. His procedure corresponds to that in Euclid's *Liber divisionum*, which was translated by Gerard of Cremona from the Arabic[43], but Leonardo does not mention this source. He also treats the division of quadrilaterals, polygons and circles in two or more equal or unequal parts.

In this part Leonardo cites only Euclid. In three places Euclid is mentioned by name. Once he mentions the number of the proposition: VI.15[44]. In the other two citations he only gives the book numbers: Book 3 and 12, respectively[45].

* * *

Section 5 (Boncompagni, 'Scritti 2', pp. 148–158) is on determination of cube roots. For this purpose, besides the arithmetic procedure Leonardo describes a geometrical method. Here the problem is reduced to finding two mean proportions between two given quantities. In antiquity this procedure was used to solve the problem of the duplication

---

[41] Boncompagni, 'Scritti 2', pp. 40f. The Latin text of the theorem and its proof in the *Verba filiorum* of the Banū Mūsā is edited and translated into English in Clagett, 'Archimedes 1', pp. 278–289.

[42] *solutio ... mihi proposita a quodam veronense* (Boncompagni, 'Scritti 2', p. 50, second third).

[43] Both Gerard's translation and the Greek text are lost.

[44] *trigona ... habent unum angulum comunem; quare circa ipsum angulum comunia patiuntur latera; hoc est quod sunt mutuae proportionis, ut in quinto decimo theoremate sexti libri Euclidis habetur* (Boncompagni, 'Scritti 2', p. 111, lines 8–6 from the bottom).

[45] *Et quoniam equalis est recta .tk. recte .ae., ipse due recte ab eodem circulo equales sectiones auferunt, ut in tertio libro Euclidis panditur* (Boncompagni, 'Scritti 2', p. 146 middle; the text refers to Prop. III.28); *sunt enim circuli ad se invicem sicuti ad quadrata dyametrorum; ut in duodecimo Euclidis monstratum est* (Boncompagni, 'Scritti 2', p. 147 middle; the text refers to Prop. XII.2).

of the cube; but it is also convenient for finding cube roots geometrically. Leonardo gives three methods of solution. The first (Boncompagni, 'Scritti 2', pp. 153f.) corresponds to the solution which is attributed to Archytas and is presented in Prop. 16 of the *Verba filiorum* of the Banū Mūsā. The second solution (Boncompagni, 'Scritti 2', p. 154) is also to be found in a treatise by Philo of Byzantium. The third solution (Boncompagni, 'Scritti 2', pp. 154f.) is Prop. 17 of the *Verba filiorum*, slightly altered[46]. Leonardo does not state his sources, but only says that for the second proof the penultimate proposition of Euclid's Book 3, i.e. about the length of a tangent from a given point to a given circle (III.36), is necessary[47].

* * *

Section 6 (Boncompagni, 'Scritti 2', pp. 158–202) is on the determination of the volumes of solids (right parallelepipeds, pyramids, cones, spheres, polyhedra). In this chapter, too, Leonardo made full use of the *Verba filiorum* of the Banū Mūsā: on 10 pages of the printed edition (Boncompagni, 'Scritti 2', pp. 178–187) Leonardo gives the contents of the enunciations and proofs of Propositions 8–15[48]. The correspondence is not literal, but the texts of Leonardo and the Banū Mūsā are so similar that a dependence must be assumed, even though Leonardo does not mention his source.

In section 6 there are again many references to Euclid. In one place (Boncompagni, 'Scritti 2', p. 159 middle) Leonardo says that in Book 14 Euclid explains how octahedra, dodecahedra and icosahedra may be inscribed in a sphere[49]. In twelve places Leonardo refers to specific propositions of Euclid's *Elements*, which he needs for his own theorems; sometimes he gives the number of the book of the *Elements* which he is citing, but never the proposition number[50]. Of special interest is Leonardo's citation in section 6 of two lengthy passages from the *Elements*: on Bon-

---

[46]See Clagett, 'Archimedes 1', pp. 658–664; Clagett, 'Archimedes 3', p. 1274.

[47]*Probatur hec figura per penultimam tertii Euclidis* (Boncompagni, 'Scritti 2', p. 154, end of the second paragraph).

[48]See Clagett, 'Archimedes 3', pp. 1275–1283.

[49]This refers to propositions 14, 17, 16 of Book 13 (not Book 14).

[50]The Euclidian propositions in question are:
III.9: p. 179, lines 9f. (E)
VI.19 Por.: p. 175, lines 10–12 (C)
XI.2: p. 179, lines 2f. (EB)
XII.7: p. 169, lines 2–4 (EB)
XIII.3: p. 200, lines 7–4 from the bottom (E)
XIII.8: p. 196, lines 4–2 from the bottom (EB)

compagni, 'Scritti 2', pp. 159–162 the text of most of the enunciations of Books 11–13 and on Boncompagni, 'Scritti 2', p. 162 of most of the enunciations of Book 14. I shall later refer to these literal citations.

Besides Euclid, Leonardo also mentions Menelaus, Theodosius and Ptolemy in section 6. In a problem about the sphere Leonardo cites a deduction which can be found in the books of Menelaus and Theodosius[51]. A little later Leonardo writes that it is to be found in Menelaus and in the *Almagest* that the intersection of a sphere and plane is a circle and that the line between the centre of the sphere and the centre of the circle is perpendicular to the plane of the intersection[52].

* * *

In the short section 7 (Boncompagni, 'Scritti 2', pp. 202–206) Leonardo treats the measurements of altitudes and depths by means of the quadrant. The only author he cites in this section is Euclid: Leonardo uses a proposition from Euclid's Book 1[53].

* * *

The last section 8 (Boncompagni, 'Scritti 2', pp. 207–224) contains some special geometrical problems (*subtilitates*). In the first part (Boncompagni, 'Scritti 2', pp. 207–216) problems about pentagons and decagons are treated. The irrational roots of the quadratic equations which occur here are calculated by Leonardo with great exactitude and ex-

---

XIII.14: p. 194, lines 4–3 from the bottom (EB)
XIII.17: p. 196, lines 11–9 from the bottom (EB)
XIV.3: p. 197, lines 4f. (EB)
XIV.4: p. 199, lines 4–6 (E)
XIV.8: p. 197 middle (ZB)
XIV.9: p. 201, lines 6–10 (E).

Here, too, the letters in brackets mean: E = Euclid, cited by name; C = text given as citation; B = book number cited. All page numbers refer to Boncompagni, 'Scritti 2'.

51 *ut in libro Miles. et Teodosii probatum est* (Boncompagni, 'Scritti 2', p. 179, line 12). The reference is to Theodosius' *Sphaerica* I.8. I have not found an appropriate passage in Menelaus.

52 *ut habetur in libro Milei et in Almaiesti ...* (Boncompagni, 'Scritti 2', p. 197, line 13). The theorem cited, that the intersection of a plane and a sphere is a circle, I have not found in either Menelaus' *Sphaerica* or Ptolemy's *Almagest.* It appears, however, in Theodosius' *Sphaerica* I.1.

53 *Item enim permutatim, ut in primo libro Euclidis ostenditur ...* (Boncompagni, 'Scritti 2', p. 206 middle). This refers to Prop. I.29: *In equidistantes rectas recta incidens et permutatim angulos equales alternis facit ...* (Busard, 'Mediaeval Latin Translation', p. 43).

pressed in sexagesimal fractions. The text on pentagons and decagons is mostly taken from Abū Kāmil's treatise on the subject[54], even to the numerical examples. But Leonardo does not give his source. In fact, the only author cited by Leonardo is again Euclid: in five places he quotes propositions from Books 13 and 14 of the *Elements*[55].

In the second part of this section Leonardo proposes a problem in number theory (Boncompagni, 'Scritti 2', pp. 216–218): to find a square number which, increased by 5, is again a square number[56]. Leonardo gives two methods for solution. He does not give his source.

3. Minor works

a. *Flos*

This work begins with a problem presented to Leonardo by John of Palermo in the presence of the emperor Frederick II: to find a square number which, when increased and decreased by 5, produces square numbers (Boncompagni, 'Scritti 2', p. 227). Here Leonardo gives only the result; elsewhere, in the *Liber quadratorum*, he presents the procedure to find these numbers. As he himself says in the *Flos*, he had already begun the treatise on the square numbers[57].

Another problem was proposed to him by John of Palermo (Boncompagni, 'Scritti 2', pp. 228–234): to solve the cubic equation $x^3 + 2x^2 + 10x = 20$. Leonardo first proves that x cannot be a (positive) integer, nor a fraction, nor a square root of a rational number. Then he shows that x cannot have the form of any of the irrational quantities to be found in Euclid. Finally he gives an approximate solution in sexagesimal fractions, though without describing the procedure to find it. His result is remarkable enough: it exceeds the true value of the root only by $1\frac{1}{2}$ units of the sixth sexagesimal place. This problem is to be found, with the same numerical values, in 'Umar Khayyām[58].

---

[54] See Lorch, 'Abū Kāmil'.

[55] XIII.1: p. 215, line 2 from the bottom – 216, line 2 (E)
XIII.3: p. 215 middle (E)
XIII.8: p. 215, lines 4–2 from the bottom (E)
XIII.11: p. 207, lines 8–7 from the bottom (EB)
XIV.1: p. 215 middle (E)
XIV.3: p. 211 middle (EB).

[56] For this problem, see the remarks on the *Flos* in the next section.

[57] *quare hinc sumens materiam, libellum incepi componere ... quem libellum quadratorum intitulavi* (Boncompagni, 'Scritti 2', p. 227, line 5–4 from the bottom).

[58] See Vogel, 'Unbestimmtes Problem', p. 66.

At the beginning of this section Leonardo gives a detailed treatment of Euclid's tenth book (Boncompagni, 'Scritti 2', p. 228, first paragraph). He writes that when he had thought about this problem, it became clear to him that the solution can be expressed in terms of the quantities found in Euclid X[59]. In consequence, he had studied Book 10 intensively in order to understand the meaning and to commit the contents in memory. Because this book is more difficult than the previous and some of the following books, he had begun to write notes on it, reducing Euclid's statement about lines and areas to statements about numbers[60]. It would seem that this work on Book 10 has been lost – unless his following remarks in the *Flos* may be interpreted as his notes. For Leonardo now proceeds to list the 15 different varieties of quantity which Euclid introduces in Book 10. He gives their names, some of which reveal the Greek origin[61]. On the following pages Leonardo cites Euclid in three places: Book 10 twice and Book 6 once[62]; in addition there is at least one literal quotation from Euclid, but without mentioning his name[63].

The second part of the work (Boncompagni, 'Scritti 2', pp. 234–247) is a small collection of problems in recreational mathematics. One of the problems, in which a sum of money must be distributed among three people, was suggested to him by John of Palermo. The same problem is to be found in the third book of al-Karağī's *Algebra*. Leonardo gives the solution, which he himself describes as "exceedingly beautiful" (*nimis pulchrum modum inveni*), in the *Flos*[64]. He gives three other methods for this problem in his *Liber abbaci*[65].

---

[59] *solutionem egredi ex his que continentur in X^{o} libro Euclidis.*

[60] *Et quia difficilior est antecedentium et quorumdam sequentium librorum Euclidis, ideo ipsum X^{m} librum glosare incepi, reducens intellectum ipsius ad numerum, qui in eo per lineas et superficies demonstratur* (Boncompagni, 'Scritti 2', p. 228, top).

[61] Especially *rite* ("rational"; Boncompagni, 'Scritti 2', p. 228) and *apothami* (Boncompagni, 'Scritti 2', p. 232); see *Liber abbaci*, chapter 14.

[62] VI.14: Boncompagni, 'Scritti 2', p. 229, lines 8–6 from the bottom; X.11: Boncompagni, 'Scritti 2', p. 229, lines 5–3 from the bottom; X.73: Boncompagni, 'Scritti 2', p. 230, lines 5–7.

[63] X.22: *Et quia quod a media secundum ritim et ductum latitudinem facit ritim et incommensurabilem ei cui adiacet longitudine* (Boncompagni, 'Scritti 2', p. 230, beginning of the last third). The wording is the same as in the Greek-Latin translation, which has correctly *eductum* instead of the incorrect *et ductum* in Boncompagni's edition; see Busard, 'Mediaeval Latin Translation', p. 229.

[64] Boncompagni, 'Scritti 2', pp. 234–236; see Vogel, 'Unbestimmtes Problem', pp. 67f.

[65] See Vogel, 'Unbestimmtes Problem', pp. 70–72.

b. *Epistola ad Magistrum Theodorum*

This letter, which Leonardo wrote to a certain "magister Theodorus" (Boncompagni, 'Scritti 2', pp. 247–252), contains several problems. The first belongs to the indeterminate analysis; it is the "problem of 100 birds". Then follows a geometrical problem: in an isosceles triangle to inscribe a regular pentagon with the condition that one of the vertices is the common point of the equal sides of the triangle and the opposite side is a part of the opposite side of the triangle. At the end there is another problem, which leads to a system of five linear equations and five unknowns.

In this letter Leonardo mentions no sources.

c. *Liber quadratorum*

In the year 1225 Leonardo wrote his "Book of squares" (*Liber quadratorum*; Boncompagni, 'Scritti 2', pp. 253–283), which contains problems in indeterminate quadratic equations. One of them was already mentioned in the *Flos*, one of the problems suggested to Leonardo by master John of Palermo, the philosopher at the court of Frederick II: to determine a (rational) square number such that, if it is increased and decreased by 5, produces (rational) square numbers (Boncompagni, 'Scritti 2', p. 271). In the last problem of the *Liber quadratorum* (Boncompagni, 'Scritti 2', pp. 279–283) three numbers x, y and z must be determined so that each of the sums $x+y+z+x^2$, $x+y+z+x^2+y^2$, $x+y+z+x^2+y^2+z^2$ is a square number.

In the "Liber quadratorum" Euclid is again the only author cited: Leonardo quotes him three times, propositions in Book 8 and 10[66].

4. Which version of Euclid did Leonardo use?

As we have seen, Euclid is the most-cited author in Leonardo's writings. He knew the *Elements* in great detail, especially Book 10. Since he explicitly quotes substantial passages of the *Elements*, the question may be asked, which of the Latin versions of Euclid he used.

Which Latin Euclid texts could have been available to Leonardo? In the twelfth century the *Elements* were translated three times from Arabic into Latin. The first translation was made by Adelard of Bath, perhaps in the second quarter of the twelfth century[67]. As author of

[66] VIII.11: Boncompagni, 'Scritti 2', p. 276, 2nd paragraph (E); VIII.24: Boncompagni, 'Scritti 2', p. 254, lines 5–3 from the bottom (E); lemma 1 to X.28: Boncompagni, 'Scritti 2', p. 255, penultimate paragraph (EB).

[67] The so-called "Adelard I" text; ed. Busard, 'Adelard of Bath'.

the second translation the manuscripts give Hermann of Carinthia[68]. If Hermann really is the author of the translation, it appears that he did not translate literally – as it was usual for him. The third translation from the Arabic was made by Gerard of Cremona (d. 1187)[69]. The text as we have it appears to be a later reworking of the original.

The most influential Euclid text in the Latin West in the twelfth century was the so-called "Version II"[70]. Clagett called it "Adelard II". We now know that the author was not Adelard, but almost certainly Robert of Chester. At least, he was responsible for the text of the enunciations. For these he used different Latin texts (Adelard's and Hermann's translations; the so-called "Geometry II" of Pseudo-Boethius). Later, proofs were added to the enunciations. This addition might have been made in two stages, as Burnett has suggested[71]. The text of the enunciations must have been written shortly before 1140; a *terminus ante quem* for the complete work is 1163–1168.

Another Euclid text which might have been available at the end of the twelfth century is the so-called "Version III"[72]. It was composed after "Version II", as it seems, by John of Tinemue. The oldest manuscript (Oxford, Balliol College 257) is dated late twelfth century.

Besides the translations from the Arabic and the Latin texts based upon them, there was another twelfth century Latin text: a translation made direct from the Greek. It was composed in Sicily and, as it seems, after 1165 (ed. Busard, 'Mediaeval Latin Translation'). It belongs to a group of texts which were translated in the second part of the twelfth century either in Southern Italy or in Sicily. These texts include Ptolemy's *Almagest*, Euclid's *Elements, Data, Optics* (*De visu*) and *Catoptrics* (*Liber de speculis*); to these we may add Proclus' *Elementatio physica*.

In the twelfth century the Norman kingdom of Southern Italy and Sicily was the most important meeting-point of Greek and Latin cultures. By 1090 the island had become a Norman kingdom in which Greek, Latin, and Arabic civilizations lived side by side in peace and toleration. The three languages were in current use in the royal charters and registers, as well as in many-tongued Palermo, so that knowledge of more than one of them was a necessity for the officials of the royal court. The production

68 Ed. Busard, 'Hermann of Carinthia'.

69 Ed. Busard, 'Gerard of Cremona'.

70 Ed. Busard / Folkerts, 'Robert of Chester'.

71 Burnett, 'Influences'.

72 Called "Adelard III" by Clagett; ed. Busard, 'Johannes de Tinemue'.

of translations was encouraged by the Sicilian kings from Roger to Frederick II and Manfred as part of their efforts to foster learning. The two principal translators, Henricus Aristippus and Eugene the Emir, were members of the royal administration. Aristippus was the first translator of the *Meno* and *Phaedo* of Plato and of the fourth book of Aristotle's *Meteorology*, and his Latin rendering remained in current use during the Middle Ages and the early Renaissance. He was also instrumental in bringing manuscripts to Sicily from the library of the Emperor Manuel at Constantinople. One of these possesses special importance, a codex of Ptolemy's *Almagest*, from which the first Latin version was made by an anonymous scholar about 1160. In his preface the translator says that "as he was laboring over the study of medicine at Salerno, he learned that a copy of Ptolemy's great treatise had been brought from Constantinople to Palermo, as a present from the Greek emperor, by an ambassador of the Sicilian king. He set out to seek this emissary, by name Aristippus, and braving the terrors of Scylla and Charybdis and the fiery streams of Etna he found him at Pergusa, near the fount, engaged, not without danger, in investigating the marvels of Etna".

J. E. Murdoch, who in 1961 discovered a manuscript of the translation of the *Elements* from the Greek, argued that the anonymous translator of the *Almagest* and that of the *Elements* are identical and that these two works were not translated by the same person who translated the *Data*, the *Optics* and the *Catoptrics*[73]. But Busard, who edited the translation of the *Elements* in 1987, doubts Murdoch's assumption. He thinks that two groups should be distinguished and that the person who translated Books I–XI.33 and XV, might be different from the translator of Books XI.34–XIII.

The translation of the *Elements* from the Greek is transmitted by only two manuscripts, one of them being incomplete[74]. It should be said that the only complete manuscript has after the end of Book 13 and before the beginning of Book 15 an abbreviated version of Books 14 and 15 of the *Elements*[75]. Thus it does not contain the original text of Book 14, but two versions of Book 15: the original and an abbreviation.

* * *

[73]See Murdoch, 'Euclides Graeco-Latinus'.

[74]Paris, BnF lat. 7373, s. 13, f. 2r–175v; Florence, BNC, conv. soppr. C I 448, 104 fols., s. 14 (breaking off in X.48).

[75]Paris, BnF lat. 7373, f. 167v–172v.

After this long excursus let us return to Leonardo. Apart from citations of individual propositions from Euclid's *Elements*, Leonardo presents substantial quotations of enunciations (and definitions) from four sections of the *Elements*: from Books 1, 2, 11–13 and 14. A comparison of these passages with the Latin versions of Euclid available in the early thirteenth century might answer the question which version Leonardo used.

However, the extracts from *Book 1* of the *Elements* which Leonardo puts at the beginning of his *Practica geometriae* (Boncompagni, 'Scritti 2', pp. 1–3) show no great agreement with any of the texts of the Latin Euclid. Here Leonardo apparently used no one version or, at any rate, no one version known to us, but brought together various elements from different sources. According to Leonardo, the material he gives is necessary for those occupied with geometry[76]. The order of the material itself shows that Leonardo has written this text by compilation[77].

Book 2: In two different places Leonardo gives the content of a large part of the enunciations of Book 2: in the *Liber abbaci*[78] and later in the *Practica geometriae*[79]. If we compare the text of the five enunciations common to the two citations, we find that the formulations are quite different. This suggests that Leonardo did not copy the texts word for word, but cited the content of what he had committed to memory. This explanation would account for the lack of agreement with the known texts of Euclid[80]. But there is one exception to this pattern of citation: Leonardo's text of Proposition II.9 is, apart from minor details, the text of the direct translation from the Greek[81].

[76] This is apparently the meaning of the sentence which appears in Boncompagni's edition (Boncompagni, 'Scritti 2', pp. 2f.) in corrupt form: *Sine his omnibus, et sine radicum inventione, ipsi qui secundum vulgarem modum procedere voluerint: cum his que ostendam inferius, convenienti loco poterunt sufficienter procedere.* (A *non* appears to be missing in the last part of the sentence.)

[77] The order is: *Elements* I. def. 1–14, 19, 15–18; III. def. 6, 10; I. def. 23, 29; I.1, 9/10, 11–13, 15, addition, 33, 23, 31, 35–38; I. def. 22; I 34, 32; ax. 1–4, 4a, 5–8; addition to post. 5. We may note that after I. def. 18 not only the definition of the segment of a circle (III. def. 6), but also the definition of the sector of a circle (III. def. 10) is inserted.

[78] On Boncompagni, 'Scritti 1', p. 352 the enunciations of II.2, 1, 4, 7, 5 and 6.

[79] On Boncompagni, 'Scritti 2', pp. 13–17 the enunciations of II.2, 1, 3, 4, 7, 5, 6, 9 and 10.

[80] That is: Boethius, Adelard I, Hermann, Gerard, Version II, translation from the Greek.

[81] This proposition is quoted only in the *Practica geometriae*.

Books 11–13: As an introduction to calculations about solids in Chapter 6 of his *Practica geometriae* Leonardo quotes almost all theorems of Books 11–13 of the *Elements*[82]. Somewhat later he quotes *in extenso* the construction problems of the same books[83]. As in Books 1 and 2, there is no significant agreement with the known Latin texts of Euclid. But it is remarkable that the order of the cited propositions is the same as in the Arabic-based versions, which is quite distinct from the order in Greek and in the direct translation[84]. This ordering is a strong indication that Leonardo here was using an Arabic-based translation. This translation is either lost (or not yet found) or was one of the known Arabic-based translations whose text Leonardo reformulated to suit his own style.

Book 14: Between the two blocks of citations from Books 11–13 (see above) Leonardo added a passage with the title *Incipiunt theoremata quartidecimi libri*, which reproduces the enunciations of Book 14[85]. The wording of these citations corresponds very exactly to the formulation in the compendium of Books 14 and 15 which appears in the only complete manuscript of the direct translation from the Greek between Books 13 and 15[86]. It is therefore clear, beyond any doubt, that Leonardo knew and used this compendium. Indeed, Busard adduces arguments to support the idea that Leonardo himself wrote the compendium[87].

* * *

Now a few words about the citations of individual propositions from Euclid's *Elements*. It is not always clear in which places Leonardo reproduces the exact wording of the text he had before him and in which he expresses the content with his own words. In the majority of cases he appears to paraphrase the text; in no case have I found a definite agreement between Leonardo's text and that of an Arabic-based version. But on the other hand there are several places where Leonardo's wording is

[82]On Boncompagni, 'Scritti 2', pp. 158–162. Besides the definitions of some basic stereometric terms, he cites the following propositions: XI.1–10, 13, 15–22, 24, 25, 28–32, 34, 33, (def. 9), 35–39; XII.4, 5, 7, 9, 8, 14, 10, 12, 11, 15, 18; XIII.1–3, 5, 4, 6, 7, 12, 9, 10, 8, 11, 16, 17.

[83]On Boncompagni, 'Scritti 2', p. 162, bottom. There are: *Elements* XI.11, 12, 23, 26, 27; XII.16, 17; XIII.13–18.

[84]Above all, the interchange XII.9 – XII.8; the omission of XII.6 and XII.13–14; the order of XII.10, 12, 11 and XIII.5, 4, 6, 7, 12, 9, 10, 8, 11.

[85]Boncompagni, 'Scritti 2', p. 162. The cited material is XIV.1–7, 9 and 10 (in Busard's numbering [Busard, 'Mediaeval Latin Translation', pp. 399–407]).

[86]See above. This text has been edited in Busard, 'Mediaeval Latin Translation', pp. 399–411.

[87]See Busard, 'Mediaeval Latin Translation', pp. 19f.

very similar to the direct translation from the Greek; in some cases the agreement is so exact that it is not merely probable, but certain, that he knew and used this translation.

List of parallel texts:

I.32:

Leonardo (*Practica geometrie*; Boncompagni, 'Scritti 2', p. 31):

cum omnes tres anguli cuiuslibet trigoni duobus rectis sint equales ...

Greek-Latin translation (Busard, 'Mediaeval Latin Translation', p. 44):

Et interiores tres anguli trigoni duobus rectis sunt equales.

III.27:

Leonardo (*Practica geometrie*; Boncompagni, 'Scritti 2', p. 97):

... equales angulos super equas periferias consistere, cum ad centrum vel ad periferiam sint constituti ...

Greek-Latin translation (Busard, 'Mediaeval Latin Translation', p. 85):

In equalibus circulis qui super equales periferias consistunt anguli equales sibi invicem sunt sive ad centra sive ad periferias fuerint constituti.

VI.1:

Leonardo (*Practica geometrie*; Boncompagni, 'Scritti 2', p. 110):

sunt enim trigona sibi invicem sub eadem altitudine existentia sicut bases

Greek-Latin translation (Busard, 'Mediaeval Latin Translation', p. 125):

Trigona et parallilogramma sub eadem altitudine existentia ad se invicem sunt ut bases.

VI.14:

Leonardo (*Flos*; Boncompagni, 'Scritti 2', p. 229):

Equalium vero et unum uni equalem habentium angulum parallilogramorum contrarie potiuntur latera, que circa equales angulos subtenduntur

Greek-Latin translation (Busard, 'Mediaeval Latin Translation', p. 135):

Et equalium et unum uni equalem habentium angulum parallilogrammorum contrarie latera patiuntur que circa equales angulos.

VI.20, por. 2:

Leonardo (*Practica geometrie*; Boncompagni, 'Scritti 2', p. 175):

Et quando tres recte continue proportionales sunt, erit sicut prima ad tertiam ita figura, que est a prima, ad figuram que est ad secundam similem, et similiter descriptam

Greek-Latin translation (Busard, 'Mediaeval Latin Translation', p. 141):

... si tres recte proportionales fuerint, erit ut prima ad tertiam ita figura que a prima ad eam que a secunda similem et similiter descriptam.

X.11:

Leonardo (*Flos*; Boncompagni, 'Scritti 2', p. 229):

Et cum quatuor quidem quantitates proportionales sunt, fueritque sicut prima ad secundam, ita tertia ad quartam; et prima fuerit secunde commensurabilis, et tertia quarte commensurabilis erit

Greek-Latin translation (Busard, 'Mediaeval Latin Translation', p. 221):

Si quattuor quantitates proportionales fuerint, fueritque prima secunde commensurabilis, et tertia quarte commensurabilis erit.

X.22:

Leonardo (*Flos*; Boncompagni, 'Scritti 2', p. 230):

Et quia quod a media secundum ritim et ductum latitudinem facit ritim et incommensurabilem ei cui adiacet longitudine

Greek-Latin translation (Busard, 'Mediaeval Latin Translation', p. 229):

Quod a media secundum ritin eductum latitudinem facit ritin et incommensurabilem ei cui adiacet longitudine.

Another proof that Leonardo used the direct translation from the Greek is his use of some technical terms which are otherwise only known in the direct translation, e.g.: *epipedi* (VII. def. 17), *riti* (X. def. 3), *riton et medium potens* (X.40), *media* (X.21), *abscisio* (X.73), *apotomi* (X.92), *dodrans* and *dextans* (XIV.8)[88].

It is well known that the direct translation of the *Elements* was little used; this can be seen, for instance, from the paucity of manuscripts which survive (only two, one of which being incomplete). In general, this translation had little influence. It is, however, not surprising that Leonardo should have known it, for he had a close relationship with

[88] This was mentioned by Busard, 'Mediaeval Latin Translation', pp. 19f.

the Hohenstaufen emperor Frederick II and with the philosophers and scientists at his court. The *Flos* and the *Liber quadratorum*, both written in 1225, deal with mathematical problems given by John of Palermo in the presence of the emperor. John himself probably translated a short Arabic treatise on the hyperbola from Arabic into Latin. Michael Scot, who lived at the court in Palermo, received the revised edition of the *Liber abbaci* in 1228 from Leonardo himself. As the translation of the *Elements* from the Greek was done in Sicily some decades before Leonardo's activity, it is not surprising that Leonardo had access to it and used it.

5. Summary and final remarks

The study of the citations of Euclid has shown that Leonardo knew the text of the *Elements* very well and fully understood its mathematical content. Euclid was Leonardo's principal source for theoretical mathematics: he cites him often and, according to his own words, wrote some kind of commentary on Book 10, the longest and most difficult book of the *Elements*. As we have seen, Leonardo knew and used the direct translation of the *Elements* which was made in Sicily after 1160. He also knew a compendium of Books 14 and 15 which is transmitted together with the Greek-Latin direct translation; and it is not impossible that Leonardo himself compiled this text. Further, he was acquainted with another Euclid text which followed the Arabic order of the propositions as shown by the translations of Adelard, Hermann and Gerard.

Leonardo's writings show that for his time he possessed a comprehensive and deep knowledge of both theoretical and practical mathematics. He was also very well-read: besides Euclid, he cites Archimedes, Ptolemy, Menelaus, Theodosius, Aḥmad b. Yūsuf and the *agrimensores*. Further he knew the *Liber trium fratrum* of the Banū Mūsā and some writing of Abū Kāmil. These are the authors whom Leonardo either names himself or whom we may identify with certainty from the wording of the citation – there may be more of the second sort which have yet to be identified. Research into Leonardo's sources has hardly begun. No doubt, new results will be obtained when the content of his writings is carefully analysed and systematically compared with the works of other authors.

Bibliography

B. Boncompagni, *Scritti di Leonardo Pisano, matematico del secolo decimoterzo.* Vol. 1: *Il Liber abbaci di Leonardo Pisano pubblicato secondo la lezione del codice Magliabechiano C. I, 2616, Badia Fiorentina, n.° 73*, Rome 1857.

B. Boncompagni, *Scritti di Leonardo Pisano, matematico del secolo decimoterzo.* Vol. 2: *Leonardi Pisani Practica geometriae ed opuscoli*, Rome 1862.

Ch. Burnett, "The Latin and Arabic Influences on the Vocabulary Concerning Demonstrative Argument in the Versions of Euclid's *Elements* Associated with Adelard of Bath", in J. Hamesse (ed.), *Aux origines du lexique philosophique européen: L'influence de la* latinitas, Louvain-La-Neuve 1997, 117–135.

H. L. L. Busard, "The Translation of the *Elements* of Euclid from the Arabic into Latin by Hermann of Carinthia (?): books I–VI", *Janus* 54 (1967), 1–140; books VII–XII, Amsterdam 1977.

H. L. L. Busard, *The First Latin Translation of Euclid's* Elements *Commonly Ascribed to Adelard of Bath*, Toronto 1983.

H. L. L. Busard, *The Latin translation of the Arabic version of Euclid's* Elements *commonly ascribed to Gerard of Cremona*, Leiden 1983.

H. L. L. Busard, *The Mediaeval Latin Translation of Euclid's* Elements *Made Directly from the Greek*, Stuttgart 1987.

H. L. L. Busard / M. Folkerts, *Robert of Chester's (?) Redaction of Euclid's* Elements: *the so-called Adelard II Version*, 2 vols., Basel / Boston / Berlin 1992.

H. L. L. Busard, *Johannes de Tinemue's Redaction of Euclid's Elements, the So-Called Adelard III Version*, 2 vols., Stuttgart 2001.

M. Clagett, *Archimedes in the Middle Ages.* Vol. 1: *The Arabo-Latin Tradition*, Madison 1964.

M. Clagett, *Archimedes in the Middle Ages.* Vol. 3: *The Fate of the Medieval Archimedes 1300 to 1565*, Philadelphia 1978.

Th. L. Heath, *The Thirteen Books of Euclid's Elements Translated from the Text of Heiberg.* Vol. 2. Second Edition, Cambridge 1956.

I. L. Heiberg, *Euclidis Elementa*, vol. 2, Leipzig 1884.

R. Lorch, "Abū Kāmil on the Pentagon and Decagon", in M. Folkerts, J. P. Hogendijk (ed.), *Vestigia Mathematica. Festschrift für H. L. L. Busard*, Amsterdam 1993, 215–252.

R. Lorch, *Thābit ibn Qurra On the Sector-Figure and Related Texts. Edited with Translation and Commentary*, Frankfurt/Main 2001.

M. Levey, *The* Algebra *of Abū Kāmil, Kitāb fī al-jābr wa'l-muqābala, in a Commentary by Mordecai Finzi. Hebrew Text, Translation, and Commentary with Special Reference to the Arabic Text*, Madison / Milwaukee / London 1966.

H. Lüneburg, *Leonardi Pisani Liber Abbaci oder Lesevergnügen eines Mathematikers*, Mannheim etc. 1992.

J. E. Murdoch, "Euclides Graeco-Latinus: A Hitherto Unknown Medieval Latin Translation of the *Elements* Made Directly from the Greek", *Harvard Studies in Classical Philology* 71 (1966), 249–302.

Ptolemäus, Handbuch der Astronomie, Bd. 1. Deutsche Übersetzung und erläuternde Anmerkungen von K. Manitius. Vorwort und Berichtigungen von O. Neugebauer, Leipzig 1963.

W. R. Schrader, *The* Epistola de Proportione et Proportionalitate *of Ametus filius Iosephi*. Ph.D. Thesis, Univ. of Wisconsin, 1961.

J. Sesiano, "The Appearance of Negative Solutions in Mediaeval Mathematics", *Archive for History of Exact Sciences* 32 (1985), 105–150.

J. Sesiano, "La version latine médiévale de l'Algèbre d'Abū Kāmil", in M. Folkerts, J. P. Hogendijk (ed.), *Vestigia Mathematica. Festschrift für H. L. L. Busard*, Amsterdam 1993, 315–452.

K. Vogel, "Zur Geschichte der linearen Gleichungen mit mehreren Unbekannten", *Deutsche Mathematik* 5 (1940), 217–240.

K. Vogel, "Ein unbestimmtes Problem al-Karağīs in Rechenbüchern des Abendlandes", *Sudhoffs Archiv* 61 (1977), 66–74.

X

PIERO DELLA FRANCESCA AND EUCLID

If one reads Piero's three treatises – the *Trattato d'abaco* (here called *Trattato*)[1], the *De prospectiva pingendi*[2] and the *Libellus de quinque corporibus regularibus* (here called *Libellus*)[3] – one finds that the name «Euclid» is mentioned very often. Indeed, Euclid is the only mathematician named by Piero in the dedicatory letter at the beginning of the *Libellus*, where he says that the content of the treatise was known «to Euclid and other geometers», but now was transferred to arithmetic[4]. As Margaret Daly Davis has shown[5], Piero's citations of Euclid do not come from an intermediate source, such as Leonardo Fibonacci's *Practica geometriae* (1220), but from a genuine Euclid text. Because we do not know when and where Piero learned mathematics – it has been suggested that he came into contact with scientists like Toscanelli and Alberti in 1439 or later, when he was in Florence – we have to consult his writings themselves in our search for the text (or texts) Piero might have used. Before doing this, I should like to make some remarks about the

[1] Piero della Francesca, *Trattato d'abaco, dal Codice Ashburnhamiano 280 (359*-291*) della Biblioteca Medicea Laurenziana di Firenze*, ed. G. Arrighi, Pisa 1970.

[2] Piero della Francesca, *De prospectiva pingendi*, ed. G. Nicco Fasola, Firenze 1942.

[3] G. Mancini, *L'opera «De corporibus regularibus» di Pietro Franceschi detto della Francesca, usurpata da Fra Luca Pacioli*, in «Atti della R. Accademia dei Lincei», Memorie della Classe di Scienze morali, storiche e filologiche, s. v, XIV, [1915] 1916, pp. 441-580.

[4] «Etenim licet res apud Euclidem, et alios geometras nota sit, per ipsum tamen nuper ad arithmeticos translata est»: *ibid.*, p. 488.

[5] M. Daly Davis, *Piero della Francesca's Mathematical Treatises. The «Trattato d'abaco» and «Libellus de quinque corporibus regularibus»*, Ravenna 1977.

Latin and Italian versions of Euclid's *Elements* which were available in Piero's time.

LATIN EUCLID TEXTS; THE CAMPANUS VERSION

In the twelfth century, there were three translations into Latin of Euclid's *Elements* from the Arabic: by Adelard of Bath, Hermann of Carinthia and Gerard of Cremona. But more important than these translations was a redaction, compiled about 1140, the «Version II». Although Marshall Clagett called this text «Adelard II», it seems that its author was not Adelard of Bath, but Robert of Chester. It was a reworking based upon the translations by Adelard and Hermann, having abbreviated proofs in the earlier books. «Version II» was well-known, especially in the 12th and 13th centuries; more than sixty manuscripts of it are still extant. Moreover, this text was reworked several times. Of these the most important are: the version that Clagett called «Adelard III»; a commentary on the *Elements* which was the source for another commentary ascribed to Albertus Magnus; and the reworking by Campanus of Novara in the middle of the 13th century. The Campanus version was to become easily the best known Euclid text in the Middle Ages. I know 131 manuscripts which still survive; nearly sixty of them are from the 14th century, and as many are from the 15th. The text was first published by Erhardus Ratdolt in Venice in 1482; later editions were by Henricus Stephanus (Paris 1516), J. Hervagius (Basel 1537, 1546, 1558) and others.

Although the Campanus redaction of Euclid was the best known and most used version of the *Elements* in the Latin West from the 13th to the 16th centuries, only special aspects have been studied. John Murdoch has given some general characteristics of the Campanus version[6], and H.L.L. Busard has shown which parts of it are drawn from other texts of the 13th century, especially from Jordanus Nemorarius' *Arithmetica*[7].

[6] See *Euclid: Transmission of the Elements*, in *Dictionary of Scientific Biography*, vol. 4, New York 1971, pp. 437-459, especially pp. 446-448, 457, and *The Medieval Euclid: Salient Aspects of the Translations of the* Elements *by Adelard of Bath and Campanus of Novara*, in *XIIe Congrès International d'Histoire des Sciences, Colloques*, in «Revue de synthèse», 89, 1968, pp. 67-94.

[7] See H.L.L. Busard, *The Translation of the* Elements *of Euclid from the Arabic into Latin by Hermann of Carinthia (?), books VII, VIII and IX*, in «Janus», 59, 1972, pp. 125-187 (here:

Campanus took over the enunciations from Version II of Euclid, but not the proofs, such as they were. He tried to write a «textbook» of the *Elements* that would include material of interest to the scholastic philosophers. Particularly notable is Campanus' inclusion of contemporary and near-contemporary material, such as Nayrīzī's commentary on the *Elements* (which had been translated by Gerard of Cremona) and, above all, Jordanus' *Arithmetica*. Some excursus were added by Campanus to discuss and, if possible, make clear obscure points of the Euclidean text. Relatively well-known are his remarks on the so-called «horn angle» between the diameter and tangent of a circle[8], his notes on the star polygons[9] and his attempt to make sense of the crucial definitions of proportionality at the beginning of book V, which were not understandable as transmitted[10]. But there are other annotations and comments. Of course, this is not the place to deal with them in detail, but some notice should be made of extra propositions which will become important for our study of Piero's treatises.

To the twenty-five propositions of book V contained in Version II Campanus added a further nine (V. 26-34). Three of them are to be found in Nayrīzī's commentary on the *Elements*[11]. The proofs given by Campanus are similar to, but not identical with, those in the Nayrīzī commentary.

The most important of Campanus' sources for his redaction of Euclid was Jordanus Nemorarius' *Arithmetica*. In his commentary to V def. 16 Campanus explicitly mentions this contemporary work, which was a highlight in Western mediaeval mathematics[12]. Campanus added to the arithmetical books VII, VIII and IX of the *Ele-*

pp. 132-139), and Id., *The translation of the* Elements *of Euclid from the Arabic into Latin by Hermann of Carinthia (?), books VII-XII*, Amsterdam 1977, pp. 17-20 (Table II).

[8] At the end of III.15. (The numbering of the propositions is, if not otherwise stated, from the Campanus version). See e.g. M. Cantor, *Vorlesungen über Geschichte der Mathematik*, vol. II, Leipzig 1900, p. 104.

[9] At the end of I.32. See S. Günther, *Lo sviluppo storico della teoria dei poligoni stellati nell'antichità e nel medio evo*, in «Bollettino di Bibliografia e di Storia delle Scienze Matematiche e Fisiche», 6, 1875, pp. 313-340.

[10] See J. Murdoch, *The Medieval Language of Proportions: Elements of the Interaction With Greek Foundations and the Development of New Mathematical Techniques*, in A.C. Crombie ed., *Scientific Change*, London 1963, pp. 237-271, 334-343.

[11] Busard, *The translation*, 1977, cit., pp. 131-132.

[12] «Unde Jordanus in secundo aritmetice sue diffiniens que proportiones sunt eedem et que diverse, dicit easdem esse que eandem denominationem recipiunt. Maiorem vero que maiorem et minorem que minorem». (All citations of the Campanus text are from the 1482 edition).

ments a number of definitions (and to book VII he also added postulates and axioms). As Dr. Busard has shown, nearly all of them have been taken from Jordanus' *Arithmetica*[13]. The same is true of some annotations which Campanus made to theorems in books VII, VIII and IX of Euclid[14].

In the stereographic books of the *Elements*, i.e. the genuine books XI-XIII and books XIV-XV which were later appended, Campanus also added material to the traditional text (as transmitted in the so-called Version II). Though these additions are not less important than those in the arithmetical books VII-IX, they have been almost completely neglected by historians of mathematics. In book XII, where the method of exhaustion is applied to solids such as pyramids and cones, Campanus made some additions that concern the relationships between pyramids, parallelepipedal solids and cones[15]. Book XIII, which is devoted to the regular solids, shows how the sphere can be found which circumscribes each solid. It is crowned by theorem 18, in which the edges of the five regular solids inscribed in the same sphere are arranged according to their magnitudes, and in which is proved that there are not more than five regular solids. This book has no spectacular additions by Campanus. But in his text of book XIV there are several new theorems. This book was written by Hypsicles, but in the Middle Ages treated as an integral part of Euclid's work. It contains some interesting additional propositions about the regular solids, such as the relative dimensions of the dodecahedron and icosahedron inscribed in one and the same sphere. While the Greek edition has (in the numeration by Heath) only eight propositions, Campanus has added ten more[16]. They have to do with dividing the side of the regular hexagon cut in extreme and mean ratio and with the relations between special regular solids and the sphere circumscribing them.

Book XV is also concerned with the regular solids. It shows how

[13] Busard, *The translation*, 1972, cit., pp. 132-135 with Table II on pp. 135-139, and Id., 1977, cit., pp. 17-20 (Table II).

[14] There are striking similarities between Jordanus and Campanus' annotations to VII.17, 20; VIII.8; porism VII.33; IX.5; additions to IX.16: see Busard, *The translation*, 1972, cit., pp. 133-134; additional material in Id., 1977, cit., p. 9. A list of propositions in Jordanus' *Arithmetica* which correspond to propositions by Campanus and might have been his source is given in H.L.L. Busard, *Jordanus de Nemore, De elementis arithmetice artis. A medieval treatise on number theory*, Stuttgart 1991, pp. 36-43.

[15] XII.6 add.1-4, 7 add.1, 8 add.1-5, 10 add.1.

[16] IX.3, 10-18 in his numeration.

to inscribe some of them in others[17]; it explains how to calculate the number of edges and angles in the five solids; and it shows how to determine the angle of inclination between faces meeting in an edge of any solid. This book was also enlarged by Campanus, who added to the original five propositions eight new ones[18] which deal with inscribing certain additional regular solids or the sphere in certain others. It is not clear where the additions in books XIV and XV in Campanus' version come from. The additions to the books of the *Elements* about regular solids must have been of special interest for Piero. The question which manuscript he might have had before him will be raised later.

As I said, the Campanus version was the common Euclid text in the 15th century. It was used for the Euclid lectures given at the universities within the curriculum of the *artes liberales*. Although there were new attempts to study Euclid during Piero's lifetime – I mention Pope Nicholas V, who in the 1450s was not only responsible for the new Archimedes translation made by Jacobus Cremonensis, but also tried to assemble a Greek Euclid text[19], and Regiomontanus who tried to reconstruct the original Euclid text with the help of Bessarion's Greek manuscripts – only after Piero's death was anything new discovered and made known: in 1498 Giorgio Valla published books XIV and XV, together with commentaries, and also took over Euclidean material into his encyclopedic *De expetendis et fugiendis rebus*[20]. In the 1490s Bartolomeo Zamberti became acquainted with a Greek Euclid manuscript which he translated into Latin, but published only in 1505. This translation was ignored by Luca Pacioli, whose Euclid edition was published in the same year, 1509, in which his treatise on regular solids was printed. Pacioli tried to save the Campanus version by improving it. But this is another story.

ITALIAN EUCLID MANUSCRIPTS

It may well be asked whether Italian texts of Euclid's *Elements* were available in Piero's time. This question has not been studied

[17] Tetrahedron in cube; octahedron in tetrahedron; octahedron in cube; cube in octahedron; dodecahedron in icosahedron.

[18] XV.5, 7-13.

[19] See the notes in the introductions of the Campanus manuscripts Stuttgart, HB XI 24, and Vat. Pal. Lat. 1352.

[20] Published in Venice, 1501.

in detail. It is well-known that the first printed Italian translation was that by Tartaglia (published in 1543). There are also manuscripts of the *Elements* in Italian from the 16th century, some of which may have circulated before Tartaglia's redaction[21]. But of greater interest are two manuscripts in Italian – at least two – that transmit the complete Euclid text and were written in the 15th century: Siena, L.IV.16 and L.IV.17. Nothing has been published on one of these and only a short note has been published, by Gino Arrighi, on the other[22]. An examination of the manuscripts has shown that the two agree very well with each other, that both have the characteristics of the Campanus version[23], and that therefore they seem to be translations of the Campanus text. Thus it was possible to have access to Campanus' version of the *Elements* without using a Latin manuscript.

GENERAL CHARACTERISTICS OF PIERO'S WRITINGS

Although research into mediaeval writings on practical geometry in Italian has scarcely begun, we can say that Piero's contributions are extraordinary. This is true not only for the *De prospectiva pingendi* and the *Libellus*, but also for the geometrical part of the *Trattato*. It differs from the geometrical sections in usual *abaco* treatises by solving almost no practical problems of measurement. Instead of this, Piero deals almost exclusively with polygons and polyhedra. The *Trattato*, which was written before 1477-78, is Piero's earliest work. Between 1478 and 1482 he wrote the Italian version of *De prospectiva pingendi* for Federigo da Montefeltro, Duke of Urbino (1422-1482). It was later translated into Latin by Matteo da Borgo Sansepolcro and presented to the library of the Duke. The *Libellus* was dedicated to Federigo's son Guidubaldo da Montefeltro who

[21] E.g. Bologna, BU, 326/5 (by Bottrigai); Bologna, BU, 535 (by Sanseverini; a. 1553); Firenze, BN, II.383 (by Benedetto Varchi); Brescia, Codex Viganò (see G. Arrighi, *Über eine Übersetzung der Euklidischen «Elemente» in italienischer Sprache*, in «Sudhoffs Archiv», 63, 1979, pp. 356-366).

[22] G. Arrighi, *Il codice L.IV.16 (sec. XV) della Biblioteca degl'Intronati di Siena. Gli «Elementi» di Euclide, un vocabolario di abbreviature e gli «scemi delle botti»*, in «Accademie e biblioteche d'Italia», 38, 1970, pp. 34-41; and Id., *La prima volgarizzazione degli «Elementi» di Euclide (nel Cod. L.IV.16 della Biblioteca degl'Intronati di Siena)*, in «Bollettino U.M.I.», 4, 12, 1975, pp. 410-416.

[23] E.g. 34 propositions in book V, 18 in book XIV and 13 in book XV.

became Duke of Urbino in 1482; and therefore it must have been written after 1482.

The mathematical contents of the *Trattato* and the *Libellus* have been analyzed by Margaret Daly Davis. I should like to summarize her main results[24]. In the geometrical sections of the *Trattato*[25] Piero treats polygons and polyhedra in a remarkably precise and thorough way. His application of algebra to geometrical principles – e.g. in calculating the sections of a line divided according to the golden section, which is essential for the pentagon and the dodecahedron – has its roots in Leonardo Fibonacci's *Practica geometriae*, in which some problems similar to Piero's are treated. Although Piero does not cite Leonardo explicitly, he depends on his works, as contemporary *abaco* writers were wont to do. But Piero's studies were more exhaustive and were at least partially based upon the study of books XIII and XIV of Euclid's *Elements*. Propositions XIII.13-18, which treat the construction of the regular bodies in a sphere together with their proportional relationship to the sphere and to each other, were crucial for Piero in calculating the numerical values of the unknown magnitudes when the diameter of the sphere or an other straight line was given. He also utilized propositions XIV.5-10 in the problems concerning the dodecahedron and icosahedron[26].

The *Libellus* is closely associated with the *Trattato*. Over half of the problems given in books I-III of the *Libellus* had previously been demonstrated in the *Trattato* and were taken over, partly reworked[27]. Frequently the *Libellus* is more concise than the *Trattato* and adds passages to clarify mathematical procedures. But more interesting than these reworkings of older problems are those parts in the *Libellus* which have no equivalent in the *Trattato*. Among them are problems 1 to 13 of book III dealing with the measure of one regular body located in another[28]. All of Piero's problems, except one, relate directly to book XV of the *Elements* in the Campanus reworking, and therefore Piero must have studied this book before writing this part of the *Libellus*. While in book XV the geometrical

[24] Daly Davis, *Piero della Francesca*, cit., pp. 27-63.

[25] Ff. 80r-120v of the manuscript. In citing the problems of the *Trattato*, I follow the practice of Daly Davis numbering each of the problems in parentheses, according to the order of its appearance on the individual folio in manuscript Ashburnham 359* of the Biblioteca Laurenziana, Firenze.

[26] See Daly Davis, *Piero della Francesca*, cit., pp. 39-40.

[27] *Ibid.*, pp. 46-54, with information about which problems were taken over and which have been reworked.

[28] *Ibid.*, pp. 54-57.

proportions are given that exist among the regular bodies inscribed in one another, Piero gave numerical examples and applied the arithmetical procedures to the geometry in book XV by calculating unknown quantities. – In book IV of the *Libellus* Piero deals with irregular bodies. Some of them are derived from problems in the *Trattato*, while others have to do with the semi-regular solids already treated by Archimedes – I shall give later some ideas about Piero's sources in that part. Several solids mentioned by Piero are related to the study of architecture and have to do with perspective, because painters may have to represent them[29]. Therefore this part of the *Libellus* is linked with Piero's *De prospectiva pingendi.*

EUCLID CITATIONS IN PIERO'S WORKS

Piero cites Euclid very often by name. If I have counted correctly, the name «Euclid» (in some form) is given 12 times in the *Trattato*, 11 times in *De prospectiva pingendi* and 52 times in the *Libellus*; with the exception of 4 citations in *De prospectiva pingendi*, which are from Euclid's *Optics* (see below), all refer to the *Elements*. At two places the name «Campanus» is explicitly given[30]. If one looks at which propositions of Euclid have been cited, one finds that most of them are from books I-III and XIII-XV (see Appendix 1). Of course, this is not surprising, because the first books give the fundamentals of geometry, and the last deal with the regular solids. Very often the Pythagorean theorem (I.46) is cited, mostly in the form *per penultimam primi Euclidis*. Other important theorems for Piero were the chord theorem (III.34), the last theorem of book XIII (XIII.18), which compares the sides of the regular solids with each other, and the theorems in book XV, which give the proportions between regular bodies inscribed in one another. From the many propositions which Piero cites it is clear that he must have studied the *Elements* thoroughly. This is especially true for book XV, which was his main source for the first part of book III of the *Libellus*. From propositions III.1-13, twelve can be found in book XV of the

[29] *Ibid.*, pp. 57-63.

[30] *Libellus*, III.5: «Et si velis quod anguli cubi terminent in centro octobasium triangularium, ut inquit Campanus, erit cubus pro quolibet suo latere radix 48». *Libellus* IV.1: «Hoc corpus docet conficere Campanus in XIIII^a^ parte XII^i^ Euclidis, qui per lineas tantum suorum laterum quantitatem demonstrat; nec quae sit eius superficies ostendit, de qua quaeritur».

Campanus version[31], and when dealing with the icosahedron inscribed into a cube – a problem which is not in the Campanus Euclid – Piero writes: «Sed quia Euclides non tangit quod dictum corpus constituatur in cubo ...»[32].

Of course, Piero might have used for his writings parts of the *Elements* without citing Euclid's name. When Piero gives the definitions of the circle, its diameter and the circumference, he might have taken this from Euclid, although the wording is somewhat different[33]. In the case of the sphere, Piero gives two definitions, one from Euclid and the other from Theodosius[34]. The definition attributed to Euclid is close to the Campanus text, while the Theodosius definition seems to have come from the reworking of Theodosius' *De sphaera* attributed to Campanus[35].

Piero cites texts from Euclid in the way most mathematicians and philosophers did in the late Middle Ages, e.g. *Habes per XXX^am^ III^ii^ Euclidis ...*[36]. Sometimes this citation is given alone, and sometimes a paraphrase of the content of the proposition referred to follows. If the content of the proposition is given, in almost all cases Piero's text does not agree with the wording of Campanus. To give an example:

Libellus I.*36*: Et ita probat in x^a^ XIII^i^, quod linea quae subtenditur angulo pentagonico divisa cum proportione et medium et duo extrema habente maior pars est latus pentagoni.

[31] Listed by Daly Davis, *Piero della Francesca*, cit., p. 57. See Appendix 2.

[32] *Libellus*, III.4.

[33] *Ibid.*, before I.43: «Circulus est quaedam superficies unica linea comprehensa quae dicitur circumferentia. Et ea linea, quae in illo maior fiat, diameter appellatur, qui circulum ac superficiem aeque ac bipartite dividit. Punctum vero, quod medium tenet, centrum dicitur». Campanus has (I. Def.15-17): «Circulus est figura plana una quidem linea contenta, que circumferentia nominatur, in cuius medio punctus est: a quo omnes linee recte ad circumferentiam exeuntes sibi invicem sunt equales. Et hic quidem punctus centrum circuli dicitur. Diameter circuli est linea recta que super eius centrum transiens extremitatesque suas circumferentie applicans circulum in duo media dividit».

[34] *Ibid.*, after III.13: «Est igitur spera, auctore Euclide, transitus circumferentiae dimidii circuli, quotiens fixa diametro, quousque ad locum suum redeat circumducitur. A Theodosio autem sic deffinitur: Spera est solidum corpus una superficie contentum, in cuius medio punctus est a quo omnes lineae ductae ad circumferentiam sunt aequales. Punctus ille dicitur centrum sperae. Linea autem transiens recta per centrum sperae applicans extremitates suas ad circumferentiam ex utraque parte, dicitur axis sperae».

[35] The definition of the sphere in this version runs: «Spera est figura solida una tantum superficie contenta in cuius medio punctus est a quo omnes linee recte ad eius superficiem exeuntes sibi invicem sunt equales. Et hic quidem punctus spere centrum dicitur. Diameter spere est linea recta per centrum spere transiens extremitatesque suas superficiei spere applicans ex utraque parte». (Information kindly supplied by Richard Lorch).

[36] *Ibid.*, I.46.

Campanus XIII.11[37]: Si duobus propinquis angulis pentagoni equilateri intra circulum descripti a terminis suorum laterum due recte linee subtendantur utraque alteram secundum proportionem habentem medium duoque extrema secabit maiorque ipsius portio lateri ipsius pentagoni equalis erit.

One should not conclude from the different wording that Piero did not use the Campanus text. This he undoubtedly had before him, because only here were the additional propositions in book XV to be found, and because he cites Campanus' name twice, in both cases for the Euclid redaction[38]. Most scholars agree that the *Libellus* was originally written in Italian and later translated into Latin, and therefore the Latin translation might have induced small changes in meaning. But this does not suffice to explain the differences between Piero and Campanus, because in the *Trattato*, of which we have the original, the Euclid citations differ from the Campanus text, too. In general, the style in which both the *Libellus* and the *Trattato* are written widely differs from that in the Latin Euclid tradition. Many terms used in the *Libellus* are totally un-Euclidean: e.g. the names of the regular solids (*quadribasis triangularis aequilatera* = tetrahedron[39], *corpus octobasium triangulare et aequilaterum* = octohedron[40]), *quadribasis triangularis* = general pyramid[41], *suum posse* = its [sc. a line's] square[42], *vis* = square[43], *linea plana* = a plane cutting a sphere into two parts[44]. In the second part of XIV.8 Campanus uses the Roman names for fractions, such as *dodrans* for 3/4 and *dextans* for 5/6, while Piero has *tres quarte* and *quinque sextas*[45]. This difference between Piero's style and the Euclid tradition is not surprising, because Piero came from the *abaco* tradition and seems at first not to have been familiar with the style used in scientific mathematical works.

One question still remains: did Piero use more than one Euclid text, and did he change his procedure of citation during the long time he must have worked on the three treatises? I think there is

[37] As to the different numbering, see the following chapter.
[38] See note 30.
[39] *Ibid.*, II.1ff.
[40] *Ibid.*, II.21ff.
[41] *Ibid.*, II.9ff.
[42] *Ibid.*, I.42.
[43] E.g. *Ibid.*, II.30.
[44] Proposition before II.1, Mancini, *L'opera*, cit., p. 520: «Linea plana est illa quae duas in partes speram resecat, facitque superficiem circularem».
[45] *Libellus*, II.29.

no reason to assume this. The style, at any rate of the *Trattato* and the *Libellus*, is consistent, even in the passages taken from Euclid. True, there is more liberal citation of Euclid in the later work: of the problems which exist in both treatises there are some in which Piero cites Euclid in the *Libellus*, but does not in the *Trattato*[46], while the opposite does not happen. Therefore, it seems that Piero, when writing his *Libellus*, added in some cases the source which he used. This agrees with the fact that the *Libellus* as a whole is more elaborate than the *Trattato*.

WHICH COPY OF CAMPANUS DID PIERO USE?

As I have already said, the first edition of Campanus' version of Euclid's *Elements* was printed in 1482. Because Piero wrote his *Trattato* before this time, he cannot have used this edition for it. Theoretically this could have been the case for his latest work, the *Libellus*. But one characteristic of the printed Campanus text is that at the end of book I an additional proposition (no. 48) is to be found. In the *Libellus*, Piero cites the Pythagorean theorem (I.46) always as the penultimate of the first book of Euclid and therefore cannot have been citing the printed text.

But which manuscript copy did he use? Of course, one is tempted to consult the book collection in Urbino, which in the time of Federigo da Montefeltro was one of the most famous humanist collections in the world; Vespasiano da Bisticci, who describes, among others, this collection in his *Vite di Uomini Illustri del Secolo XV*[47], states that Federigo acquired the books of «all the Latin writers on astrology, geometry, arithmetic, architecture, and military science»[48]. And, indeed, the old catalogue of this collection, made before 1487 by the librarian Agapito[49], lists three manuscripts of Euclid's *Elements*: one Greek[50] and two Latin[51]. Both Latin codices are now part of the Urbino collection in the Vatican Library (Urb.

[46] *Ibid.*, I.14, 34, 39, 42, 46; II.4, 6, 10, 15, 27, 30; III.23, 25, 27, 28.

[47] Ed. P. d'Ancona, E. Aeschlimann, Milano 1951.

[48] See P.L. Rose, *The Italian Renaissance of Mathematics*, Genève 1975, pp. 54f.

[49] See the *Index vetus* written before 1487. It was edited from Vat. Urb. Lat. 167 by C. Stornajolo, *Codices Urbinates Graeci ...*, Romae 1895, pp. 1v ff. A concordance with other catalogues and the manuscripts now in the Vatican Library is in C. Stornajolo, *Codices Urbinates Latini*, vol. III, Roma 1902-21.

[50] *Index vetus*, no. 126, which is not in the *fondo Urbinate Greco*.

[51] *Ibid.*, nos. 283 and 284.

Lat. 506-507)[52]. They transmit the usual Campanus text. Since these manuscripts were already in Urbino at the time of Piero, he might have consulted them. But they were at least not the main source for his knowledge of Euclid, because in neither are the propositions numbered, while Piero always gives exact numbers of them in both the *Trattato* and the *Libellus*.

In most cases the citation-numbers in both texts correspond to each other and to the usual numbering in the Campanus text. But there is one important exception: Piero cites some enunciations in book XIII by numbers which differ by one from the usual numeration in the Campanus text: XIII.8 instead of XIII.9[53], XIII.9 instead of XIII.10[54], XIII.10 instead of XIII.11[55], XIII.13 instead of XIII.14[56]. Because he always cites them in this way[57], this cannot be a coincidence: it must be concluded that Piero's text was a manuscript in which these four propositions were numbered as XIII.8, 9, 10, and 13. I have checked all Campanus manuscripts except three, which were not available to me[58]. Of the 131 known manuscripts of the Campanus text[59], 64 transmit these propositions. 37 of them have them numbered in the usual way, i.e. as XIII.9, 10, 11, 14. In 21 manuscripts the propositions are not numbered, but the four propositions have their correct places: 9, 10, 11, 14. To these belong both Urbino manuscripts I have just mentioned. Three manuscripts deserve special mention, because the order of propositions XIII.9 ff. is disturbed. In one of them, Paris, Bibliothèque Nationale lat. 16197, from the 13th century, with no numbering, there follow, after the end of the proof of XIII.7, the enunciation of XIII.8 and the proof of XIII.9. The proof of XIII.8 and the enunciation of XIII.9, which originally were left out by the copyist, were later supplemented on a special folio[60]. The whole proposition XIII.8 (enunciation and proof) is missing in two manuscripts: New York, Columbia Un-

[52] Urb. Lat. 506 = *Index vetus*, no. 284; Urb. Lat. 507 = *Index vetus*, no. 283.

[53] *Trattato*, f. 90r(1); *Libellus*, I.28 and I.36.

[54] *Trattato*, f. 90r(1); *Libellus*, I.28 and I.36.

[55] *Trattato*, f. 90r(1); *Libellus*, I.36.

[56] *Trattato*, f. 110r(1).

[57] Except *Libellus*, IV.3, where he cites Euclid XIII.9, giving the correct number. I think this is not a counter-argument, because it is only one out of ten instances, and scribal errors like «VIIII» instead of «VIII» could easily have occurred: there are other miswritings of numbers in the manuscript of the *Libellus*.

[58] Glasgow, UL, Gen. 1115; Pommersfelden 230; Salamanca 2624.

[59] Listed by M. Folkerts, *Euclid in medieval Europe*, Winnipeg 1989 (Questio II de rerum natura), pp. 38-43. The manuscript Princeton, University Library, Garrett 104 should be added.

[60] F. 114r (f. 114v being empty).

iversity, Plimpton 165 (end of 13th century), and Jena, Universitäts-bibliothek, El. fol. 67 (14th century). In the Plimpton manuscript the missing proposition was added on an inserted folio[61], while in the Jena codex only the figure pertaining to XIII.8 is given in the margin. From there we can conclude:

1. There were Campanus manuscripts in which XIII.8 was missing, so that propositions XIII.9 ff. took the places of 8 ff. It is evident that Piero had such a manuscript before him.
2. Neither the two Euclid manuscripts in Urbino (now Vat. Urb. Lat. 506 and 507) nor the Italian translation in Siena, L.IV.16 and 17 omit proposition XIII.8 or another one at the beginning of book XIII. Therefore they were not the copies Piero worked with.
3. Although the Campanus manuscripts now in New York and in Jena leave out proposition XIII.8, Piero cannot have used them, because they are not numbered, and therefore it seems not very likely that Piero took from them the numbering 8-10, 13 instead of 9-11, 14.
4. Hence it must unfortunately be concluded that Piero worked with a Campanus manuscript which is no longer extant, but was similar to the Plimpton or the Jena manuscript.

OTHER CITATIONS IN PIERO'S WRITINGS

In his three treatises Piero mentions only a few persons by name. In his dedicatory letter at the beginning of the *Libellus*, which is a paraphrase of the introduction to book III of Vitruvius' *De architectura*[62], and also at the beginning of book III of his *De prospectiva pingendi*[63] he gives the names of some Greek artists. Apart from Euclid's *Elements* and Campanus' reworking of it, he cites in the geometrical parts only Euclid's *Optics*, Archimedes, Theodosius, Ptolemy; all these names are given much less frequently than citations of Euclid's *Elements*. I should like to make some remarks about them.

Euclid's *Optics* is cited only in the *De prospectiva pingendi*, and

[61] On f. 126r the text of proposition 8 is written, numbered in the margin as «9ª». At the end the word *interposita* is written in the margin. The *verso* of the folio is blank. Between XIII.7 and XIII.9 (on f. 123r) we find in the margin the note *hic intrat* «8ª». Propositions XIII.1 to 7 are numbered by «1ª» to «7ª», XIII.9 by «9ª», while the following propositions are not numbered.

[62] Daly Davis, *Piero della Francesca*, cit., pp. 44-45.

[63] Piero della Francesca, *De prospectiva pingendi*, ed. Nicco Fasola, cit., p. 129.

there four times[64]. Piero always gives its title as *De aspectuum diversitate.* Although the most common version of Euclid's *Optics* in the West was the direct translation from the Greek (usually known as *De visu*), this citation refers to a translation from the Arabic. There were at least two of them (known as *Liber de aspectibus* and *Liber de radiis visualibus*) and an additional version that might stem from Theon of Alexandria's recension[65]. It is very likely that Piero refers to this last text: there is one and only one manuscript of this version, and this belonged to the Urbino library (now Vat. Urb. Lat. 1329). It was written in 1458 and has exactly the same title that Piero gives.

Now to Theodosius, Ptolemy and Archimedes: Theodosius is cited only once, when Piero defines the sphere[66]. Ptolemy's name is given in three places[67], always as a witness for the formula concerning the relation between the side of a regular pentagon and the diameter of the circumscribed circle[68]. This formula is proved in Ptolemy's *Almagest*[69], but it seems more likely that Piero knew it from a text derived from the *Almagest*, such as the *Almagestum parvum*, which was well-known in the Latin West[70]. Archimedes' name is cited more frequently: once in the *Libellus*[71] and eight times in the *Trattato*[72]. All these instances have to do with either the circle or the sphere (including surface and volume of spherical segments), and accordingly the citations refer to *De mensura circuli* or *De sphaera et cylindro*. Clagett gives some arguments[73] that Piero had not read the Archimedes text itself, when writing the *Trattato*, but some mediaeval handbook such as Leonardo Fibonacci's *Practica geometriae* and/or John de Muris' *De arte mensurandi*, which con-

[64] I.6, 12, 13; III.10: *ibid.*, pp. 69, 75, 76, 211.

[65] The manuscripts are listed in D. Lindberg, *A Catalogue of Medieval and Renaissance Optical Manuscripts*, Toronto 1975, nos. 79 E-L (*De visu*), 79 A and C (translations from the Arabic), 79 B (Theon's recension?). The different versions are discussed in W.R. Theisen, *The Mediaeval Tradition of Euclid's* Optics, Ph.D. diss., University of Wisconsin, 1972, especially pp. 11-17 and 321-335. On p. 324, note 10 Theisen gives arguments for attributing the last version to Theon's recension.

[66] *Libellus*, after III.13; see above.

[67] *Trattato*, f. 98r(1) and f. 110v(1); *Libellus*, II.29.

[68] $s_5 = d/4\sqrt{10 - \sqrt{20}}$. Piero always has $d = 4$.

[69] I.10; pp. 32-34 in Heiberg's edition.

[70] See R. Lorch, *Some remarks on the Almagestum parvum*, in *Amphora. Festschrift für Hans Wussing*, Basel 1992, pp. 407-437.

[71] F. 106v(2).

[72] III.15, 16, 20, 23, 26; IV.10 (2x), 11.

[73] M. Clagett, *Archimedes in the Middle Ages*, vol. III, Philadelphia 1978, pp. 383-415.

tain Archimedean material. Later, Piero became acquainted with the Cremonensis translation of Archimedes, and in all likelihood he used the manuscript of it which was in the Ducal Library of Urbino (now Vat. Urb. Lat. 261). One of the strongest arguments is that Piero cites some propositions by numbers which differ from the usual numbering, but agree with that in the Urbino manuscript.

PIERO'S SOURCES ON THE SEMI-REGULAR SOLIDS

Margaret Daly Davis rightly emphasizes that Piero's *Libellus* does not belong to a pre-existing genre and is original in conception and content[74]. But it should be noted that the study of regular solids was common in the Middle Ages, because they formed part of the *Elements* and had a special place in the Campanus reworking. Leonardo Fibonacci deals with them in his *Practica geometriae*; and in Thomas Bradwardine's well-known *Geometria speculativa*, written in the first half of the 14th century, there is, too, a long chapter on the regular bodies[75]. More remarkable is that Piero also deals with semi-regular or Archimedean polyhedra. They belong to the first type of semi-regular solids in which the requirement of congruence of the faces has been abandoned. We know from Pappus that Archimedes discovered thirteen of them, starting from the five regular solids and dividing the edges of them successively into two equal parts, into three equal parts, and into three parts the middle of which being to each of the two outermost as the diagonal of a face to the side[76]. Six of Archimedes' thirteen semi-regular solids are to be found in Piero's writings: two (P_1 and P_2) in his *Trattato*[77] and five (P_8, P_9, P_3, P_4, P_1) in his *Libellus*[78]. It has been asked what Piero's sources were – Clagett was profoundly puzzled by this question – for Pappus' *Collectio* seems not have been known in the

[74] Daly Davis, *Piero della Francesca*, cit., p. 18.

[75] Chapter 4.2 in G. Molland's ed., *Thomas Bradwardine, Geometria speculativa. Latin text and English translation with an introduction and a commentary*, Stuttgart 1989. See Id., *An Examination of Bradwardine's Geometry*, in «Archive for History of Exact Sciences», 19, 1978, pp. 113-175, especially p. 130.

[76] T.L. Heath, *A History of Greek mathematics*, vol. I, Oxford 1925, p. 295; vol. II, Oxford 1925, pp. 98-101.

[77] P_1 (4 triangles, 4 hexagons) and P_2 (8 triangles, 6 squares): Piero della Francesca, *Trattato d'abaco*, ed. Arrighi, cit., pp. 230-233.

[78] P_8 (12 pentagons, 20 hexagons), P_9 (20 triangles, 12 decagons), P_3 (6 squares, 8 hexagons), P_4 (8 triangles, 6 octagons), P_1 (4 triangles, 4 hexagons): chapters IV.2-IV.6.

Middle Ages in the West[79]. Some suggestions may be given: The semi-regular solid P_9 (20 triangles, 12 decagons) is treated in Euclid XIII.9, and this proposition is explicitly cited by Piero (IV.3). Another author on regular and semi-regular solids has been ignored up to now: John de Muris. In chapter 11 of his *De arte mensurandi*, which was probably completed before 1345, he treats the Archimedean solids P_2, P_7 and P_8, generating them from regular solids in a way similar to that of Archimedes (according to Pappus)[80]. John de Muris explicitly states that a stone-cutter made models of the regular solids for him and under his guidance[81] – as Luca Pacioli did in 1489. It is possible that John de Muris' treatise was available to Piero – at least one manuscript of it still exists in Florence[82]. But John cannot be the only source for Piero, because Piero also deals with semi-regular solids missing in the *De arte mensurandi*. Since John de Muris was a mediocre mathematician, it must be assumed that he took his information from texts which are now lost. We know that there were some Arabic treatises on semi-regular solids (e.g. by Thābit ibn Qurra[83], Abū'l-Wafā[84] and al-Kāshī), and it is possible that something of them was known in the Latin West, though no translation into Latin is known. It can be assumed that Piero had access to some of the material on semi-regular solids which undoubtedly existed in the West in the time of John de Muris.

FINAL REMARKS

Perhaps this paper has raised more questions than it has answered. But at least it has become clear that Euclid's *Elements* were the most important scientific book for Piero, that he used the Cam-

[79] Clagett, *Archimedes*, cit., pp. 398-406.

[80] The following information is from Busard who prepares an edition of this text.

[81] Capitulum XI, prop. 5: «Fortassis autem talium corporum figuram imaginatio non capit satis prompte, igitur consulo corpora 5 regularia in materia sensibili fabricari, sicut et ego manu propria lineavi et postmodum per artificem latomorum in materia lapidis feci predicta corpora preparari me presente, ne in errorem propter linearum multitudinem latomus verteretur. Et tunc sensus et intellectus in die illa concordes facti sunt pariter et amici».

[82] Biblioteca Nazionale, Magl. XI.2 (15th c.).

[83] F. Sezgin, *Geschichte des Arabischen Schrifttums*, vol. 5, Leiden 1974, p. 269. See E. Bessel-Hagen, O. Spies, *Ṯābit b. Qurra's Abhandlung über einen halbregelmäßigen Vierzehnflächner*, in «Quellen und Studien zur Geschichte der Mathematik, Astronomie, Physik», Abt. B: Studien, 2, 2, 1932, pp. 186-198.

[84] Sezgin, *Geschichte*, cit., p. 324.

panus text, and that this version was already well-known to him when he started his *Trattato*. The copy he used does not still exist. Later, when preparing his *Libellus*, he seems to have studied Campanus' text in more detail, especially book XV, which became the base for the first part of book III of his *Libellus*. But Piero did not simply take over the Campanus text. He used it for calculating plane figures and solids, applying to them and taking for granted the methods of algebra, which he had explained in the first part of his *Trattato*. He performs the calculations, which are sometimes very complicated, with a facility that exceeds by far the abilities of the «common» *maestro d'abaco*. To introduce such a mass of theoretical mathematics as is given in Euclid's *Elements* into the practical mathematics of the *abaco* treatises, and to apply it to problems which were nearly useless for practicioners, is an achievement by Piero. But even more important are some contributions which – according to our present knowledge – seem to be original, or at least have no known equivalent in the West. I have mentioned one proposition on regular solids inscribed in each other of which Piero himself said that he did not find it in the Euclid text. The treatment of the semi-regular solids may also be – at least in part – original. Another example is Piero's treatment of the geometry of the cross vault of which no mediaeval or early Renaissance Latin equivalent is known[85]. Perhaps Piero acquired formulas for them from discussions with architects or mathematicians, or he may have discovered it for himself by analogy[86]. Another contribution, which in all likelihood is original to Piero, is the drawing of geometrical diagrams in correct perspective: in all mathematical texts that I know from the 15th century (and even in many from the 16th century) the figures which illustrate the text are drawn without any perspective. This is particularly true of all of the manuscripts of the Campanus Euclid. The difference between Piero and the Campanus text (both manuscripts and printed edition) is obvious (see figures).

All of this shows that Piero was not only a well-known painter, but also a mathematician who ranked high among the practicioners of his time; in the words of Morris Kline[87]: «He was the painter-

[85] *Libellus*, IV.10, concerning the volume of the common segment of two perpendicularly intersection cylinders, and IV.11, which makes use of the surface area of that common segment.

[86] See Clagett, *Archimedes*, cit., p. 415.

[87] *Mathematical Thought from Early to Modern Times*, New York 1972, p. 233.

mathematician and the scientific artist par excellence, and his contemporaries so regarded him». It is no exaggeration to say that with him a new period of cooperation between artists and theoretical mathematicians began (figg. 1-3).

APPENDIX I: THEOREMS OF EUCLID'S *ELEMENTS* CITED BY PIERO

I.def.23: *Trattato*, before f. 87v(1)[1]
I.24: *Prospectiva* I.3, 7
I.34: *Prospectiva* I.9
I.35: *Trattato* f. 87v(1)[2]
I.37: *Libellus* IV.13
I.41: *Libellus* I.2 (2x)
I.46: *Trattato* f. 104v(1), 106r(1), 112r(1), 112v(1); *Libellus* I.1, 13, 14, 34, 39, 42, before II.1, II.4, 6, 10, 15, 31, 32, III.1, 2, 9, 23, 25, 26, 27, 28, IV.12, 18
II.4: *Trattato* f. 103r(2)[3]; *Libellus* I.25[3]
III.17: *Libellus* IV.14
III.30: *Libellus* I.42 (?)[4]
III.34: *Libellus* I.46[5], III.23, 25
VI.8: *Libellus* III.27
VI.21: *Prospectiva* I.8, 10 (?)[6]
VI.22: *Trattato* f. 104r(1)[7]; *Prospectiva* I.11 (?)[8], 16[8]
XI.def.14: *Libellus*, after III.13[9]
XII.5: *Libellus* IV.13 (?)[10]
XII.9: *Libellus* II.24
XII.14: *Libellus* IV.1
XIII.9: *Trattato* f. 90r(1)[11]; *Libellus* I.28[11], 36[11], IV.3
XIII.10: *Trattato* f. 90r(1)[12]; *Libellus* I.28[12], 36[12]
XIII.11: *Trattato* f. 90r(1)[13]; *Libellus* I.36[13]
XIII.14: *Trattato* f. 110r(1)[14]
XIII.15: *Libellus* II.22
XIII.18: *Trattato* f. 112r(1); *Libellus* II.27, 30, 31 (2x), addition to II.37 (2x)
XIV.8, 2: *Trattato* f. 110v(1)[15]; *Libellus* I.36[15], II.29[15]
XV.1: *Libellus* III.2, 12
XV.3: *Libellus* III.3
XV.4: *Libellus* III.5[16]
XV.5: *Libellus* III.1
XV.9: *Libellus* III.9
XV.10: *Libellus* III.8
Euclid: *Libellus* III.4

[1] No number of definition is given, but only the name *simile almuachin*.
[2] Erroneously given as I.15.
[3] No number of proposition given.
[4] Erroneously given as III.21.
[5] Erroneously given as III.30.
[6] Erroneously given as 30 (without indicating the book number).
[7] Erroneously given as VI.24.
[8] Erroneously given as VI.25.
[9] No number of definition given.
[10] Erroneously given as XI.19.
[11] Given as XIII.8.
[12] Given as XIII.9.
[13] Given as XIII.10.
[14] Given as XIII.13.
[15] Given as XIV.9.
[16] Neither the number of proposition nor the name *Euclides* given, but the name *Campanus*.

APPENDIX 2: CONCORDANCE *LIBELLUS* III.1-13 - CAMPANUS XV

Libellus		Campanus
III.1	Octahedron in tetrahedron	XV.2
III.2	Tetrahedron in cube	XV.1
III.3	Octahedron in cube	XV.3
III.4	Icosahedron in cube	–
III.5	Cube in octahedron	XV.4
III.6	Tetrahedron in octahedron	XV.5
III.7	Cube in dodecahedron	XV.8
III.8	Tetrahedron in dodecahedron	XV.10
III.9	Octahedron in dodecahedron	XV.9
III.10	Icosahedron in dodecahedron	XV.7
III.11	Cube in icosahedron	XV.11
III.12	Tetrahedron in icosahedron	XV.12
III.13	Dodecahedron in icosahedron	XV.6

Addenda to X

[pp. 297f.] Information about the early Euclid manuscripts in Italian is to be found in an article by P. Pagli: see item III, pp. 57f., with a list of such manuscripts.

[p. 304, note 59] An up-to-date list of all Western medieval manuscripts of Euclid's *Elements* is given in item III, pp. 31–58.

[p. 308, note 80] Busard's edition has appeared in 1998 (H. L. L. Busard (ed.), *Johannes de Muris,* De arte mensurandi. *A Geometrical Handbook of the Fourteenth Century*, Stuttgart 1998).

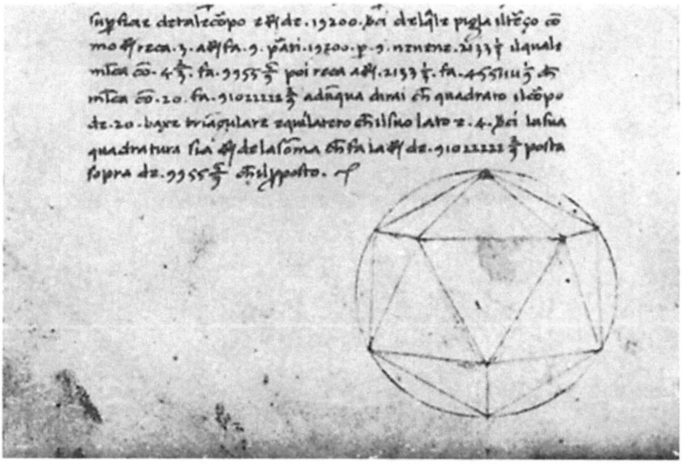

Fig. 1. Drawing of an icosahedron in Piero della Francesca's *Trattato d'abaco* (Florence, Biblioteca Medicea Laurenziana, Ashburnham 280 (359*), f. 113v).

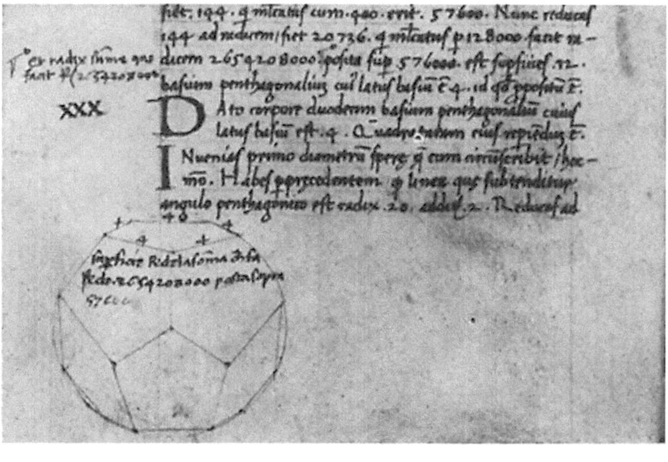

Fig. 2. Drawing of a dodecahedron in Piero della Francesca's *Libellus de quinque corporibus regularibus* (Biblioteca Apostolica Vaticana, Urb. Lat. 632, f. 33v).

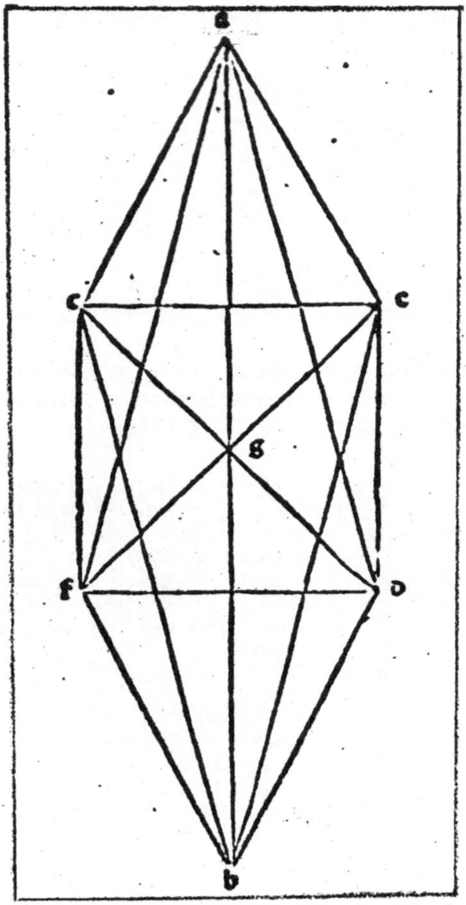

Fig. 3. Drawing of an octahedron in the printed edition of Campanus' Euclid (Venice 1482), XIV. 16. The lack of perspective should be noticed.

XI

LUCA PACIOLI AND EUCLID

Luca Pacioli's importance in the history of mathematics lies principally in his *Summa de arithmetica, geometria, proportioni e proportionalita,* published 500 years ago (1494). It is well known that the work is not original, but a compilation of the mathematics known in his time. Of course, Euclid was included in this. How high Pacioli estimated Euclid, can be seen from the famous painting presenting Pacioli with a Euclid text and a diagram from it. In this paper Euclid's rôle in Pacioli's oeuvres will be discussed, with respect not to his treatment of the regular solids, but to his translation of the *Elements* into Italian and his edition in Latin[1].

1. Pacioli's knowledge of Euclid

Whether Pacioli as a young man in Sansepolcro knew the *Elements* is not known. It has been supposed that he was taught there by Piero della Francesca (ca. 1410-1492). Even if this is so, it is unlikely that Pacioli learned from him much about the *Elements*, for Piero himself apparently did not read the work at all extensively until he wrote his *Trattato d'abaco* (before 1477). But when Pacioli went to Venice about 1464, there was a greater chance of his reading theoretical mathematical works[2]. Whether he read Euclid with Domenico

[1] L. PACIOLI, *Euclidis megarensis philosophi acutissimi mathematicorumque omnium sine controuersia principis opera a Campano interprete fidissimo tralata...*, Venice, 1509.

[2] For the importance of Venice for the development of mathematics in the 15th and early 16th centuries see P.L. ROSE, *The Italian renaissance of mathematics. Studies on humanists and mathematicians from Petrarch to Galileo*, Geneva, 1975, pp. 44-54.

Bragadino, who taught him mathematics in Venice[3], is not known. Also unknown is whether he was in contact with Cardinal Bessarion and his extensive library of Greek classics (including Euclid) and Latin translations. But it should be noted that the library was permanently in Venice only after Bessarion's death in 1472. Another opportunity to read Euclid was after his meeting with Alberti in Rome (1470-1471).

At the very latest he must have studied the *Elements* about the time of his entry into the Fransiscan Order, some time between 1470 and 1477. It is unthinkable that he gave his lectures at Perugia on arithmetic and geometry (1477-1480)[4] without knowing Euclid, although in the arithmetical treatise, which Pacioli wrote 1477-78 in Perugia[5], he did not deal with Euclid. Between 1480 and 1487, at the behest of his Order, Pacioli travelled to various places in Italy and Dalmatia. In Zara he wrote a treatise, now lost, on arithmetic (1481). Later (1487-88) he returned to Perugia to give lectures on mathematical topics, «ad docendum abicum». In 1489, when he was in Rome, he had models of the five regular and some of the semi-regular solids made by a craftsman. His interest in these solids does not mean that he had read the relevant part of Euclid – i.e. books XIII-XV – , but only that he knew Piero's *Libellus de quinque corporibus regularibus*. This text was written some time after 1482 and printed in Italian in 1509 as the third part of Pacioli's *Divina proportione*. In Rome Pacioli also gave lectures in mathematics[6]. Some time before 1494 he was in Naples; here he expounded Euclid to a certain Camillo Vitelli for three years – as he himself says in the introduction to his treatise on architecture[7].

[3] See L. PACIOLI, *Summa de Arithmetica Geometria Proportioni et Proportionalita*, Venice, 1494, I, f. 67.

[4] In the records of the magistrate of Perugia we find that on October 14, 1477, Pacioli received 30 florins «ad docendum artem arismetricam». When his salary was raised by 20 florins at January 11, 1478, we read: «cum iam legisset dictam artem [i.e. doctrinam arismetrice seu geometrie] per duos menses et optimam de se experientiam ostendisset»: see B. BONCOMPAGNI, *Intorno alle vite inedite di tre matematici (Giovanni Danck di Sassonia, Giovanni de Lineriis e Fra Luca Pacioli da Borgo San Sepolcro) scritte da Bernardino Baldi*, in «Bullettino di bibliografia e di storia delle scienze matematiche e fisiche», 12 (1879), pp. 352-438, 863-872; here p. 432.

[5] Transmitted in manuscript Vat. lat. 3129.

[6] L. PACIOLI, *Summa*..., cit., II, f. 74*v*: «Hauenga che nel 1489 nella cita de roma doue publice legiauamo...».

[7] «Con la Illustre Signoria miser Giouaniacomo traulzi con lo degno oratore del Dominio Fiorentino alora Pier vetori con presentia del Pontano nel palazzo del conte de Sarno in Napoli. E non manco con lo Magnifico e degno condoctiero Signor Camillo vitelli dela cita de castello legendoli Jo per anni tre el sublime volume del nostra Euclide»: FRA L. PACIOLI, *Divina proportione. Die Lehre vom Goldenen Schnitt. Nach der venezianischen Ausgabe vom Jahre 1509 neu herausgegeben, übersetzt und erläutert von Constantin Winterberg*, Vienna, 1889, p. 128. Because Pacioli mentions his stay in Naples in the dedicatory letter of his *Summa* to the Duke of Urbino, he must have been there before 1494.

Pacioli began his opus magnum, the *Summa*, during his second stay in Perugia, i.e. 1487-88[8]. He finished it in 1494 and went to Venice to supervise its publication. This work, dedicated to the young Duke of Urbino, Guidobaldo da Montefeltro (1472-1508), is the first clear indication that he had intensively studied the *Elements*.

Between 1497 and 1499 Pacioli was at the court of Duke Ludovico Sforza in Milan. Among the scholars and artists gathered at this court was Leonardo da Vinci. We know that Leonardo asked Pacioli questions about mathematics and that in return he drew the figures for the *Divina proportione*. Pacioli had written the first part in his time in Milan, finishing it in 1498[9], and he also held lectures there on Euclid before large audiences. In the preface to the *Divina proportione*, which was dedicated to Ludovico Sforza, Pacioli writes[10]:

And may this serve first as a recommendation of the mathematical sciences, the number of which in your renowned city increases daily, thanks to your ducal encouragement and because of the public lectures newly introduced by you for the benefit of the excellent audiences, for whom I, by the grace vouchsafed to me by Your Highness, expound, clearly and with great care (as they judge), the eminent volume of the aforementioned Euclid in the sciences of arithmetic, geometry and ratio and proportion. I have already come to the very worthy end of ten of his books. To his theory I always accommodate our practice, to make it more useful and more completely understandable, and I spend the rest of the time in carrying it out.

After the French took Milan (1499), Pacioli spent most of the time up to 1506 in Florence. In this time he deepened his knowledge of the *Elements* and lectured on the subject at the University of Pisa, which at that time functioned in Florence – sometimes on all the books and sometimes only on book VIII[11]. During his time in Florence he was for one year (1501-02) «lector ad mathematicam» at the University of Bologna together with Scipione del Ferro.

Because Pacioli had a special interest in the theory of proportions – he had already treated it in detail in the *Summa* –, book V of

[8] See L. PACIOLI, *Summa...*, cit., I, f. 67*v* and 98*v*.

[9] Of this version two manuscripts exist: Milan, Bibl. Ambr., & 170 sup. (edited in transcription, L. PACIOLI, *De divina proportione*, preface by Carlo Castiglioni, Milan, «Fontes Ambrosiani» 31, 1956, and in facsimile, L. PACIOLI, *De divina proportione*, with introduction by Augusto Marinoni, Milan, «Fontes Ambrosiani» 72, 1982) and Geneva, Univ. Libr., m.l.e. 210.

[10] FRA L. PACIOLI, *Divina proportione*, (1889), cit., 39f.

[11] The entries in the records of Florence are printed in B. BONCOMPAGNI, *Intorno alle vite...*, cit., pp. 438 and 864-869.

the *Elements* was of particular importance for him. In 1508 he gave public lectures in Venice on this book. From his edition of Euclid[12] we know that these lectures began on 11th August 1508 in the church of San Bartolomeo. Pacioli himself reports that there were more than five hundred in the audience and gives the names of the most important[13]. A little later he obtained a fifteen-year permission from the Doge to publish a series of books, inter alia his Latin Euclid, which was printed in Venice in 1509[14]. After this he appears to have done nothing more on Euclid. Until his death in 1517 he made few journeys outside Sansepolcro.

2. Pacioli's Italian translation of Euclid; the excerpts in the *Summa*

In his biography of Pacioli, Bernardino Baldi says that Pacioli had prepared an Italian translation of the *Elements*[15]:

He wrote many works. He considered more important than many other things a translation of Euclid's *Elements* for the young people of Perugia in «lingua volgare», a success like the other works, as he says in the dedication to Piero Soderini. He made use of Campano's translation, which was never very far from his meaning.

This passage gives the impression that the translation was about one of Pacioli's early works. But closer inspection reveals that the situation is more complicated. Baldi obtained his information from Pacioli's writings. In fact Pacioli himself refers three times to his translation of Euclid's *Elements*: in his letters of dedication to the *De viribus quantitatis*[16], to the *Divina proportione*[17] and to his Latin edi-

[12] L. PACIOLI, *Euclidis megarensis...*, cit., f. 30*r*-31*v*.

[13] He ends with: «Aliique plurimi quorum nomina sigillatim referre ad quingentos et amplius operosum nimis foret florem tantum hominum decerpsi», L. PACIOLI, *Euclidis megarensis...*, cit., f. 31*r*.

[14] The privilege is printed in B. BONCOMPAGNI, *Intorno alle vite...*, cit., 431f.

[15] «Scrisse molte opere, e prima ad instanza per quanto mi credo de la giouentù di Perugia trasferì in lingua uolgare gli Elementi di Euclide, il che gli successe (come egli afferma in una dedicatoria à Pietro Soderini) assai felicemente. Seruissi de la traduttione del Campano non discostandosi punto dal opinione di lui». See *Ibidem*, p. 422.

[16] «Ma ormai a proximandosi de mia uita lultimi giorni acio le durate fatighe et assidue uigilie non douesino al tutto anichilarsi como e ditto ali non mediocri affani posta gia la extrema mano con la egregia per noi similmente traductione de latino in uulgare de uerbo ad uerbum del maximo Monarcha dele Mathematici discipline megarense Euclide...». See *Ibidem*, pp. 418 and 430, from manuscript Bologna, BU 250.

[17] L. PACIOLI, *Divina proportione*, Venice, 1509, f. Aiir: «Accessit nunc ad eam curam: vt confluente studiosorum copia Megarensis Euclidis elementa lingua patria donare coactus sim:

tion of Euclid[18]. In addition, Daniele Gaetani mentions the translation in a letter to Andrea Mocenigo, which was also printed with the *Divina proportione*[19]. From these letters we gather that the translation was taken from the Campanus redaction, which Pacioli rated very highly[20], and that it was made before the *Divina proportione* was printed, i.e. before 1509[21]. The date of writing may be more precisely estimated from Pacioli's dedicatory letter to Pietro Soderini in the *Divina proportione* (1509): immediately after mentioning his Euclid translation, Pacioli says that, encouraged by the interest of the audience, he shortly afterwards wrote his *Divina proportione* (i.e. the first part, finished in 1497) and dedicated it to Ludovico Sforza[22]. It is accordingly very likely that his translation of the *Elements* was made before 1497.

No manuscript of Pacioli's Italian translation of Euclid has yet been found. On the other hand there are two Euclid manuscripts in Italian written before 1500: Siena, L.IV.16 and L.IV.17. Both transmit the complete text that Campanus of Novara compiled shortly before 1260. The translator is not mentioned in either manuscript. But there is not only no «a priori» reason for supposing that the translation is Pacioli's, but the text of these manuscripts does not match the wording of the Euclid excerpts in Pacioli's *Summa*.

As far as I know, it has not been realized by historians of mathematics that there are substantial parts from the *Elements* in Pacioli's *Summa*. There he quotes texts from books I, II, III, V, VI and XI. The

cessit id diis bene iuuantibus faelicissime». The letter of dedication to Pietro Soderini is dated: Venice, 13th June 1509.

[18] L. PACIOLI, *Euclidis megarensis...*, cit., f. 2*v*: «Leges qua soles frontis serenitate vernacula lingua per me donatum Euclidem: vt quod opus ad vtilitatem nostrum omnium conscriptum est maximasque humano generi vtilitates parit: Ita omnibus obuium sit: vt cum lingua patria nostra et in his disciplinis locupletetur». The letter of dedication is addressed to Pietro Soderini's brother Francesco, cardinal in Volterra, and is dated Venice, 13th June 1509.

[19] L. PACIOLI, *Divina proportione*, (1509), cit., f. Aiiv: «Ad hoc etiam illud quod subiectum certe formidandum tanta facilitate prosequitur vt a promptae et planae disputationis communione ne idiotarum quidem aut imperitorum siscitatio (!) repudietur: quemadmodum in Euclide cernere est quem de rhomano Vernaculum fecit nihil ab opinione Castigatissima domini Campani declinans quem summopere probat et sequitur». Gaetani's letter, dated Padova, 15th May 1509, follows Pacioli's letter of dedication.

[20] At the end of his letter to Mocenigo (see note 19), Gaetani writes: «nihil ab opinione Castigatissima domini Campani declinans quem summopere probat et sequitur».

[21] The exact date of *De viribus quantitatis* is unknown. We only know that it was written between 1496 and 1508, because in its dedication the year 1496 is mentioned and because Pacioli obtained in 1508 a printing privilege for it.

[22] L. PACIOLI, *Divina proportione*, (1509), cit., f. Aii r: «Nec vero multo post spe animos alentes libellum cui de diuina proportione titulus est: Ludouico Sphorciae Duci mediolanensi nuncupaui».

extracts from book V, on the theory of proportions, are included in the first, arithmetic, section; and the excerpts from the other books are integrated into the second, geometrical, section.

On the passages from book V I need say nothing, since they were treated in detail in 1990[23]. The extracts from books I-III, VI and XI are to be found in the second part of the *Summa*, the *Tractatus geometriae*, which, in Pacioli's own words, are «about continuous quantities, i.e. about geometry, [mostly] in practical applications and also the theory of all operations, always with their foundations». Clearly, Euclid's *Elements* are among these foundations. It is therefore not surprising that the *Elements* is, alongside Fibonacci's *Practica geometriae*, the principal source for the geometrical part of the *Summa*.

In the first *distinctio*, which treats the fundamental constructions and areas of figures, geometrical algebra and the theory of similarity, Pacioli took over all of the definitions, postulates, axioms and enunciations of books I, II and VI (except enunciations II.1-10) from Euclid[24]. The fourth *distinctio*, on circles, begins with the reproduction of all the definitions and enunciations of book III[25]. There follows the calculation of the areas of the circle and of its various parts, a subject not treated by Euclid. The sixth *distinctio* is on the calculation of volumes, i.e. on stereometry. Here again Pacioli begins with Euclid (book XI)[26] and then goes on to solve particular problems involving calculation.

There is no doubt that Pacioli took these Euclid extracts from Campanus' redaction. This may be seen in the order of the quotations, which follow the order of Campanus[27], and in the additions to the definitions in books II and VI, which are to be found only in Campanus. There are also Arabisms that are to be found in Campanus, for instance «mutekefia» (VI.14 and 15 in Heiberg). Pacioli also uses the nicknames for particular diagrams, which occur in the Arabic transmission and in some Latin texts dependent on it, among others Cam-

[23] M. Bartolozzi, R. Franci, *La teoria delle proporzioni nella matematica dell'abaco da Leonardo Pisano a Luca Pacioli*, in «Bollettino di Storia delle Scienze Matematiche», vol. 10 (1990), pp. 3-28 (here: pp. 17-27).

[24] Book I, on f. 1*r*-3*v*, forms chapters 1 and 2. Book II is on f. 3*v*-4*r* (chapter 3) and book VI on f. 4*r*-6*v* (chapter 4).

[25] On f. 26*r*-30*r* (chapter 1).

[26] L. Pacioli, *Summa*..., cit., f. 44 (chapter 1).

[27] The order is (in Heiberg's numbering of the Greek text): VI. 1-8, 13, 11, 9-10, 14-17, 19-20, 18, 21-22, 24, 26, 23, 25, 27-30, 32, 31, 33.

panus: III.7 is «campa docha», i.e. «pes anseris»; III.8 is «coda di pavone», i.e. «cauda pavonis»[28].

For the great majority of the Euclid extracts Pacioli's Italian follows Campanus' Latin almost word for word. This is true for the postulates and axioms of book I, the definitions of books II, III and VI and the enunciations in books I-III and VI.

In the definitions from book I and all the extracts from book XI the similarity to Campanus' redaction is not so striking. Even here it is recognizable that the text given by Pacioli is based upon Campanus' text[29], but the order of the material in general was changed, numerous additions to the definitions were made, and several enunciations in book XI were left out. The additions to the definitions of books I and XI are mostly about special lines, angles, figures and solids important in practical mathematics, but not in Euclid. Some were taken over from Leonardo Fibonacci's *Practica geometriae*[30]. We see that the compiler took trouble to be more practical than Euclid.

It is striking that in Pacioli's text there is no clear distinction between definitions and enunciations. The definitions in book III are called «conclusiones», but the same term is used for the propositions. Sometimes we find theorems within the definitions[31] and definitions within the theorems[32]. The compiler intended to write a mathematical guide for non-specialists, not a substitute for Euclid.

The excerpts from Euclid always come at the beginning of a section. They are there to give the theoretical justification for the following calculations. It is therefore understandable that no proofs are given (with three exceptions)[33]. In most cases there are only enunciations and explanations of them by reference to diagrams[34], though when the enunciation is about a construction, details are usually given. This shows the practical inclination: for a practical mathematician constructions are important, but proofs are superfluous.

It is tempting to suppose that the extracts from Euclid books I-III and VI were taken from Pacioli's own translation of Campanus.

[28] See P. Kunitzsch, «*The Peacock's Tail*»: *On the names of some theorems of Euclid's Elements,* in *Vestigia mathematica. Studies in medieval and early modern mathematics in honour of H. L. L. Busard,* ed. by M. Folkerts and J. P. Hogendijk, Amsterdam / Atlanta, 1993, pp. 205-214 (here: pp. 209f.).

[29] Striking similarities are in I.def.17, 18, III.def.6 and I.def.23.

[30] Not only some additions, but also Pacioli's order of definitions in book I agree with Leonardo (I.def.14, 19; III.def.6, 10; I.def.23; I.29).

[31] E.g. after I.def.23 the equality of alternate angles is stated.

[32] E.g. after XI.31 comes the definition of similar solids.

[33] III.30 P(acioli) = III.31 H(eiberg), III.31 P. = 32 H.; III.35 P. = 36 H.

[34] Not at the beginning of book I and in book XI.

But this assumption is incorrect, because the same Euclid excerpts exist in the same order and almost word for word in an earlier manuscript in a different dialect: Palatino 577 of the Biblioteca Nazionale in Florence. The geometrical section of Pacioli's *Summa* agrees in the other parts, too, with that Florence manuscript[35]. Palatino 577 seems to have been written about 1460, and therefore we must assume that Pacioli took over the Euclid excerpts in his *Summa* from this manuscript or a similar one.

3. The Euclid edition (1509)

The clearest evidence for Pacioli's interest in the *Elements* is provided by his Latin edition of it, which appeared in Venice in 1509. There is no reason to suppose that the work had already been printed in 1489 as is stated by Fabricius and the bibliographers who copied him[36]. The purpose of the edition is given on the title-page: that Campanus had translated the text very accurately, that it had been so corrupted at the hands of the scribes as to be scarcely recognized as Euclid's text; accordingly, Pacioli had found and corrected the mistakes, had correctly redrawn 129 faulty diagrams and had added numerous other, necessary, diagrams; finally, he had provided clear and learned comments to very many difficult passages – and in this he had been helped by Scipio Vegius of Milan[37].

To understand this note we must bear in mind that four years before Pacioli's edition, i.e. in 1505, an edition of Euclid was published, again in Venice, which was not based on Arabic versions, as Campanus' text was, but was translated from a Greek manuscript. Although the editor, the lawyer and humanist Bartolomeo Zamberti

[35] E. PICUTTI, *Sui plagi matematici di frate Luca Pacioli*, in «Le Scienze», (1989), pp. 72-79 (here: pp. 76f.).

[36] J.A. FABRICIUS, *Bibliothecae Graecae... liber III*, Hamburg, 1707, p. 374; J.C. HEILBRONNER, *Historia matheseos universae...*, Leipzig, 1742, p. 159. Boncompagni's opinion, that the year 1489 is an oversight, may be correct: B. BONCOMPAGNI, *Intorno alle vite...*, cit., 412f.

[37] «Euclidis megarensis philosophi acutissimi mathematicorumque omnium sine controuersia principis opera a Campano interprete fidissimo tralata Que cum antea librariorum detestanda culpa mendis fedissimis adeo deformia essent: vt vix Euclidem ipsum agnosceremus. Lucas paciolus theologus insignis: altissima Mathematicarum disciplinarum scientia rarissimus iudicio castigatissimo detersit: emendauit. Figuras centum et vndetriginta que in alijs codicibus inuerse et deformate erant: ad rectam symmetriam concinnauit: et multas necessarias addidit. Eundem quoque plurimis locis intellectu difficilim commentariolis sane luculentis et eruditissimis aperuit: enarrauit: illustrauit. Ad hec vt elimatior exiret Scipio vegius mediol. vir vtraque lingua: arte medica: sublimioribusque studijs clarissimus diligentiam: et censuram suam prestitit».

(1473 - after 1539)[38], did not use a good original[39], he made available for the first time a relatively pure text, which at any rate did not contain the obscurities of the Arabic-Latin translations. Zamberti was aware of this: on the very title-page he declared that there was much that was in the right form in his edition, but in the wrong order or wrongly presented in Campanus[40]. In the preface he referred to Campanus as «ille interpres barbarissimus», who used such barbaric expressions as «helmuain» for rhombus and who wrongly translated the definitions of book V, which were of central importance for Eudoxus' theory of proportions. In the course of the edition there are numerous animadversions by Zamberti against Campanus' text. These remarks are introduced by the word «Interpres», but whether Zamberti intended «translator» or «commentator» by this term is uncertain.

Pacioli's edition of 1509 is to be understood as an attempt to defend Campanus against Zamberti's criticisms. According to Pacioli, the mistakes in the printed Campanus text (Pacioli refers to the 1482 edition) are not to be ascribed to Campanus, but to careless copyists, and it should suffice to point these errors out. As Rose rightly remarks, «Pacioli portrays himself as a castigator of a different sort from Zamberti, whom he implicitly chides for ignorantly showing off at the expense of Campanus»[41].

Independently of the quarrel between Zamberti and Pacioli it may be asked how far Pacioli's Euclid edition of 1509 differs from the conventional Campanus text printed in 1482 (also in Venice). A comparison shows that the differences are not very important; and this is true of the figures as well as the text. Most of Pacioli's diagrams agree completely with those of the 1482 edition, though now and then they are slightly altered; sometimes Pacioli puts numerical values on the lines. In contrast to the 1482 edition, the diagrams are labelled with the corresponding proposition-numbers. As in Pacioli's Italian translation of the *Elements* the words «Pes anseris» and «Cauda pauonis» appear under the diagrams for III.7 and III.8. Few diagrams were added by Pacioli and these were mostly in cases in which he had extended the text with a commentary. In the stereometric books XI-XIII and in books XIV-XV Pacioli's figures are as lacking in perspective

[38] For Zamberti and his edition of Euclid see P.L. ROSE, *The Italian Renaissance...*, cit., pp. 50-52 and H. WEISSENBORN, *Die Übersetzungen des Euklid durch Campano und Zamberti. Eine mathematisch-historische Studie*. Halle/Saale, 1882, pp. 12-28.

[39] See P.L. ROSE, *The Italian renaissance...*, cit., p. 73, note 247.

[40] «...quibus multa quae deerant ex lectione graeca sumpta addita sub nec non plurima subuersa et prepostere: voluta in Campani interpretatione: ordinata digesta et castigata sunt».

[41] P.L. ROSE, *The Italian Renaissance...*, cit., p. 52.

as those in the 1482 edition. If Pacioli really corrected 129 figures as he maintains, he must have counted every minute change as a correction.

The text of the *Elements* in Pacioli's edition appears, from numerous samples, to be identical in all parts to the 1482 edition. Pacioli's claim on the title-page that he had corrected mistakes in Campanus' text is therefore not correct. But in many places he did add comments, mostly at the ends of propositions; in almost all cases they were prefaced by the word «Castigator». The following list shows that these additions are unevenly distributed in the work:

Book	Number of additions (in brackets: additions over 10 lines long)	
I	14	(5)
II	6	(1)
III	14	(5)
IV	11	(3)
V	22	(5)
VI	14	(3)
VII	5	(2)
VIII	—	
IX	2	(1)
X	28	(13)
XI	8	(1)
XII	4	(-)
XIII	2	(-)
XIV	3	(3)
XV	3	(-)

This survey shows that most of the additions are relatively short. It is noticeable that in books VII-IX, on the theory of proportion going back to the Pythagoreans and on the theory of numbers, and in the stereometical books XI-XV there are few additions, but Pacioli commented more liberally on the geometrical books I, III-IV and VI and also on books V and X, which treat, respectively, the Eudoxan theory of proportion and irrationals.

More persuasive than the above statistics is, of course, an analysis of the material added by Pacioli. I can give here only a few examples. In general, Pacioli confines himself to explaining terms or individual steps within a proof or construction. In such cases he indicates his comments by raised numbers placed after the relevant word. Thus in a comment to I.1 he says that «to construct a figure on a given line» means «to lay down the line as one of the sides of the figure that is thus to be described on it»[42]. In his additions Pacioli often makes remarks that are not immediately necessary for the understanding of the theorem, but are suggested by it. Here are some examples[43]:

- Prop. I.5 says that in an isosceles triangle the angles at the base are equal to one another, and, if the equal lines be produced further, the angles under the base will be equal to one another. Pacioli adds here, «This fifth theorem is called "the flight of the wretched", because those of wretched intellect take flight when they reach it. But, so that there is no occasion for flight, I shall demonstrate it briefly and with an easy proof that should suffice for the beginner»[44]. There follows a proof, which differs from that of Campanus. The name «fuga miserorum» is not to be found in Campanus but in other manuscripts of the Arabic-Latin tradition, *inter alia* in a codex in the possession of Cardinal Bessarion (now Venice, BN, f.a. 332)[45]. This shows that Pacioli knew not only the printed Campanus edition, but Latin manuscripts as well – perhaps the Bessarion codex just mentioned.
- In another remark on the same theorem Pacioli explains by many words that an angle in a triangle is normally denoted by three letters and that the first and third can be interchanged, though in general the alphabetical order is chosen[46].
- For I.39 Pacioli gives three further proofs.
- In an addition to Pythagoras' theorem (I.46) Pacioli points out that in an isosceles right-angled triangle the square on the hypotenuse is double the square on one of the sides containing the right angle. He uses here for «hypotenuse» and «side containing the right angle» «diameter» and «costa», terms that are to be found in texts of the High Middle Ages. Later Pacioli uses this theorem in the treatment of irrationals in book X.

[42] L. PACIOLI, *Euclidis megarensis*..., cit., f. 5*r-v*: «Constituere siue formare figuram aliquam super datam lineam est ipsam ponere vnum ex lateribus figure sic describende super eam».

[43] Further examples, mostly from book I, are given by H. WEISSENBORN, *Die Übersetzungen*..., cit., pp. 38-51.

[44] L. PACIOLI, *Euclidis megarensis*..., cit., f. 6*r*: «Hec 5. vocatur fuga miserorum quoniam miseri ingenii cum ad eam perueniunt fugam aripiunt. sed ne fuge detur occasio ostendam eam breuiter et leui ostensione quae sufficiet adiscenti».

[45] See P. KUNITZSCH, «*The Peacock's Tail*»..., cit., pp. 206-208.

[46] L. PACIOLI, *Euclidis megarensis*..., cit., f. 6*r*: «Nota quando dicit angulum a.b.c. intelligit angulum b. qui designatur per litteram mediam et tali modo semper vtitur in triangulis etc. et si diceret angulum b.a.c. intelligeretur angulus a. et dicendo a.c.b. vel b.c.a. intelligitur angulus c. sed communiter in proferendo seruatur ordo alphabeti videlicet literarum a.b.c.d. etc.».

– To the definitions of book III Pacioli attaches the remark that the circle is the most perfect figure, emphasizing that of all isoperimetric figures it contains the greatest area. In this connection he uses physical arguments and cites Aristotle[47].

– Pacioli is particularly interested in the theory of proportion, which is also given detailed treatment in his *Summa*. Therefore, it is no surprise that he makes large insertions in book V. Before presenting the text of the definitions he informs the reader of his lecture on book V that he had given in Venice in 1508[48]. He goes into the importance of proportion in various scientific disciplines (philosophy, theology, medicine, astronomy, cosmography, architecture, painting, music, rhetoric, grammar, jurisprudence) and enumerates the most important members of the audience at his lecture. He also gives lengthy additions to the definitions of book V. The longest concerns definition 5 (in Campanus' numbering), which in the transmitted form was incomprehensible[49]. However, Pacioli does not go into the problems of the definition but treats such general questions as the classification of proportions.

From these few examples it is clear that in his edition of Euclid Pacioli – contrary to what is promised in the title – did not trace and correct the errors of Campanus' text introduced by inaccurate copyists, but simply reproduced the text printed in Venice in 1482 practically unchanged. Rather, his contribution consisted in making the transmitted text in some passages clearer and more understandable. We recognize here that Pacioli applied the mathematical knowledge that he had acquired to enliven the often dry material of the *Elements* and to show the importance of the mathematical content for other disciplines. We may assume that the edition of Euclid contained elements from Pacioli's mathematical lectures.

The rivalry between the Campanus text and Zamberti's Latin translation, aggravated by the appearance of Pacioli's edition, was continued in the following decades. In 1516 the humanist Jacques Lefèvre d'Etaples published them as a parallel-text; and further editions of the double text appeared until 1558. Even the Greek «editio

[47] L. PACIOLI, *Euclidis megarensis*..., cit., f. 16*r*: «Circularis figura inter alias optinet principatum quia multe conditiones sibi competunt que nulli alii figurarum ysoperimetrarum nam ipsa est prima figurarum et perfectissima et simplicissima et regularissima et capacissima omnium ysoperimetarum et pulchrerima omnium ut constat ex electione summi opificis et sic addere potes ipsa quod est maxime apta motui ut inde physico auditu dixit aristoteles».

[48] *Ibidem*, f. 30*r*-31*r*.

[49] «Quantitates que dicuntur continuam habere proportionalitatem: sunt quarum eque multiplicia: aut equa sunt: aut eque sibi sine interruptione addunt aut minuunt». Pacioli's remarks are on ff. 33*v*-34*r*.

princeps», published by Simon Grynaeus in 1533, did not at first alter the situation. Only in 1572, when Commandino brought out a new Latin edition, did the versions of Campanus and Zamberti – and with them Pacioli's endeavours – fade into oblivion.

4. Concluding remarks

The American historian of mathematics, D. E. Smith, said of Pacioli[50]: «His edition of Euclid (1509), interesting as an illustration of the typography of the time, made no advance in geometry and exerted little influence.» From the modern point of view this is certainly true. But, considered relative to his time, this estimate is too negative: in his edition of Euclid and other writings Pacioli endeavoured to make Euclid's theoretical mathematics accessible to readers who, like Pacioli himself, were also interested in practical questions. To this end he used the wide mathematical knowledge that he had inherited. Pacioli was a compiler; even in his treatments of Euclid he showed no originality. But by his writings he contributed to making Euclid's thought accessible to those not exclusively interested in theoretical mathematics.

[50] S. MORRISON, *Fra Luca de Pacioli of Borgo S. Sepolcro*, New York, 1933, p. 5.

XII

Algebra in Germany in the Fifteenth Century

The symbols that we use today in mathematics in algebraic equations were mostly invented in the fifteenth and sixteenth centuries. The abbreviated representation of equations and most algebraic symbols – the plus and minus signs, the sign x for the unknown and the square root sign – go back to mathematicians of Southern Germany. Some of these mathematicians were teachers of practical mathematics – the so-called *Rechenmeister* –; others were university teachers.

The *Rechenmeister* must not be seen as a specifically German phenomenon. They follow the pattern of the Italian *maestri d'abbaco*, who had been active since the end of the thirteenth century. The *maestri* taught principally the practical mathematics used by merchants, but they also concerned themselves with recreational mathematics, with geometrical questions and with algebra. The *maestri d'abbaco* contributed much to the solution of the general cubic and quartic equations; and the German *Rechenmeister* created modern algebraic symbolism.

The development of algebra in Southern Germany in the fifteenth century has been investigated since the end of the nineteenth century by German historians of mathematics, especially by Hermann Emil Wappler, Peter Treutlein, Maximilian Curtze, Kurt Vogel and Wolfgang Kaunzner. They made the principal sources available. But their publications were printed only in German and have therefore remained largely unknown outside German-speaking countries. Further, these writers underestimated or overlooked the importance of Johannes Regiomontanus in developing algebraic symbolism.

Much more recent is our knowledge of algebra in Italy from the thirteenth to the fifteenth centuries. Although the most important printed works, the *Summa* of Luca Pacioli (1494) and G. Cardano's *Ars magna* (1545), were well known and long ago found their proper place in the history of mathematics, the vast bulk of unprinted sources, after Leonardo Fibonacci's *Liber abbaci* (1202/1228) and before Pacioli, were gathering dust in various libraries. But in the last few decades considerable work on this material has been done. It started with Gino Arrighi, who edited from the 1960s some works of the Italian *maestri d'abbaco*. The fundamental work in listing the sources is Van Egmond's 'Catalog of Italian Abbacus Manuscripts' (1980). At the same time a group headed by Raffaella Franci and Laura Toti Rigatelli at the University of Siena started the series *Quaderni del Centro Studi della Matematica Medioevale*, in which a number of Italian texts on mathematics have been published[1].

In the following we will first present a summary of the most important contributions to algebra by the Italian *maestri d'abbaco*[2]. After this comes a sketch of the development of algebra in Germany in the fifteenth and early sixteenth centuries. I shall point to some hints about the extent to which the German texts were influenced by the Italians. My paper will also contain some new information about the development of algebraic symbolism in Southern Germany.

Starting with the *Liber abbaci* of Leonardo of Pisa, the Italian *maestri d'abbaco* took over the new Hindu-Arabic numerals and methods and often added a section on algebra or (later) wrote separate algebraic treatises.

The earliest surviving example of an Italian *abbacus* treatise, where algebra is explained, is the *Libro di ragioni* written by the Florentine Paolo Gerardi in Montpellier in 1328[3]. Gerardi was the first to try to solve third degree equations not equivalent to second degree ones. His rules were wrong, but they became well known and can be found in many manuscripts of the fourteenth and fifteenth centuries.

In 1344 a Magister Dardi of Pisa wrote a treatise called *Aliabraa–Argibra*, which was the earliest vernacular treatise entirely devoted to algebra[4]. He presents a list of 194 equations. Among them there are four equations of the third and fourth degree coming from problems of capital and interest. Dardi gives a formula which is valid for a whole class of cubic equations arising for the calculation of compound interest. The same formulas can be found in many manuscripts of the fourteenth and fifteenth centuries.

In the fourteenth century Antonio de'Mazzinghi (1353–1383) of Florence wrote a treatise on algebra. He introduced two unknowns for solving difficult problems. He was the first vernacular author to use two unknowns with different names (*cosa* and *quantità*).

---

[1]Up to now, 26 volumes have appeared.

[2]It is based on Franci / Toti Rigatelli, 'Towards a History of Algebra'.

[3]The algebraic part of the text was edited by Van Egmond, 'Earliest Vernacular Treatment'.

[4]See Van Egmond, 'Algebra of Master Dardi'.

In two anonymous manuscripts of the fourteenth century there is the interesting idea of solving the general third-degree equation by completing one side of the equation to a perfect cube[5]. This may have influenced Scipione del Ferro and N. Tartaglia in the sixteenth century in finding the general solution of a third-degree equation.

In the fifteenth century there were several famous algebraic schools in Florence[6]. Furthermore, the well-known painter Piero della Francesca (c. 1420–1492) contributed not only to perspective, but also to algebra. For these and other reasons Luca Pacioli, whose *Summa de arithmetica geometria proportioni e proportionalità* was printed in Venice in 1494, has enjoyed unmerited fame, for his algebra contains nothing new of any value.

In the German-speaking countries algebra made a late start, not before the middle of the fourteenth century. It seems that it started in Vienna. The University of Vienna was founded in either 1365 or 1385, by Albert of Saxony (c. 1316–1394) and Henry of Hesse – both of whom came from Paris and had a good knowledge of Parisian mathematics. Among the works they are supposed to have known are the writings of Johannes de Muris, and in particular his *Quadripartitum numerorum*, which shows his acquaintance with Arabic algebra.

But it was not before 1450 that algebra was visibly cultivated – in Vienna in the 1450s, in Regensburg in the 1460s, in Leipzig in the 1480s and in Erfurt perhaps in the 1460s. Since that time writings wholly or partly on algebra have come down to us. The relationship between these writings is not always clear, but they afford evidence of algebraic activities in the German *Kulturgebiet.* In the following I give an account of our present knowledge of these texts and their authors.

Johannes Regiomontanus (1436–1476), the most important German mathematician of the fifteenth century, was central for the transmission of Italian ideas about algebra to Central Europe. This is not the place to detail the mathematical achievements of Regiomontanus[7]. Here I shall confine myself to his contributions to algebra. These belong to the

---

[5]See Franci / Toti Rigatelli, 'Towards a History of Algebra', pp. 45–48.

[6]See Franci / Toti Rigatelli, 'Towards a History of Algebra', pp. 51–56. The fundamental work on the *abbacus* schools in Florence is Ulivi, 'Benedetto da Firenze', especially pp. 195–209.

[7]For this, see Folkerts, 'Regiomontanus als Mathematiker' and 'Die mathematischen Studien'. For special aspects see Folkerts, 'Transmission and transformation'; 'Regiomontanus' Approach to Euclid'; 'Transmission of Mathematical Problems'.

years he spent in Vienna (1450–1461) and in Italy (1461–1467)[8]. At this time he studied the Greek mathematical classics (Euclid, Archimedes, Apollonius, Diophantus)[9] and also some medieval writers (al-Khwārizmī, Jordanus de Nemore, Johannes de Muris)[10]. Of these writers especially important for algebra are Diophantus, al-Khwārizmī and Johannes de Muris. Regiomontanus' copies of al-Khwārizmī's *Algebra* and of Johannes de Muris' *Quadripartitum numerorum* have survived and show – by his annotations – how thoroughly he studied these works[11].

Regiomontanus made important contributions to algebra. Although no special work on the subject by Regiomontanus has come down to us, it is possible to guess something from two manuscripts which belonged to him (New York, Columbia University, Plimpton 188, dated 1456; the autograph of his *De triangulis*, now in St. Petersburg, first half of the 1460s) and from his correspondence with Giovanni Bianchini (1463–64), Jakob von Speyer (1465) and Christian Roder (1471) (autograph now in Nuremberg). In these writings we find many problems in practical and theoretical arithmetic, geometry and – important for the present discussion – also in algebra. Since I have dealt with them elsewhere[12], it may suffice here to say that some problems lead to cubic equations and that Regiomontanus solves one of them by using a three-dimensional *gnomon*[13]. This may perhaps be seen as a step towards the solution of the general cubic found by the Italians in the first half of the sixteenth century. But it should be said that similar problems had already been solved by Master Dardi of Pisa in the fourteenth century and in other Italian texts of the fifteenth century[14]. Accordingly, it seems that Regiomontanus had knowledge of some of these texts.

A collection of problems written by Regiomontanus, presumably in 1456 (New York, Columbia University, Plimpton 188, f. 82v–96r), demon-

---

[8]For his life, see Folkerts, 'Transmission and transformation', pp. 89f., and Folkerts, 'Transmission of Mathematical Problems', pp. 411f.

[9]See Folkerts, 'Transmission and transformation', pp. 92–101. For Euclid, see especially Folkerts, 'Regiomontanus' Approach to Euclid'.

[10]See Folkerts, 'Transmission and transformation', pp. 101–103; Folkerts, 'Transmission of Mathematical Problems', pp. 413f.

[11]See Folkerts, 'Transmission and transformation', pp. 102f.

[12]Folkerts, 'Die mathematischen Studien', pp. 197–209; Folkerts, 'Transmission and transformation', pp. 103–110; Folkerts, 'Transmission of Mathematical Problems', pp. 415–426.

[13]See Folkerts, 'Transmission and transformation', pp. 104–106; Folkerts, 'Transmission of Mathematical Problems', pp. 415f.

[14]See Folkerts, 'Transmission and transformation', p. 105.

strates that Regiomontanus was a central figure in the transmission of mathematical knowledge from Italy to Germany: it can be shown that many problems in this collection are identical with those found in Latin and vernacular texts from Italy, and some of Regiomontanus' examples are to be found in later algebraic texts written in Germany.

To show Regiomontanus' dependence on Italian sources we may point to several problems in this collection that treat standard problems in recreational mathematics, many of them being similar to problems found in Leonardo's *Liber abbaci*[15]. Again, other problems[16] serve as examples for the six types of equations mentioned by al-Khwārizmī. Many of them can be found in Italian texts, especially in Paolo Gerardi's *Libro di Ragioni*[17].

To illustrate Regiomontanus' influence on later writers in Germany we may note the six forms of linear and quadratic equations with the characteristic title *Regule de cosa et censo sex sunt capitula, per que omnis computatio solet calculari* found in the Plimpton manuscript[18]. This was plainly the source of examples given by Fridericus Amann five years later to illustrate the six types of equations[19]; I shall mention Fridericus later. Even more impressive is a passage on f. 90v of the Plimpton manuscript. Here Regiomontanus gives a list of 18 types of equation, up to the fourth degree; its order differs from that given in previous Italian texts, but with one exception agrees with the list in the "German algebra" of 1481 (see p. 9 below).

Of special interest is the symbolism which Regiomontanus used in his algebraic texts. Abbreviations for the constant terms, for the unknown and for its powers were used in several older treatises, beginning with Diophantus, in Indian and Arabic texts and also in a Latin manuscript written about 1380; these abbreviations were normally the first letter – and at least in one case (the Arabic writer al-Qalaṣādī, d. 1486) the last letter – of the respective name of the unknown and its powers[20]. The Arabic words were: *shay'* = "thing" or *jidhr* = "root" (for x), *māl* = "wealth" (for x^2), and *ka'b* = "cube" (for x^3). Accordingly, in the

---

[15] They are on f. 82v–84v and f. 88r–90v of Plimpton 188.

[16] On f. 85r–87v.

[17] Problem 17 (to solve $(10 - x) : x = 5$); problem 19 (to solve $(x - \frac{x}{3} - \frac{x}{4}) = 16$); problem 20 ($x : y = 2 : 3, x \times y = x + y$); problem 22 (a problem in compound interest). See Van Egmond, 'Earliest Vernacular Treatment', p. 166, no. 1 and 2; p. 167, no. 3; pp. 167f., no. 4.

[18] On f. 85r.

[19] In the Munich manuscript Clm 14908.

[20] See Tropfke, 'Elementarmathematik', pp. 281 and 374–378.

Latin algebraic texts, which were based upon Arabic writings, we find *res* (*shay'*), *radix* (*jidhr*), *census* (*māl*) and *cubus* (*ka'b*); the constant term (in Arabic *dirham*, *dīnār* or *fals*) was reproduced in Latin as *dragma*, *denarius* or *numerus*. In Italian texts the term *res* for the unknown x was translated as *cosa* and therefore a book on algebra was called "Liber de la cosa"; from this came the German word *Coss* for the unknown, and the name *Cossisten* for persons engaged in algebra was used in the late fifteenth and the early sixteenth centuries.

Against this background Regiomontanus' symbols must be seen. He used raised signs for the unknown (x) and for its square x^2, the sign for x being an r with added abbreviation stroke (so = *res*) or an R with added stroke (so = *radix*) and the sign for x^2 a c with stroke (so = *census*). This may be considered traditional. What is new is the long line for the "equals" sign and the abbreviation for *minus*: not *mi* plus the abbreviation for *-us* and the horizontal stroke above (for the *n* in *minus*), but *i* with the *-us* mark and the horizontal stroke. These abbreviations are already to be found in manuscript Plimpton 188, written in 1456, and also in Regiomontanus' correspondence written in 1463 and in his autograph of *De triangulis* written about the same time[21]. Therefore it is clear that Regiomontanus had his symbolism as early as 1456, five years before Fridericus Amann's symbolism of 1461 (see p. 8 below). According to our present knowledge, therefore, Regiomontanus was the first to use the symbolism that transformed algebra in the subsequent decades.

Let us leave Regiomontanus and turn to other places in Germany in which algebra was studied. One place was the Benedictine monastery of St. Emmeram near Regensburg, where Fridericus Amann lived in the middle of the fifteenth century. His activities in mathematics have to do with the so-called *Algorismus Ratisbonensis*, an arithmetical compilation which was written before 1450 by an anonymous author who apparently worked in Regensburg. The first two parts are on computations with whole numbers (*de integris*) and with fractions (*de minuciis*); and the third part comprises practical problems (*de practica*). The whole is written in Latin. Only the last part has been edited[22]. This work depends

[21] See Folkerts, 'Die mathematischen Studien', pp. 201f. On Regiomontanus' symbolism in his correspondence see Cajori, 'Mathematical Notations', pp. 95f., and Kaunzner, 'Regiomontanus als Mathematiker', pp. 134–136. In Kaunzner's article there are facsimiles from Regiomontanus' correspondence and from his autograph of *De triangulis* in tables 2–6. Samples of Regiomontanus' symbolism are reproduced at the end of this article.

[22] Vogel, 'Practica'.

mostly on the algorismus of Johannes de Sacrobosco and the *Algorismus de minuciis* of Johannes de Lineriis; the immediate sources that the compiler used are known[23].

One copy of the *Algorismus Ratisbonensis*, now in Munich (BSB, Clm 14908), was written by Fridericus Amann. We now know that he is not identical with Fridericus Gerhart, as formerly thought – Fridericus Gerhart was a fellow monk in St. Emmeram[24]. Fridericus Amann apparently studied in Leipzig from 1427 and then went to St. Emmeram. There he may well have met Nicholas of Cusa in 1451 – when Nicholas was visiting Regensburg. In 1453 Fridericus removed temporarily to the monastery of St. Peter in Salzburg and probably visited Vienna in 1456[25]. Fridericus Amann's main scholarly activities lie between 1457 and 1461. In these years he wrote or copied numerous mathematical, astronomical and geographical works. His autograph manuscripts are now kept in the Bayerische Staatsbibliothek in Munich. Fridericus not only used the numerous mathematical and astronomical manuscripts in St. Emmeram, but was also well informed about the mathematical, astronomical and geographical activities in Vienna. *Inter alia*, Fridericus enriched the *Practica* of the *Algorismus Ratisbonensis* by adding more than 280 problems, written partly in Latin and partly in German[26].

The problems that Fridericus added to the *Practica* are transmitted in the manuscript Clm 14908 of the BSB München. In the same manuscript there are many other mathematical texts written by Fridericus[27]. Many of them are algebraic[28]. The most important are an algebra in German, written by Fridericus in 1461, and examples of the different types of equation[29]. The algebra of 1461 is particularly important for our present study, since it seems to be the first text in German entirely devoted to

[23]His main source was the text in the manuscript St. Florian, XI. 619, f. 87r–94r. For the second part he also used a Latin treatise on sexagesimal fractions which largely depends on the Lineriis text (Clm 14111, f. 242r–243v), a treatise in German on computation with common fractions followed by applications to commerce (Clm 14111, f. 244r–246v) and a *Rechenbuch* in German which treats computation with whole numbers and with common fractions and in addition contains practical problems (St. Florian, XI. 619, f. 100r–117r).

[24]See Gerl, 'Fridericus astronomus'.

[25]The arguments for this are given in Durand, 'Vienna–Klosterneuburg Map Corpus', p. 75, note 5. On pp. 72–76 Durand gives information about Fridericus' life and his activities in geography and cartography.

[26]Edited by Vogel, 'Practica'.

[27]For the list of contents of the manuscript, see Vogel, 'Practica', pp. 12–19.

[28]These are edited in Curtze, 'Beitrag'.

[29]On f. 133v–134v and f. 136v–157r. Edited by Curtze, 'Beitrag', pp. 49–73.

algebra. It describes in words the solution of the six types of linear and quadratic equation treated by al-Khwārizmī.

There are striking similarities between the examples given by Fridericus in Clm 14908 and those to be found in Regiomontanus' algebraic collection in Plimpton 188: the beginning of Regiomontanus' Latin text that contains the examples of the six types of quadratic equations agrees word-for-word with the German translation written by Fridericus; and all further problems in this part of the Plimpton manuscript can also be found in Clm 14908[30]. The title *Regule delacose secundum 6 capitula*, which Fridericus uses, agrees, too, with Regiomontanus' title and reminds us of the Italian origin. Furthermore, in the examples that follow the "German algebra" Fridericus uses symbols to represent the powers of the unknown which bear a remarkable similarity to those in Regiomontanus' manuscript (Plimpton 188)[31]. This cannot be a coincidence, and we must assume that Fridericus took over this use of symbols from Regiomontanus. Since Fridericus seems to have visited Vienna in 1456, there are good reasons to assume that he met Regiomontanus there and at this meeting Fridericus learnt of his symbols.

In the second half of the fifteenth century another region in Germany became important for the development of algebra: the area of Thuringia and Saxony in Central Germany, especially the cities of Erfurt and Leipzig with their universities. The best information about the mathematical activities in these places can be obtained from a voluminous manuscript which is now kept in the Sächsische Landesbibliothek Dresden and has the shelf-mark C 80. The main part of this manuscript was written in Germany in the 1470s and 1480s, but there are many notes and additions from later times. Codex C 80 contains exclusively mathematical texts: "classical" medieval writings in Latin and "modern" texts from the end of the fifteenth century (in Latin or in German), which were used at the universities of Erfurt and Leipzig. But this manuscript was not only used by university teachers, but also by the famous *Rechenmeister* Adam Ries[32].

The manuscript Dresden, C 80, transmits, *inter alia*, several texts on algebra. The most important of them are: a lecture given by Gott-

[30]Clm 14908, f. 136v–157r, equals Plimpton 188, f. 85r–87v.

[31]For the symbols used by Fridericus and their names see Tropfke, 'Elementarmathematik', pp. 281f.

[32]For the importance of this manuscript, see Folkerts, 'C 80', where a detailed description of its contents is given. For Adam Ries, see pp. 11f. below.

fried Wolack in Erfurt in 1468, though it is not strictly algebraic[33], and two writings which are often referred to as "the Latin algebra" and "the German algebra". The anonymous "Latin algebra" gives, after an explanation of the cossistic symbols, general rules and numerical examples for the six basic forms and for 18 further forms of the quadratic equation[34]. In this text the "+" and the "–" signs (in the modern form) occur for addition and subtraction – as it seems, for the first time. The "German algebra", which was finished in 1481 by an unknown author[35], seems to depend on the "Latin algebra". It easily surpasses in length and content the short German algebra written by Fridericus Amann in St. Emmeram in 1461. In the "German algebra", as in the "Latin algebra", the number of types of equation is kept to twenty-four. The order of the types, which elsewhere varied, is the same in the "German algebra" in MS C 80 and in the Regiomontanus text in MS Plimpton 188. This cannot be a coincidence. Therefore the "German algebra" depends, directly or indirectly, on the Regiomontanus text.

From notes in the codex C 80 we know that a lecture entirely devoted to algebra was given in the year 1486 at the University of Leipzig by Johannes Widmann[36]. This was apparently the first lecture in Germany on the topic. Widmann was born in Eger, as it seems between 1460 and 1465. In 1480 he matriculated at Leipzig University, where he became *baccalaureus* in 1482 and *magister* in 1485. After this time he gave mathematical lectures. Between 1490 and 1500 Widmann left Leipzig for Annaberg, a newly-founded town in the Erzgebirge, which had become rich in a very short time from silver-mining. He died some time after 1504[37].

Johannes Widmann was for some time the owner of the manuscript C 80 and there are many marginal notes and additional texts by his hand in this codex. Widmann took over the symbols "+" and "–", which are found in the "Latin algebra" and in other texts in manuscript C 80, and used them regularly in his printed "Rechenbuch", *Behende vnd hubsche Rechenung auff allen kauffmanschafft*, which was published in 1489. After

---

[33] On f. 301v–303r. For this, see p. 12 below.

[34] On f. 350r–364r. Edited by Wappler, 'Zur Geschichte der deutschen Algebra', pp. 11–30. Author and date of this treatise are unknown.

[35] On f. 368–378v. Edited by Vogel, 'Die erste deutsche Algebra'.

[36] For details about this lecture, see Gärtner, 'Behende vnd hubsche Rechenung', pp. 6 and 33–36. A student's copy of this lecture made at the time is to be found in Leipzig, UB, Ms. 1470, f. 479r–493v.

[37] For his life, see Gärtner, 'Behende vnd hubsche Rechenung', pp. 3–11, and Kaunzner, 'Johannes Widmann'.

the publication of this book these symbols gradually came into use, at least in the German-speaking part of Europe[38].

Somewhat later than Widmann, another teacher worked at Leipzig University, who, too, was an important figure for the development of algebra: Andreas Alexander[39]. We know that he was a pupil of one Aquinas, a monk of the Order or Preachers. Aquinas seems to have been an important figure in the development of algebra in Germany in the second half of the fifteenth century, but we know very little about him. He was one of the "Wandermathematiker", who travelled around and taught mathematics for money. We do know that Regiomontanus learnt a great deal from Aquinas[40]. Andreas Alexander originated from Regensburg. In 1485 he matriculated in the Faculty of Arts at the University of Cologne and, having gained his B.A., from 1493 he studied in Leipzig. We do not know when he started his teaching activities at Leipzig University or when he died. What we know is that between 1502 and 1504 he gave lectures on "arithmetica communis", "mathematica", "perspectiva communis" and books I and III of Euclid's *Elements* and that he published two books in Leipzig in 1504: the *Mathemalogium*, which explains mathematical terms and problems found in the logical writings of Aristotle, and an edition of John Pecham's *Perspectiva communis*.

From later sources, especially from notes given by Adam Ries in his *Coss* (see p. 12 below), we know that Andreas Alexander also wrote an *Introductorium* to the first book of Euclid's *Elements* and two algebraic treatises, one in Latin and one in German; eleven problems given by him were cited by Adam Ries[41]. There are convincing arguments that Andreas Alexander's Latin algebra and his introduction to Euclid are both contained in manuscript 1696 of the University Library in Leipzig[42].

There is some connection between the algebraic work in the manuscript Leipzig 1696 and another algebraic treatise which was published under the name "Initius Algebras" more than hundred years ago by Maximilian Curtze[43]. Curtze thought that "Initius" was part of the author's name, but it is much more probable that "initium Algebras"

---

38 For details, see Cajori, 'Mathematical Notations', pp. 231–236.

39 What we know about his life and his works is given in Folkerts, 'Andreas Alexander'.

40 According to Regiomontanus's own words in a letter to Christian Roder of Erfurt, written in 1471 (Curtze, 'Briefwechsel', p. 325).

41 Folkerts, 'Andreas Alexander', pp. 55–57.

42 On f. 71r–126v and f. 127r–151r; see Folkerts, 'Andreas Alexander', p. 57. These texts have not yet been edited. The manuscript was written about 1500.

43 Curtze, 'Initius Algebras'.

means: "introduction into algebra" or "the introductory work of Algebras", where "Algebras" is the (fictitious) name of an Arabic writer who "invented" this discipline[44]. This work is mainly in German, but much of the text comprises a copious German commentary on Latin extracts. The whole has many similarities to Andreas Alexander's algebra in manuscript Leipzig 1696: both texts treat the same types of equation; the Latin rules to solve them are almost identical; and many examples in the second and third part of the "(Initium) Algebras" are also be found in the Leipzig manuscript. Thus it might be possible that the German parts of the "(Initium) Algebras" are also from Andreas Alexander and that this text is the algebraic treatise in German which it has been argued that Alexander wrote. But this is only a hypothesis.

The text of the "(Initium) Algebras" was edited by Curtze from a late manuscript. He did not recognize that there was another, older, copy which was written by Adam Ries[45]. Therefore, Adam Ries, too, had something to do with this text. It is even possible that he was the author.

Adam Ries is the best-known German *Rechenmeister*[46]. He was born in Staffelstein near Bamberg in 1492. From 1518 until 1522/23 he lived in Erfurt and founded there a *Rechenschule.* In Erfurt Ries came into contact with Georg Stortz, who had studied medicine and later became rector of Erfurt University. Stortz encouraged Ries to publish *Rechenbücher* to be used by ordinary people. The first came out in 1518, the second in 1522 and the third *Rechenbuch* was published in 1550. Adam Ries became famous with his second *Rechenbuch*, which had more than 100 editions. At that time the manuscript C 80 was in the possession of Stortz. Ries was allowed to use this manuscript and, like Widmann, added marginal notes to the codex. The algebraic texts in C 80 seem to have induced Ries to study algebra. He started writing his own treatise on algebra, the *Coss.* In 1522 or 1523 he moved to Annaberg, where he lived until his death in 1559. In Annaberg Ries had official positions in the mining authority and also taught at his own *Rechenschule.*

Of Adam Ries's algebraic work little has been printed. But a facsimile of his autograph, which is now kept in the Erzgebirgsmuseum in Annaberg, was printed in 1992[47], and some important parts were pub-

[44]This plausible interpretation is given in Hellmann, 'Von Pythagoras zum Liber de cosa', pp. 88–91.

[45]Manuscript Dresden, C 349.

[46]For Ries's life and work see Wußing, 'Adam Ries', who refers to the large amount of secondary literature on Ries.

[47]Kaunzner / Wußing, 'Adam Ries: Coß'.

lished in 1994[48]. Ries's autograph of the *Coss* consists of two parts, here called *Coss 1* and *Coss 2*: the first was finished in 1524, the second – which is a later redaction of the *Coss 1* – was completed by Ries himself not before 1544. It is not the place to analyze Ries's *Coss* here[49]. Suffice it to say that the first part shows some original ideas, e.g. that Ries combines the algebraic symbols of the first three powers of the unknown (*radix* $= x$, *zensus* $= x^2$, *cubus* $= x^3$) to denote powers of the unknown up to degree 9[50]. In *Coss 1* Ries formulates 24 types of equation and gives examples for each of them. *Coss 2* is more traditional; it treats topics in algebra that are also to be found in manuscript C 80. But this part was also influenced by algebraic treatises which had been published since *Coss 1*, e.g. by Christoph Rudolff's *Coß* (1525) and Michael Stifel's *Arithmetica integra* (1544), which marks a final point in the development of early algebra in Germany.

I should like to mention another figure in Southern German mathematics, who lived about 1500, who was completely unknown until a few years ago: Conrad Landvogt. Almost all that we know about him comes from four manuscripts which belonged to him and were partly written by him[51]. Landvogt was a Cistercian in a monastery in the diocese of Mainz, not far from Frankfurt. In 1468 he became a student in Erfurt, the university of the archbishopric of Mainz.

The University of Erfurt, founded in 1392, was famous in the fifteenth century for its mathematical studies[52]. At this university Christian Roder taught from 1449 until his death in 1478. He was a famous mathematician and astronomer and corresponded with Regiomontanus on these subjects. Another important mathematician in Erfurt was Gottfried Wolack, who became rector in 1462. In 1468 – in the same year as Landvogt matriculated in Erfurt – Wolack gave a mathematical lecture, of which several copies are known[53]. It contains numerous problems which are solved by the *regula de tri*. The text is not a systematic treatment of algebra, but procedures are applied that may be considered algebraic. On these grounds Wolack's lecture in Erfurt has been characterized as the first

[48] Gebhardt, 'Einblicke'.

[49] For this, see, for instance, W. Kaunzner's article in Gebhardt, 'Einblicke', pp. 171–183, and Wußing, 'Adam Ries', pp. 80–96.

[50] On p. 109.

[51] Now Uppsala University Library, MS C 652, C 656, C 657, C 658. For details about his life and his works see Folkerts, 'Conrad Landvogt'.

[52] For this, see Folkerts, 'Conrad Landvogt', pp. 232–235.

[53] Edited by Wappler, 'Zur Geschichte der Mathematik'.

lecture on algebra in Germany. At any rate, it influenced Johannes Widmann, who, as has been said, gave an unambiguously algebraic lecture in Leipzig in 1486.

Conrad Landvogt profited as a student in Erfurt from the good mathematical teaching there. This can be seen from his mathematical writings. Among them are three introductions or commentaries to Euclid's *Elements*. His main writings are two different algebraic works entitled *Cabala algebraica*. They were written shortly after 1500. In these writings Landvogt used the usual cossistic symbols for the constant (x^0) and the three first powers (x^1, x^2, x^3) of the unknown. Through combination of these symbols higher powers can be represented, e.g. x^5 may be represented by x^3 and x^2. Another cossistic notation is a dot before a symbol to mean the square root of the quantity represented by the symbol; two, three, four ... dots mean the fourth, eighth, sixteenth ... root. Compared with other texts of this time, in which one dot meant a square root, two dots meant a fourth root, but three dots meant a cube root[54], Landvogt's symbolism represented a progress in consistency.

Landvogt begins with the six traditional types of linear and quadratic equations. There follow types 7–18, in which the left-hand side of the equation is always a constant (*De numeris*) and the right-hand side is a combination of powers of the unknown with coefficients which are either integers or square or fourth roots of integers. Then come further groups of types in which the left-hand side is a single term and the right-hand side is the sum or difference of powers of the unknown up to the fourth degree, with rational or irrational coefficients. If they are irrational, they are classified according to Euclid's Book 10[55]. In the shorter version of the *Cabala algebraica* Conrad Landvogt treats 247 types of equation; and in the longer 638 types[56].

Landvogt's contributions to mathematics were notable for his time, especially his idea of connecting the different types of irrationalities, which are in Euclid's book 10, with the algebra of the cossists. His works, which were not printed, seem to have had little direct influence, but they show what individuals at this time were capable of.

* * *

[54]E.g. in the manuscript Dresden C 80, which seems to be the earliest manuscript to have the notation of dots meaning roots. For details, see Tropfke, 'Elementarmathematik', pp. 295f.

[55]The beginning of the list, which contains the types up to no. 54, is reproduced at the end of this article.

[56]For details see Folkerts, 'Conrad Landvogt', pp. 245–254.

To sum up: in the period between 1450 and 1550 a surprisingly large number of algebraic texts were written in Germany. These texts, for which the name "Deutsche Coß" has been coined, are dependent on Italian texts written by the *maestri d'abbaco*. In the second half of the fifteenth century only a few mathematicians were active in this field and all their treatises seem to stem from only four centres: Vienna, Regensburg, Leipzig and Erfurt. This activity started with Johannes Regiomontanus, who had access to Italian algebraic knowledge as early as 1456, during his time in Vienna; we have a collection of algebraic problems written by him in this year and other texts from the 1460s, when he was in Italy. Fridericus Amann from St. Emmeram near Regensburg, who wrote an algebra in German in 1461, apparently had his knowledge from Regiomontanus, whom he may have met in Vienna in 1456. From the 1480s on, the University of Leipzig became a centre of algebra: in Leipzig in 1481 a "German algebra" was written which was dependent on a detailed "Latin algebra" and is in the tradition of Regiomontanus; and in 1486 Johannes Widmann gave the first university lecture devoted to algebra, which was based upon the "Latin algebra" and the "German algebra". About 1500 Andreas Alexander taught at Leipzig University and is reported as writing a long algebraic treatise in Latin and another in German. His German treatise may have been the "(Initium) Algebras" – and if not, was the basis of it. One copy of it was written by the German *Rechenmeister* Adam Ries. Ries also wrote two treatises on algebra, the first being dependent on Andreas Alexander's work. In Erfurt, too, algebraic problems were studied from the 1460s and 1470s, especially by Gottfried Wolack and Christian Roder, with whom Regiomontanus was in contact. A singular person in the development of early algebra in Germany was Conrad Landvogt, who had studied in Erfurt and later wrote several treatises in which the "German algebra" was combined with the different types of irrationalities found in Euclid's *Elements*.

In contrast to Italy, algebra in Germany was from the beginning not exclusively a domain of the *Rechenmeister*, but belonged rather to scholars who had studied at universities. Nevertheless, as the example of Adam Ries shows, *Rechenmeister* were also involved. The "German Cossists" did not succeed in solving the general cubic equation – this was done in Italy –, but from the modern point of view they were important because they developed the algebraic symbolism which is still used today.

Bibliography

Adam Ries: Coß. Herausgegeben und kommentiert von Wolfgang Kaunzner, Regensburg, und Hans Wußing, Leipzig, Stuttgart / Leipzig 1992.

F. Cajori, *A History of Mathematical Notations*, vol. 1, Chicago 1928.

M. Curtze, "Ein Beitrag zur Geschichte der Algebra im fünfzehnten Jahrhundert", *Abhandlungen zur Geschichte der Mathematik* 7 (1895), 31-74.

M. Curtze (ed.), "Der Briefwechsel Regiomontan's mit Giovanni Bianchini, Jacob von Speier und Christian Roder", in *Urkunden zur Geschichte der Mathematik im Mittelalter und der Renaissance, Theil 2*, Leipzig 1902, pp. 185–336.

M. Curtze (ed.), "Die Algebra des Initius Algebras ad Ylem geometram magistrum suum", in *Urkunden zur Geschichte der Mathematik im Mittelalter und der Renaissance, Theil 4*, Leipzig 1902, pp. 435–611.

D. B. Durand, *The Vienna-Klosterneuburg Map Corpus of the Fifteenth Century*, Leiden 1952.

M. Folkerts, "Regiomontanus als Mathematiker", *Centaurus* 21 (1977), 214–245.

M. Folkerts, "Die mathematischen Studien Regiomontans in seiner Wiener Zeit", in G. Hamann (ed.), *Regiomontanus-Studien*, Vienna 1980, pp. 175–209. (Reprinted in this volume, item V.)

M. Folkerts, "Conrad Landvogt, ein bisher unbekannter Algebraiker um 1500", in S. Demidov, M. Folkerts, D. Rowe, C. J. Scriba (ed.), *Amphora. Festschrift für Hans Wussing*, Basel etc. 1992, pp. 229–259.

M. Folkerts, "Regiomontanus' role in the transmission and transformation of Greek mathematics", in F. J. Ragep, S. P. Ragep, S. Livesey (ed.), *Tradition, Transmission, Transformation. Proceedings of two conferences on pre-modern science held at the University of Oklahoma*, Leiden, New York, Cologne 1996, pp. 89–113. (Reprinted in this volume, item VI.)

M. Folkerts, "Andreas Alexander – Leipziger Universitätslehrer und Cossist?", in R. Gebhardt (ed.), *Rechenmeister und Cossisten der frühen Neuzeit*, Annaberg-Buchholz 1996, pp. 53–61.

M. Folkerts, "Regiomontanus' Role in the Transmission of Mathematical Problems", in Y. Dold-Samplonius, J. W. Dauben, M. Folkerts, B. van Dalen (ed.), *From China to Paris: 2000 Years Transmission of Mathematical Ideas*, Stuttgart 2002, pp. 411–428. (Reprinted in this volume, item VIII.)

M. Folkerts, "Die Handschrift Dresden, C 80, als Quelle der Mathematikgeschichte", in R. Gebhardt (ed.), *Verfasser und Herausgeber mathematischer Texte der frühen Neuzeit*, Annaberg-Buchholz 2002, pp. 353–378.

M. Folkerts, "Regiomontanus' Approach to Euclid", this volume, item VII.

R. Franci / L. Toti Rigatelli, "Towards a History of Algebra from Leonardo of Pisa to Luca Pacioli", *Janus* 72 (1985), 17–82.

B. Gärtner, *Johannes Widmanns "Behende vnd hubsche Rechenung"*, Tübingen 2000.

R. Gebhardt, *Einblicke in die Coß von Adam Ries*, Stuttgart / Leipzig / Zürich 1994.

A. Gerl, "Fridericus astronomus", in *Die deutsche Literatur des Mittelalters. Verfasserlexikon*, vol. 11 (2004), col. 466–468.

M. Hellmann, "Von Pythagoras zum Liber de cosa. Bemerkungen von Nicolaus Matz über die Geschichte der Rechenkunst", in W. Schmitz (ed.), *Bewahren und Erforschen. Beiträge aus der Nicolaus-Matz-Bibliothek (Kirchenbibliothek) Michelstadt*, Michelstadt 2003, pp. 75–112.

W. Kaunzner, "Über Regiomontanus als Mathematiker", in G. Hamann (ed.), *Regiomontanus-Studien*, Vienna 1980, pp. 125–145.

W. Kaunzner, "Johannes Widmann, Cossist und Verfasser des ersten großen deutschen Rechenbuches", in R. Gebhardt (ed.), *Rechenmeister und Cossisten der frühen Neuzeit*, Annaberg-Buchholz 1996, pp. 37–51.

J. Tropfke, *Geschichte der Elementarmathematik. 4. Auflage. Band 1: Arithmetik und Algebra. Vollständig neu bearbeitet von K. Vogel, K. Reich, H. Gericke*, Berlin / New York 1980.

E. Ulivi, *Benedetto da Firenze (1429–1479), un maestro d'abaco del XV secolo. Con documenti inediti e con un'Appendice su abacisti e scuole d'abaco a Firenze nei secoli XIII–XVI*, Pisa / Rome 2002 (= *Bollettino di Storia delle Scienze Matematiche* 22 (2002), fasc. 1).

W. Van Egmond, "The Earliest Vernacular Treatment of Algebra: The *Libro di Ragioni* of Paolo Gerardi (1328)", *Physis* 20 (1978), 155–189.

W. Van Egmond, *Practical Mathematics in the Italian Renaissance. A Catalog of Italian Abbacus Manuscripts and Printed Books to 1600*, Florence 1980.

W. Van Egmond, "The Algebra of Master Dardi of Pisa", *Historia Mathematica* 10 (1983), 399–421.

K. Vogel, *Die Practica des Algorismus Ratisbonensis*, Munich 1954.

K. Vogel, "Die erste deutsche Algebra aus dem Jahre 1481", in *Abhandlungen der Bayerischen Akademie der Wissenschaften, Mathematisch-naturwissenschaftliche Klasse*, N.F. 160, Munich 1981.

H. Wappler, *Zur Geschichte der deutschen Algebra im 15. Jahrhundert*, Zwickau 1887.

E. Wappler, "Zur Geschichte der Mathematik im 15. Jahrhundert", *Zeitschrift für Mathematik und Physik, Historisch-litterarische Abteilung* 45 (1900), 47–56.

H. Wußing, *Adam Ries*, 2nd ed., Stuttgart / Leipzig / Zürich 1992.

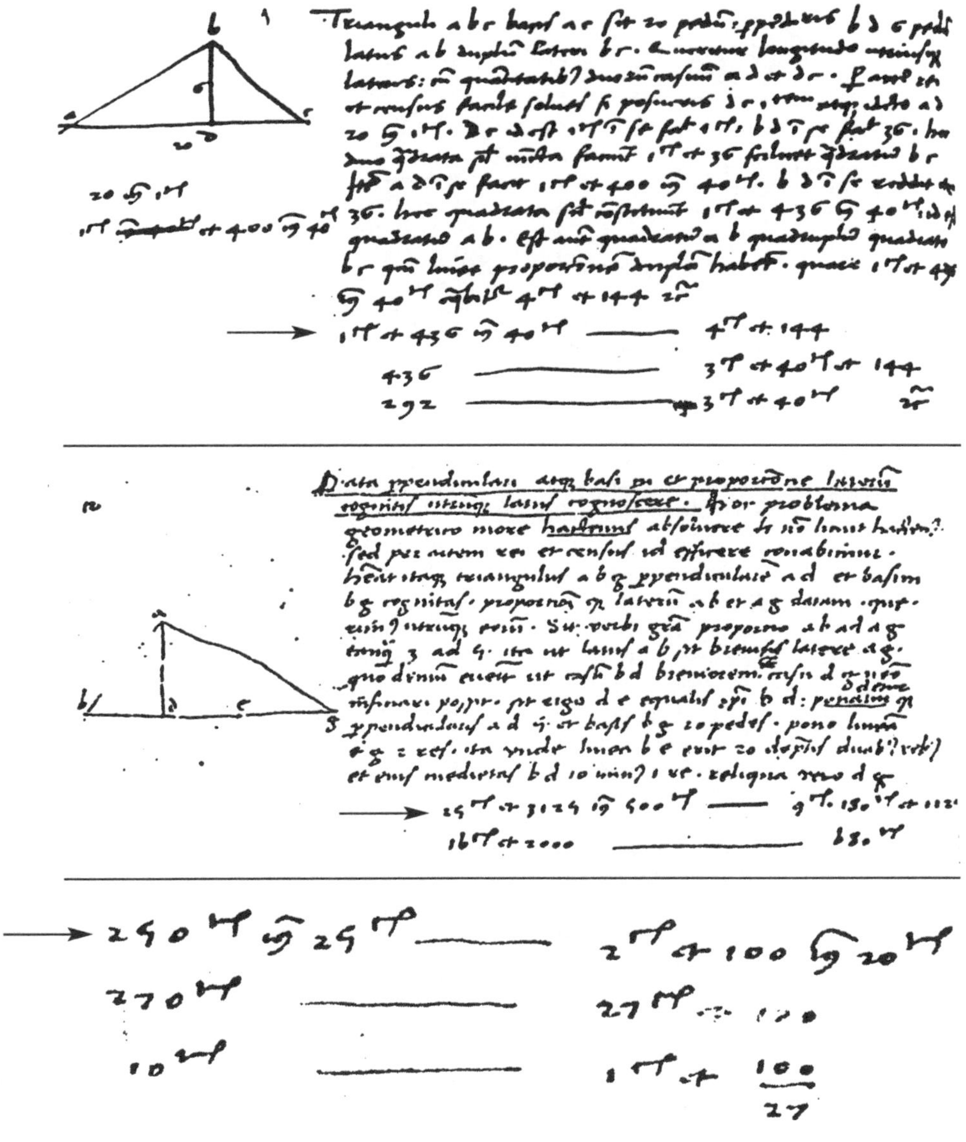

Extracts from Regiomontanus' works showing his algebraic symbolism (see p. 6 above). The first equations in the three samples are:

$1x^2 + 436 - 40x = 4x^2 + 144$ (Plimpton 188, f. 94r);

$25x^2 + 3125 - 500x = 9x^2 + 180x + 1125$ (*De triangulis*, autograph, f. 40v);

$250x - 25x^2 = 2x^2 + 100 - 20x$ (Nuremberg, Cent. V app. 56c, f. 13v).

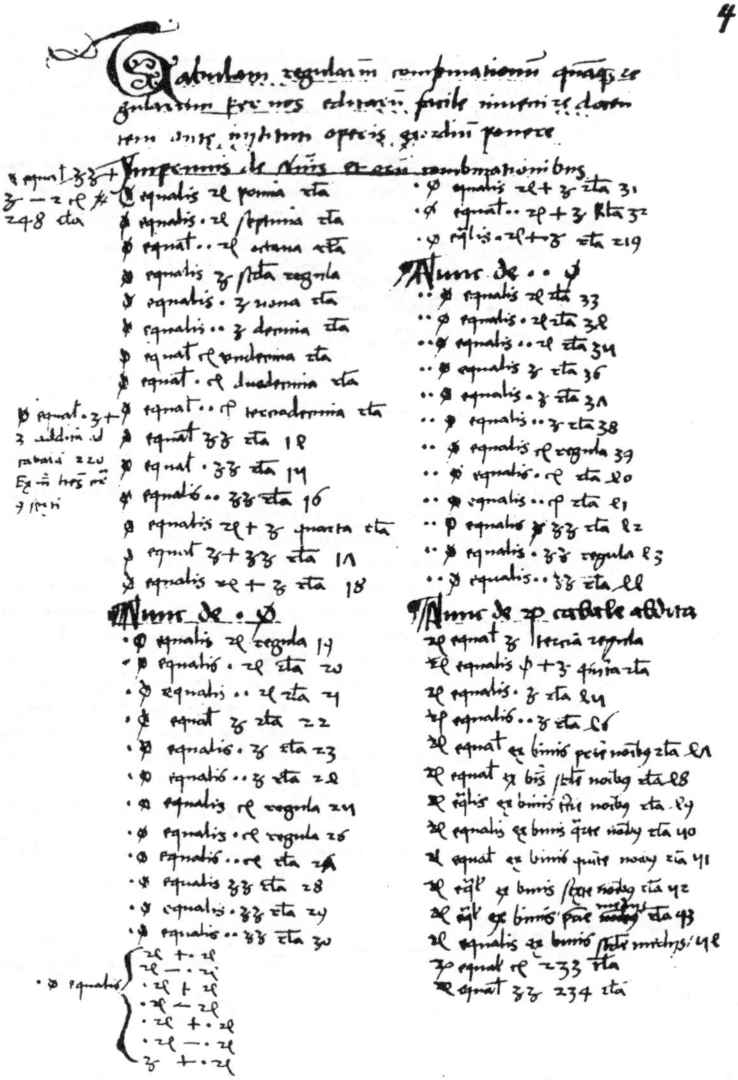

The first 54 types of Conrad Landvogt's equations (Uppsala, C 656. f. 4r; see p. 13 above).

Types 1–6: $a = bx$, $a = bx^2$, $ax = bx^2$, $a = bx + cx^2$, $ax = b + cx^2$, $ax^2 = b + cx$.

Types 7–18: $a = \sqrt{b}x$, $a = \sqrt[4]{b}x$, $a = \sqrt{b}x^2$, $a = \sqrt[4]{b}x^2$, $a = bx^3$, $a = \sqrt{b}x^3$, $a = \sqrt[4]{b}x^3$, $a = bx^4$, $a = \sqrt{b}x^4$, $a = \sqrt[4]{b}x^4$, $a = bx^2 + cx^4$, $a = \sqrt{b}x + cx^2$.

Types 19–71: The term on the left-hand side is $\sqrt{a}$ (no. 19–32); $\sqrt[4]{a}$ (no. 33–44); ax (no. 45–71).

Index of Names

This index contains all personal names (and the titles of some anonymous writings) found in the text and footnotes of this volume, including the authors of all referenced literature. Excluded are names mentioned in the titles of bibliographical items and names of publishers. Generally, Greek names have been printed in their Latin form (e. g., Menelaus for Menelaos). Names of medieval persons in the Latin West (up to ca. 1475) are usually listed under their first names, all other persons under their surnames or family names. In Arabic names the article *al-* is ignored in the alphabetical order; in genealogical connections the element *ibn* ("son [of ...]") is abbreviated as "b." Cross references are supplied in all cases where multiple forms of a name are in use.

Index of Manuscripts

This index contains page references to all manuscripts mentioned in this book. Pages on which a manuscript is cited by sigla and not by shelfmark are also indicated. These sigla are given in brackets after the shelfmark of the manuscript.